AF789910

Applications of Mass Spectrometry for the Provision of Forensic Intelligence

State-of-the-art and Perspectives

New Developments in Mass Spectrometry

Editor-in-chief:
Frank Sobott, *University of Leeds, UK*

Series editors:
Chérine Bechara, *University of Montpellier, France*
Marek Domin, *Boston College, USA*
Zhongping Yao, *The Hong Kong Polytechnic University, China*

Titles in the Series:
1: Quantitative Proteomics
2: Ambient Ionization Mass Spectrometry
3: Sector Field Mass Spectrometry for Elemental and Isotopic Analysis
4: Tandem Mass Spectrometry of Lipids: Molecular Analysis of Complex Lipids
5: Proteome Informatics
6: Capillary Electrophoresis–Mass Spectrometry for Metabolomics
7: Lipidomics: Current and Emerging Techniques
8: Processing Metabolomics and Proteomics Data with Open Software: A Practical Guide
9: Advanced Fragmentation Methods in Biomolecular Mass Spectrometry: Probing Primary and Higher-order Structure with Electrons, Photons and Surfaces
10: Advanced Mass Spectrometry-based Analytical Separation Techniques for Probing the Polar Metabolome
11: Ion Mobility–Mass Spectrometry: Fundamentals and Applications
12: MALDI Mass Spectrometry Imaging: From Fundamentals to Spatial Omics
13: Detection and Analysis of Microorganisms by Mass Spectrometry
14: Applications of Mass Spectrometry for the Provision of Forensic Intelligence: State-of-the-art and Perspectives

How to obtain future titles on publication:
A standing order plan is available for this series. A standing order will bring delivery of each new volume immediately on publication.

For further information please contact:
Book Sales Department, Royal Society of Chemistry, Thomas Graham House, Science Park, Milton Road, Cambridge, CB4 0WF, UK
Telephone: +44 (0)1223 420066, Fax: +44 (0)1223 420247
Email: booksales@rsc.org
Visit our website at books.rsc.org

Applications of Mass Spectrometry for the Provision of Forensic Intelligence
State-of-the-art and Perspectives

Edited by

Simona Francese
Sheffield Hallam University, UK
Email: S.Francese@shu.ac.uk

and

Stephen Bleay
London South Bank University, UK
Email: bleays@lsbu.ac.uk

ROYAL SOCIETY
OF **CHEMISTRY**

New Developments in Mass Spectrometry No. 14

Print ISBN: 978-1-83916-811-6
PDF ISBN: 978-1-83767-192-2
EPUB ISBN: 978-1-83767-193-9
Print ISSN: 2045-7545
Electronic ISSN: 2045-7553

A catalogue record for this book is available from the British Library

© The Royal Society of Chemistry 2024

All rights reserved

Apart from fair dealing for the purposes of research for non-commercial purposes or for private study, criticism or review, as permitted under the Copyright, Designs and Patents Act 1988 and the Copyright and Related Rights Regulations 2003, this publication may not be reproduced, stored or transmitted, in any form or by any means, without the prior permission in writing of The Royal Society of Chemistry or the copyright owner, or in the case of reproduction in accordance with the terms of licences issued by the Copyright Licensing Agency in the UK, or in accordance with the terms of the licences issued by the appropriate Reproduction Rights Organization outside the UK. Enquiries concerning reproduction outside the terms stated here should be sent to The Royal Society of Chemistry at the address printed on this page.

Whilst this material has been produced with all due care, The Royal Society of Chemistry cannot be held responsible or liable for its accuracy and completeness, nor for any consequences arising from any errors or the use of the information contained in this publication. The publication of advertisements does not constitute any endorsement by The Royal Society of Chemistry or Authors of any products advertised. The views and opinions advanced by contributors do not necessarily reflect those of The Royal Society of Chemistry which shall not be liable for any resulting loss or damage arising as a result of reliance upon this material.

The Royal Society of Chemistry is a charity, registered in England and Wales, Number 207890, and a company incorporated in England by Royal Charter (Registered No. RC000524), registered office: Burlington House, Piccadilly, London W1J 0BA, UK, Telephone: +44 (0)20 7437 8656.

For further information see our web site at www.rsc.org

Printed in the United Kingdom by CPI Group (UK) Ltd, Croydon, CR0 4YY, UK

Preface

Mass spectrometry is a powerful analytical tool that has been contributing to generating knowledge in a vast number of disciplines, from pharmaceutical analysis to environmental science and pathology, particularly in the last decade. Some examples of practical implementation stand out in clinical diagnostics, with mass spectrometers being used in hospitals for microbiological identification, antibiotic resistance detection and intra-operative cancer resection. However, there is only minimal awareness of the contribution that mass spectrometry makes to forensic science. Partly, this is due to viewing mass spectrometry as "just" an enabler for other disciplines; however the slow pace of innovation implementation and insufficient dissemination to other potential users within the community, also play a role.

Mass spectrometry has globally permeated forensic science, addressing investigative hypotheses as well as prosecution/defence scenarios. While some techniques are considered "emerging" and are progressively increasing in technology readiness level, as formally recognised in the UK by the Fingermark Visualisation Manual (Home Office/Dstl, UK), others have already been implemented to support both investigations and judicial debates.

The aim of this book is to disseminate awareness of both "mass spectrometry techniques of the future" and those already integrated within the forensic workflow for the analysis of specific types of evidence. It is important that forensic providers and police investigators recognise the forensic opportunities enabled by these techniques, as well as where to find knowledge around their deployment and application for the provision of forensic intelligence.

Indeed, due to different legislations, policies and research being conducted in silos, these techniques may be implemented in some countries but

New Developments in Mass Spectrometry No. 14
Applications of Mass Spectrometry for the Provision of Forensic Intelligence: State-of-the-art and Perspectives
Edited by Simona Francese and Stephen Bleay
© The Royal Society of Chemistry 2024
Published by the Royal Society of Chemistry, www.rsc.org

not in others. For these reasons, academics, industry sector-specific specialists and end users from different parts of the world were invited to contribute to different chapters of this book.

For each technique portrayed, this book aims to inform the reader of the type of forensic query and evidence analysed, the answers sought and provided, the limitations, the applications within the legal system and the instrumentation (and its specifications) enabling the specific type of intelligence that is being sought.

Finally, the last three chapters cover the use of mass spectrometry for the provision of intelligence that may not be directly exploited by the legal and forensic community but still has potential relevance to legal proceedings. This content choice stems from the potential transferability of these methods to support forensic investigations and court cases. After all, forensic science has previously borrowed and adapted the discovery of DNA profiling for the identification of perpetrators (or elimination of suspects).

The editors very much hope that this book will not only be useful reading for investigators and practitioners in all relevant environments, but also for postgraduates and early career investigators, as potential forensic scientists of the future, further developing the number and types of forensic (and more wide ranging) scientific queries that mass spectrometry can address.

Stephen Bleay
London South Bank University, UK

Simona Francese
Sheffield Hallam University, UK

Abbreviations

4-NT	4-Nitrotoluene
5-IT	5-(2-Aminopropyl)indole
6:2 FTAB	6:2 Fluorotelomer sulphonamide alkylbetaine
6-MAM	6-Monoacetylmorphine
ABS	Acrylonitrile butadiene styrene
ACIC	Australian Criminal Intelligence Commission
aDNA	Ancient deoxyribonucleic acid
Ag-LDI	Silver-assisted laser desorption ionisation
AMELX	Amelogenin-X
AMELY	Amelogenin-Y
AMS	Accelerator mass spectrometry
AMT	Alpha-methyltryptamine
ANOR	Alternate non-aqueous organic ratio
AP	Atmospheric pressure
APCI	Atmospheric pressure chemical ionization
ATS	Amphetamine-type substances
BOD	Biochemical oxygen demand
BSTFA	*N*,*O*-Bis(trimethylsilyl)trifluoroacetamide
BV3	Basic violet 3
BY40	Basic yellow 40
BZE	Benzoylecgonine
CAF	Cyanoacrylate fuming
CAR	Central African Republic
CBD	Cannabidiol
CFSRE	Centre for Forensic Science Research and Education
CI	Chemical ionisation
CID	Collision-induced dissociation

New Developments in Mass Spectrometry No. 14
Applications of Mass Spectrometry for the Provision of Forensic Intelligence: State-of-the-art and Perspectives
Edited by Simona Francese and Stephen Bleay
© The Royal Society of Chemistry 2024
Published by the Royal Society of Chemistry, www.rsc.org

CNS	Central nervous system
COA	Certificate of authenticity
CRM	Certified reference material
CSIs	Crime scene investigators
CWS	Canada's wastewater survey
Da	Dalton
DADP	Diacetone diperoxide
DART	Direct analysis in real time
dc	Direct current
DDA	Data-dependent acquisition
DDD	Defined daily doses
DEA	Drug Enforcement Agency
DESI	Desorption electrospray ionisation
DI	Desorption ionisation
DIA	Data-independent acquisition
DIMS	Direct infusion mass spectrometry
DMDNB	2,3-Dimethyl-2,3-dinitrobutane
DNA	Deoxyribonucleic acid
DPA	Diphenylamine
DTRs	Drug target residues
DUID	Driving under the influence of drugs
EA-IRMS	Elemental analysis-isotopic ratio mass spectrometry
EDTA	Etheylenediaminetetraacetic acid
EGDN	Ethylene glycol dinitrate
EI	Electron ionisation
EMCDDA	European Monitoring Centre for Drugs and Drug Addiction
ENFSI	European Network of Forensic Science Institutes
ESI	Electrospray ionization
ETDs	Explosive trace detectors
EtFOSAA	*N*-Ethyl-*N*-[(heptadecafluorooctyl)sulphonyl]glycine
ETN	Erythritol tetranitrate
FDR	False discovery rate
FET	Fingermark enhancement techniques
FID	Flame ionisation detection
FIRMS	Forensic isotopic ratio mass spectrometry network
FORS	Fiber optic reflectance spectroscopy
FPBSA	2-Phenylbenzimidazole-5-sulphonic acid
FRG	Fingermark Research Group
FSP	Forensic science providers
FSR	Forensic science regulator
FSS	Forensic science service
FTIR	Fourier transform infrared
FTIR	Fourier transform infrared spectroscopy
FTSA	Fluorotelomer sulphonic acid
FVM	Fingermark Visualisation Manual

FWHM	Full-width-half-maximum
GBL	Gamma-butyrolactone
GC–MS	Gas chromatography–mass spectrometry
GBH	Grievous bodily harm
GHB	Gamma-hydroxybutyric acid
GSR	Gunshot residue
GVPs	Genetically variant peptides
HBVs	Haemoglobin variants
HLB	Hydrophilic–lipophilic balanced
HME	Home-made explosives
HMTD	Hexamethylene triperoxide diamine
HMX	High melting explosive (or octogen)
HPLC	High performance liquid chromatography
HRAM	High-resolution accurate mass
HR-MC-ICPMS	High-resolution multi-collector ICPMS
HRMS	High resolution mass spectrometry
IAEA	International Atomic Energy Agency
ICC	International Criminal Court
IC–MS	Ion chromatography–mass spectrometry
ICPMS	Inductively coupled plasma mass spectrometry
ICP–MS	Inductively coupled plasma–mass spectrometry
IEDs	Improvised explosive devices
IMS	Ion mobility spectrometry
IP	Ionisation potential
IRMS	Isotopic ratio mass spectrometry
IRMS	Isotope ratio mass spectrometry
is CID	In-source collision induced dissociation spectra
KPCS	Kimberley Process and Certification Scheme
LA	Laser ablation
LA–ICP–MS	Laser ablation–inductively coupled plasma–mass spectrometry
LC–MS	Liquid chromatography–mass spectrometry
LC–MS/MS	Liquid chromatography–mass spectrometry/mass spectrometry
LDA	Linear discriminant analysis
LDI	Laser desorption ionisation
LLE	Liquid–liquid extraction
LOQ	Limits of quantitation
LOQ	Limit of quantitation
m/*z*	Mass-to-charge ratio
Ma	Million years
MA	Methamphetamine
MALDI	Matrix assisted laser desorption ionisation
MBTFA	*N*-Methyl-bis(trifluoroacetamide)
MCX	Strong cation exchangers
MDMA	3,4-Methylenedioxymethamphetamine

MetA-SIMS	Metal-assisted SIMS
MHN	Mannitol hexanitrate
MKEP	Methyl ethyl ketone peroxide
MOD	Ministry of Defence
MP	Microplastic
MRM	Multiple reaction monitoring
mRNA	Messenger ribonucleic acid
MS	Mass spectrometry
MSI	Mass spectrometry imaging
MSP	Mass spectrometry profiling
MVA	Multivariate analysis
MW	Molecular weight
NCPs	Non-collagenous proteins
NG	Nitroglycerine
ng	Nanogram
NIST	National Institute of Standards and Technology
NM	Nitromethane
NMR	Nuclear magnetic resonance
NPS	National Police Service
NPS	Novel psychoactive substances
NSAF	National Stolen Art File
nsSNPs	Non-synonymous single nucleotide polymorphisms
PCA	Principal component analysis
PDMS	Plasma desorption mass spectrometry
PE	Population equivalent
PEG	Poly-ethylene glycol
PETG	Polyethylene terephthalate glycol
PETN	Pentaerythritol tetranitrate
PFAS	Perfluorinated or polyfluorinated alkyl substances
PFCA	Perfluoroalkyl carboxylate, perfluoroalkyl carboxylic acid
PFDS	Perfluorodecane sulphonate
PFECSH	Perfluoroethylcyclohexane sulphonate
PFOA	Perfluorooctanoic acid
PFOS	Parent sulphonic acid
pg	Picrogram
PIXE	Particle induced X-ray emission
PLA	Poly-lactic acid
PLS-DA	Partial least squares-discriminant analysis
PMF	Peptide mass fingerprint
PMI	Post mortem interval
PNL	Normalised daily load
POCIS	Polar organic chemical integrative samplers
POPs	Persistent organic pollutants
ppm	Parts-per-million
PRM	Parallel reaction monitoring
PS	Powder suspensions

PTMs	Post-translational modifications
Py–GC–MS	Pyrolysis–gas chromatography–mass spectrometry
qPCR	Quantitative polymerase chain reaction
QQQ	Triple quadrupole
RDX	Hexahydro-1,3,5-trinitro-1,3,5-triazine
rf	Radiofrequency
R_s	Uptake rate
RT	Retention time
SALDI	Surface assisted laser desorption ionisation
SAPCI	Secondary atmospheric pressure chemical ionization
SAPs	Single amino acid polymorphisms
SCORE	Sewage Analysis Core Group
SCRAs	Synthetic cannabinoid receptor agonists
SEM-EDS	Scanning electron microscopy–energy dispersive X-ray spectrometry
SEM-EDX	Scanning electron microscopy–energy dispersive X-ray
SESI	Secondary electrospray ionization
SHU	Sheffield Hallam University
SIL-IS	Stable isotope-labelled internal standards
SIMS	Secondary ion mass spectrometry
SOPs	Standard operating procedures
SPE	Solid-phase extraction
SPME	Solid-phase micro-extraction
SRM	Selected reaction monitoring
SWATH-MS	Sequential window acquisition of all theoretical mass spectra
SWGDRUG	Scientific Working Group for the Analysis of Seized Drugs
TATP	Triacetone triperoxide
TD-IMS	Thermal desorption ion mobility spectrometry
TFA	Trifluoroacetamide
THC	Tetrahydrocannabinol
THC-COOH	11-Nor-9-carboxy-Δ9-tetrahydrocannabinol
TLC	Thin layer chromatography
TMS	Trimethylsilyl
TNT	Trinitrotoluene
TOF	Time-of-flight
ToF-SIMS	Time-of-flight secondary ion mass spectrometry
TRL	Technology readiness level
TSD	Time since deposition
UKAS	United Kingdom Accreditation Service
UKIAFT	United Kingdom and Ireland Association of Forensic Toxicologists
UPLC	Ultra-pressure liquid chromatography
UPS	Uninterrupted power supply
USE	Universal Spectrum Explorer
UV	Ultraviolet

VAMAS	Versailles Project on Advanced Materials and Standards
VDFS	Virginia Department of Forensic Science
VMD	Vacuum metal deposition
WBE	Wastewater-based epidemiology
WHO	World Health Organisation
WWTP	Wastewater treatment plant
WYP	West Yorkshire Police
XRF	X-Ray fluorescence
YHRSSS	Yorkshire and Humber Regional Scientific Support Services
ZooMS	Zooarchaeology by mass spectrometry

Contents

New Developments in Mass Spectrometry No. 14
Applications of Mass Spectrometry for the Provision of Forensic Intelligence: State-of-the-art and Perspectives
Edited by Simona Francese and Stephen Bleay
© The Royal Society of Chemistry 2024
Published by the Royal Society of Chemistry, www.rsc.org

CHAPTER 1

The Application of Mass Spectrometry to Explosive Casework: Opportunities and Challenges

B. CABOT,*[a] R. PARROTT II,[b] M. BROCHU,[c] H. W. GOH[d] AND G. FANG[e]

[a] Australian Federal Police, 47 Kings Ave., Canberra ACT 2601, Australia; [b] Federal Bureau of Investigation, 935 Pennsylvania Ave., NW Washington, D.C. 20535-0001, USA; [c] Royal Canadian Mounted Police, 73 Leikin Dr., Ottawa, Ontario K1A 0R2, Canada; [d] Home Team Science and Technology Agency, 1 Stars Ave #12-01, Singapore 138507, Singapore; [e] Health Sciences Authority, 11 Outram Rd, Singapore 169078, Singapore

*Email: ben.cabot@afp.gov.au

Abstract

Mass spectrometry is an essential technique employed in forensic investigations to identity trace explosive materials and is increasingly relied upon to analyse larger samples for the provision of forensic intelligence. Only in an operational setting are forensic intelligence efforts truly realised, and it is through this lens that readers and researchers must focus their efforts. This chapter outlines current and emerging techniques used in operational forensic laboratories in Australia, Canada, Singapore and the U.S., providing case studies as exemplars. Importantly, the challenges and limitations of such work are discussed, as well as the opportunities moving forward.

New Developments in Mass Spectrometry No. 14
Applications of Mass Spectrometry for the Provision of Forensic Intelligence: State-of-the-art and Perspectives
Edited by Simona Francese and Stephen Bleay
© The Royal Society of Chemistry 2024
Published by the Royal Society of Chemistry, www.rsc.org

1.1 Forensic Intelligence in Law Enforcement

Intelligence, in a law enforcement context, is the collection and assessment of information for further use as a decision-making tool at, predominantly, three levels: strategic, operational and tactical. Each type of intelligence requires different information to be collected and can range from a broad to a targeted number of sources. The intelligence is then used to direct resources as part of a wider organisational policy or for a specific operation/investigation, including applying for warrants or engaging specific capabilities. It is within this context that forensic intelligence finds its place and significance.

Ribaux *et al.* define forensic intelligence as, "the accurate, timely and useful product of logical processing (analysis) of forensic case data (information) for investigation and/or intelligence purposes".[1] Timeliness and accuracy of any type of intelligence are important with accurate intelligence provided too late being pointless, and inaccurate intelligence can be potentially disastrous. Disciplines such as DNA and fingerprints are easily recognised for providing forensic intelligence by identifying potential suspects or linking individual crimes based on profiles obtained from scenes. However, forensic intelligence is not only limited to these specific fields. Technical and physical attributes of devices have been exploited for forensic intelligence for many years in explosive device investigations. A recent highly publicised example is Malaysian Airlines flight, MH17, which was shot down over eastern Ukraine in 2014 with distinctive bowtie-shaped fragments recovered from the cockpit and captain's body, providing the first forensic intelligence to support the involvement of military hardware in the incident.[2] An additional critical piece of technical information to generate forensic intelligence from explosive devices is the identification and analysis of the devices' explosive materials. This is typically produced based on the information gathered in two distinct phases: (1) rapid analyses in the field; and (2) in-depth analysis in the laboratory. Here the timeliness of the information plays a crucial role in determining whether what is being provided is intelligence or not. Any indication, or identification, of the explosive associated with the device that can be provided rapidly in the field is intelligence as it is timely and useful. At the same time, there is some latitude in how accurate this intelligence needs to be. While the presence or absence of an explosive must be as accurate as possible, the type and exact nature of it demand less accuracy. Once the forensic investigation moves into the laboratory, accuracy is paramount. Timeliness continues to be important but what would be considered useful intelligence changes. Efforts to confirm or refute field intelligence are critical at the commencement of laboratory examinations and need to be expedited, but, beyond that, given the time and the techniques available, for anything to be considered useful intelligence, it must be far more insightful (see Figure 1.1).

The development of portable detection technology for the real time analysis of samples at the scene has led to this type of intelligence becoming a crucial component in all bombing related incidents. The Bali bombings of

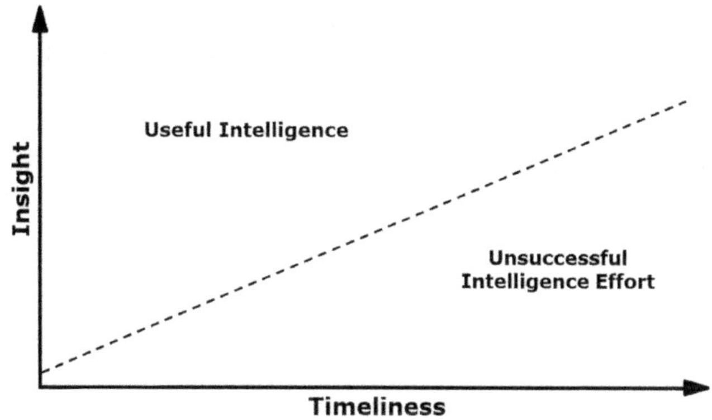

Figure 1.1 The relationship between timeliness and insight in relation to intelligence.

12 October 2002 provided a terrific example of the benefits of deploying portable detectors, as a "mobile laboratory".[3] Deploying portable equipment can assist in avoiding delays that may arise from the movement and transport of evidence from the scene to the laboratory, particularly from remote or distant locations, and it also provides an opportunity to triage evidence and prioritise, or exclude, certain items for further laboratory examination, saving the investigation valuable resources. Critically, although, rapid analysis in the field allows a forensic chemist to provide intelligence to the investigative leads at the scene. In Bali, the use of ion mobility spectrometry (IMS) in the mobile laboratory enabled detection of TNT in samples taken from two of the three scenes. This was later confirmed in the laboratory and corroborated by confessions. The success of the Bali bombing investigations, as well as numerous incidents since, was possible, despite portable mass spectrometers being unavailable. Recent advances in this area provide additional opportunities covered later in the chapter.

Sophisticated laboratory equipment, separately and holistically, plays crucial roles in providing information for forensic intelligence purposes. Laboratory equipment is generally considered to have greater sensitivity (*i.e.*, able to detect smaller amounts) and more selectivity (*i.e.*, able to detect more chemicals with greater confidence) than portable systems, offering the potential to produce confirmatory rather than presumptive results. These attributes are utilised to interrogate different characteristics of each explosive sample at a very low level to determine similarities and differences between other samples and/or to identify specific features that may indicate the sample's background (*i.e.*, production method).

Explosive sample attributes can be collated to create databases and while some databases exist, they tend to either be held within individual organisations or are more generic and publicly available over the internet. Databases held within specific law enforcement agencies are likely to be of high quality and enriched with, or linked to, broader case details including scenes and persons of interest; however, they can only be populated with samples analysed by that laboratory or from data shared directly with them. Thus, they are likely to be relatively small (depending on the jurisdiction of the agency and the number of seizures that are submitted). The need to maintain the operational capability and information security of the agencies necessitates heavily restricted access of such databases. Conversely, easily accessible databases contain characterised products procured from within the available market of the creator/s, an example being the online Smokeless Powder Database, which includes records of the manufacturer information, physical characteristics, explosives and additives present (as detected by GC–MS) of approximately 900 types of smokeless powder.[4] Although these databases are likely to contain more entries, they are unlikely to be as well characterised as secure databases. As with all commercial product databases, ensuring that the database is up to date with the market, especially in situations where there are changes in formulations, is very difficult. As somewhat of a compromise, a small number of databases accessible *via* the internet, albeit with restricted access, are hosted by various agencies/organisations and available for interrogation once an application for access has been approved.[5] Again, however, they are not exhaustive and tend to be limited to case samples analysed by the specific laboratory/group or those sampled from the marketplace. In each case, unless the specific case sample is being compared to a laboratory/user's own database with additional case context, an external database can only provide limited information.

1.2 Portable Mass Spectrometry and its Use in the Field

Field explosive detectors can be an important tool to provide forensic intelligence with critical evidence gathered from post-blast scenes, providing crucial leads as discussed previously, or by identifying linkages to other similar incidents or an organisation potentially responsible. Just as importantly, some very volatile explosives, such as triacetone triperoxide (TATP), and their decomposition products can be quickly lost. Early response to a post-blast scene with reliable portable trace detection capabilities increases the likelihood of identifying an explosive before it is lost, capturing key evidence. With time, the residual amount diminishes, posing challenges to even high sensitivity techniques within the laboratory.

A number of portable explosive detectors can be deployed to a scene to deliver this capability.[6] Presently, the most commonly deployed explosive trace detectors (ETDs) are based on thermal desorption ion mobility spectrometry (TD-IMS). This section will focus specifically on portable mass spectrometers and compare their performance against the commonly used ETDs. Current ETDs are manufactured and sold as specific products using specific configurations, although some settings are user-editable, allowing for useful comparisons between different systems. Portable mass spectrometers are relatively new and strive to achieve lower false alarm rates with higher specificity, compared to the traditional ETDs. Other operational considerations also include (i) ease of use, (ii) device response time, and (iii) the sensitivity of the device and the degree of sample preparation required.

Broadly, there are two classes of portable mass spectrometers (Table 1.1). One type utilises gas chromatography in-tandem with a quadrupole mass analyser or toroidal ion trap, while the other works based solely on ion trap technology. The former has the advantage of better resolution of different chemical species and easy reference to the NIST MS commercial database, but at the expense of weight and size. In contrast, the latter works in a higher-pressure range (*i.e.*, with a less stringent requirement for a pump to achieve a lower vacuum level), allowing them to be manufactured in a way that makes them more mobile.[7]

There are also transportable or "vehicle-portable" mass spectrometers, which (along with their accessories) are too bulky to be carried by operators but can be, relatively easily, retrofitted in vehicles or mobile laboratories. One such example is the Waters Radian™ ASAP. This mass spectrometer has been developed for the detection of drugs, but its library can be customised to include some explosives. Nonetheless, it is not an explosive trace detector and is only capable of detecting explosives in solution at tens of parts-per-million (ppm) concentration.

When considering the design of systems to utilise for the purposes of providing forensic intelligence at the earliest stages of an investigation, it is important to understand the performance of each system being considered.

For example, the comparison of two ionisation techniques, namely, secondary atmospheric pressure chemical ionisation (SAPCI) and secondary electrospray ionisation (SESI), and their performance in terms of sensitivity, selectivity and repeatability. Burns *et al.* conducted a study using a modified Waters Acquity QDa™ single quadruple mass spectrometer, to detect five explosives, namely, TNT, RDX, PETN, Tetryl and HMTD, on swabs using each of the ionisation techniques. Through electrospraying solvents during SESI, gaseous solvent ions generated interacted with the analyte droplets to ionise and subsequently measure the molecules of interest. In contrast, SAPCI allowed the ionisation of analytes, in either solid, liquid or solution form, by rapidly heating the sample using hot nitrogen gas to 400–500 °C and subsequently subjecting the evolved molecules to high voltage (3–6 kV), negating the need for the presence of a solvent.

Both techniques were able to the detect the five explosives, with SAPCI showing higher sensitivity and better specificity, especially in the absence of dopants. SAPCI is highly suitable for molecules with moderate polarity and a mass below 1000 amu, a level most explosives fall below. SESI, on the other hand, was more repeatable and its calibration curve is more linear over a wider concentration range.[8] It is performance factors such as these which must be considered before selecting a system for use in a mobile laboratory. The following question should be considered: What is the overall purpose of the forward deployment of capabilities? Accuracy in quantitative properties, or the

Table 1.1 Comparison of various battery-operated portable mass spectrometers.

Portable mass spectrometer	FLIR Griffin G510 GC/MS	Perkin Elmer Torion T-9 GCMS	INFICON HAPSITE ER identification system	MX908
Atomic mass unit range	18–510	43–500	41–300	55–470
GC programmable range	40–300 °C	50–300 °C	45–200 °C	No GC component
Mobile phase of GC	He (1 cylinder)	He (1 cylinder)	N_2 (2 cylinders)	NA
Operating hours (on battery)	1.5 hours (full GCMS mode) 2 hours (survey)	2.5 hours	2–3 hours	Up to 6 hours (2 spare batteries for hot swapping provided)
Time to recharge battery	2 hours (with external supplied charger)	Information not available	15 hours	2 hours (with external supplied charger)
Weight (with battery and gas cylinder)	16.4 kg	14.5 kg	19 kg	3.9 kg
Sampling	Heated air probe Direct liquid sampling *via* syringe	SPME or needle trap	Air probe	Use of fluorinated swab or vapour mode
Remarks	Optional accessory for direct introduction of solid sample vapours can be purchased from FLIR	Not equipped with vapour probe and so cannot perform survey analyses	Two optional sampling accessories, the Hapsite HeadSpace Sampling System for headspace sampling of volatile	Its aero attachment allows vapour to be drawn continuously (3 L min^{-1}) into MX908 to operate in survey mode.

		Uses retention time indices and an on-board spectral library (NIST) to identify compounds of interest	organic compounds in water, soil and solids, and the Hapsite SituProbe™ for water analysis	Operates in one of 3 modes for optimal performance: explosives, CWA and drugs
Column	DB-5MS	MXT-5	DB-1MS	Not applicable
Length (m)	15	5	15	
Film thickness (μm)	0.25	0.4	1	
Inner diameter (mm)	0.18	0.1	0.25	
Stationary phase	Phenyl arylene polymer (GLC)	Crossbond diphenyl dimethyl polysiloxane (GLC)	100% dimethyl-polysiloxane (GLC)	
Ion source	Electron ionisation (EI)	EI with dynamic ionisation	EI	Not applicable
GC/MS interface	Direct	Direct	Membrane	Not applicable
Mass analyser	Quadruple	Toroidal ion trap	Quadruple	Microscale ion trap
Operating pressure (Torr)	10^{-6}	10^{-4}	10^{-6} or lower	1–10
Vacuum pump	Turbomolecular (80 L s^{-1})	Turbomolecular (10 L s^{-1})	NEG	Multi-stage scroll pump (3 L min^{-1})

ability to detect as little as possible with as much confidence as possible? From a forensic intelligence perspective, detection capability would always be the preference. The ability to achieve this at a high confidence is of even greater value, in line with the definition provided by Ribaux *et al.*[1]

Weighing at approximately 27 kg, the 1st Detect Tracer 1000™ is another benchtop linear ion trap mass spectrometer, offering sensitivity down to picogram levels. The equipment has been developed to detect RDX/HMX, TNT, nitro-esters (PETN, ETN, NG, and EGDN), nitrates (*e.g.*, ammonium and urea), TATP, HMTD, tetryl, picric acid, R-salt and smokeless powders. Although the detector generally operates from an A.C. supply, it can be customised to be integrated with an uninterrupted power supply (UPS), so that the set can be made transportable and deployed in fields, on trolleys or unmanned ground platforms. Table 1.2 compares the 1st Detect Tracer 1000™ with the IonScan™ 500DT, IonScan™ 600 and NucTech™ TR2000DC, as analogous benchtop ETDs utilising IMS technology (Table 1.2). It is noteworthy that the new generation of benchtop mass spectrometers is aiming for a lower false alarm rate and the detection of a larger number of explosive substances.

It should be noted here that all the available portable products focus predominately on the detection of organic explosives. Although some detection capabilities of nitrate-based explosives do exist, these are limited to ammonium nitrate and urea nitrate chemicals. Generally speaking, inorganic explosives (see inorganic explosives further in this chapter) are not reliably detected with either IMS or portable MS technologies. Some systems, including GreyScan's ETD-100™, are entering the market to address this; however, these do not rely on mass spectrometry and hence will not be included in the chapter.

To understand the applications of mass spectrometry in the provision of forensic intelligence from explosives, both organic and inorganic, an understanding of their chemistry is required. The following sections provide an outline as well as the techniques currently being used to interrogate samples for intelligence purposes.

1.3 Organic Explosives

Organic explosives are mono-molecular explosives containing a hydrocarbon backbone upon which various functional groups can be introduced to provide a structure that emits significant energy when its chemical bonds are broken. These explosives can be categorized into family types (for the most encountered species): nitro-alkanes (R-NO_2), nitro-aromatics (Ar-NO_2), nitramines (R-N-NO_2), nitrate esters (R-O-NO_2), and peroxides (R-O-O-R) (Scheme 1.1). While these are the most common families of organic explosives, others do exist and occasionally are observed in forensic casework (*e.g.*, R-salt; an *N*-nitrosamine, and R–N–NO). Due to their structural similarities, species within a family often have similar properties in terms of

Table 1.2 Comparison of a benchtop mass spectrometer with conventional IMS-based ETDs.[a]

	1st Detector Tracer 1000	IonScan 500DT	IonScan 600	NucTech TR2000DC
Technology principle	Mass spectrometry	Ion mobility spectrometry (IMS)		
Ionisation technology	Non-radioactive	Ni-63	Non-radioactive	
Detectable explosives	Military, commercial-grade, HMEs, including RDX/HMX, TNT, PETN/ETN/NG/EGDN, AN, UN, HMTD, TATP, Tetryl, picric acid, R-salt, smokeless powders, *etc.*	Military explosives and HMEs including RDX, HMX, PETN, TNT, Tetryl, NG, EDGN, TATP, HMTD and others	Military explosives and HMEs including RDX, HMX, PETN, TNT, Tetryl, NG, EGDN, AN, UN, HMTD, TATP and others	RDX, HMX, PETN, TNT, Tetryl, NG, EGDN, AN, UN, HMTD, TATP
Sampling method	Particle sampling			
Sensitivity/limit of detection	μg–pg (dependent on chemical)	pg	Not available	ηg
False alarm	<0.5%	Not available	Not available	<1%
Warm-up time	20 min	<30 min	<10 min	<20 min
Analysis time	10–15 s	5–8 s	6–10 s	<8 s
Clear-down time	15 s (average)	Not available	Not available	<10 s (typical)
Weight	27.2 kg	19 kg	11.5 kg (with integrated printer)	<12 kg
Battery endurance	Requires an external power source	Requires an external power source	1 hour	Requires an external power source

[a] RDX: cyclotrimethylene trinitramine, HMX: cyclotetramethylene tetranitramine, TNT: trinitrotoluene, PETN: pentaerythritol tetranitrate, ETN: erythritol tetranitrate, NG: nitroglycerine, EGDN: ethylene glycol dinitrate, AN: ammonium nitrate, HMTD: hexamethylene triperoxide diamine, TATP: triacetone triperoxide, UN: urea nitrate, and R-salt: cyclotrimethylene trinitrosamine

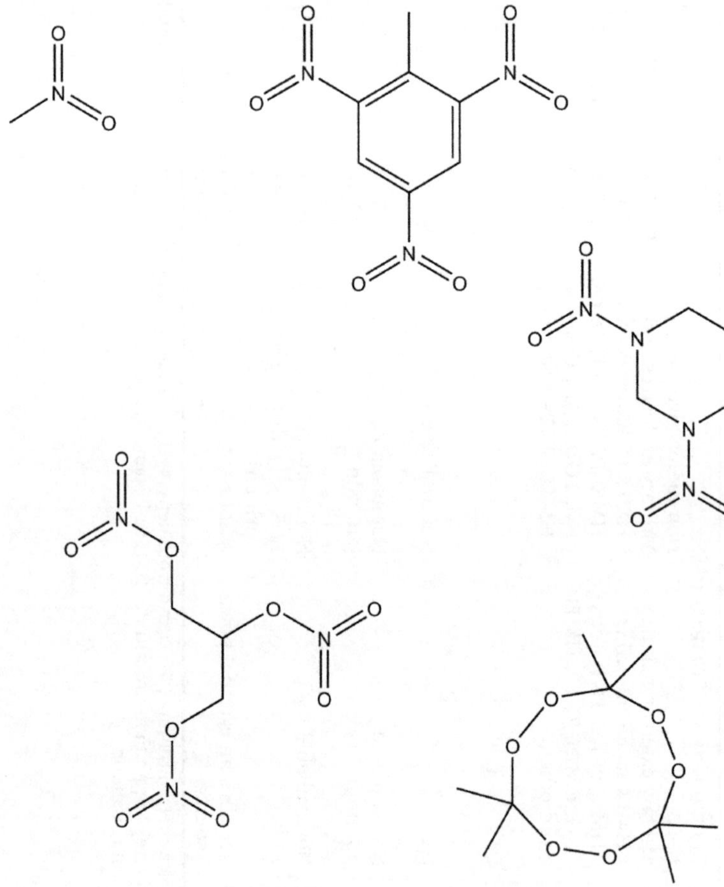

Scheme 1.1 Organic explosives. From left to right: nitromethane (a nitro-alkane), TNT (a nitro-aromatic), RDX (a nitramine), nitroglycerin (a nitrate ester), and TATP (an organic peroxide).

sensitivity, thermal lability, and detection ability based on the instrument type.

In addition to mono-molecular explosives, other organic chemicals can be utilised as fuels for blended explosives. Perhaps the most common examples of these would include fuel oil and sugar/sugar alcohols, such as erythritol. Some commercial products may also contain organic additives to provide stability or specific physical properties to an explosive product. Examples of these additives include acid scavengers utilised in smokeless powders (*e.g.*, ethyl centralite and diphenylamine) and plasticizers utilised in plastic explosives and smokeless powders (*e.g.*, bis(2-ethylhexyl) sebacate, and dibutyl phthalate).

The analysis of organic explosives, fuels, and additives can often be accomplished with spectroscopic techniques, such as Fourier transform infrared (FTIR) spectroscopy or Raman spectroscopy, when bulk material is available for interrogation. The utility of mass spectrometric techniques becomes more powerful when conducting trace residue analyses of these materials. Common exceptions to this could include a finer, detailed analysis for discriminating between different commercial explosive products. This is often done when inter-comparing between smokeless powders, plastic explosives, and other products containing organic additives, taggants, and explosives; Table 1.3 reports a list of commonly encountered explosives, fuels, and additives with their accurate masses).

The analysis of organic explosives and common organic additives to explosive products is not only significant from a law enforcement perspective but also for environmental and pollution monitoring, transportation screening, and for defensive/security applications. Given the wide range of

Table 1.3 Accurate masses of common organic compounds of interest related to explosives.

Mono-molecular explosives			
Diacetone diperoxide (DADP)	148.07356	Ethylene glycol dinitrate (EGDN)	152.00694
Triacetone triperoxide (TATP)	222.11034	Nitroglycerin (NG)	227.00258
Methyl ethyl ketone peroxide (MEKP)	210.11034	Pentaerythritol tetranitrate (PETN)	316.01388
Hexamethylene triperoxide diamine (HMTD)	208.06954	Erythritol tetranitrate (ETN)	301.99823
Cyclotrimethylene trinitramine (RDX)	222.03488	Mannitol hexanitrate (MHN)	451.98951
Cyclotetramethylene tetranitramine (HMX)	296.04651	Sorbitol hexanitrate (SHN)	451.98951
Nitroguanidine (NQ)	104.03343	Dimethyl dinitro-butane (DMDNB)	176.07971
Cyclotrimethylene trinitrosamine (R-salt)	174.05014	Nitromethane (NM)	61.01638
Trinitrotoluene (TNT)	227.01784	Nitrotoluene (NT)	137.0468
Dinitrotoluene (DNT)	182.03276	Dinitrophenol (DNP)	184.01202
Tetryl	287.01382	Picric acid (PA)	228.99710
Diazoditrophenol (DDNP)	210.00252	Hexanitrostilbene (HNS)	450.00437
Plasticisers and smokeless powder additives			
Dibutyl phthalate	278.15181	Diphenylamine (DPA)	169.08915
Diethyl phthalate	222.08921	Methyl centralite (MC)	240.12626
Dipentyl phthalate	306.18311	Ethyl centralite (EC)	268.15756
Dioctyl phthalate	390.27701	Akardite I	212.09496
Bis(2-ethylhexyl) adipate	370.30831	Akardite II	226.11061
Bis(2-ethylhexyl) sebacate	426.37091	Akardite III	240.12626
Sugars and sugar alcohols			
Sucrose	342.11622	Erythritol	122.05791
Glucose	180.06339	Sorbitol	182.07904
Fructose	180.06339	Mannitol	182.07904
Xylose	150.05282	Xylitol	152.06848

compounds that are of interest, a number of different laboratory-based techniques have been developed which are outlined below.

1.3.1 GC–MS

Due to the labile nature of many organic explosives, the utility of gas chromatography is limited, even beyond the analysis of vapour phases. However, there are specific areas of explosives analysis where GC–MS is routine and extremely beneficial. One such application is in the analysis of smokeless powders and other nitrocellulose-based propellants.[9–11] Due to the existence of international standards for the analysis of smokeless powders, this type of chemical discrimination between samples has become commonplace in many laboratories. Combined with the physical properties (*e.g.*, grain shape, grain size distribution, and marker grains) of a smokeless powder, the chemical constituents can provide an extremely useful comparison point. The identification of nitrocellulose can also be achieved *via* GC–MS through trimethylsilyl (TMS) derivatives.[12] These types of analyses can be conducted from a residual perspective and used for comparisons with samples of known origin for explosives or investigations related to organic gunshot residues. Complete identification of a smokeless powder can be limited due to similar formulations and physical attributes.

In addition to propellant analysis, extracts of plastic explosives for the discrimination of plasticizing agents have been proven to be useful. The plasticizers confer different physical properties to the final explosive products, but the identification of these plasticizers and some chemical taggants can help narrow down the field of possible products when analysing an unknown plastic explosive material.[13] Chemical taggants are an important part of plastic explosive history. In 1991, a convention on the marking of plastic explosives was held in Montreal, Canada, during which several countries agreed to add chemical taggants to manufactured plastic explosives to enhance their detectability. The agreement became effective in 1998. Currently, approximately 155 countries worldwide have signed the agreement with the most common chemical taggants being 2,3-dimethyl-2,3-dinitrobutane (DMDNB), 4-nitrotoluene (4-NT), and ethylene glycol dinitrate (EGDN).

Sugar and sugar alcohol characterization can also be accomplished through analysis on GC–MS after derivatization. Although this process is more complex and less straightforward compared to a simple extraction, it can provide useful information to the forensic investigator. A common derivatizing agent that has been utilised for the analysis of sugar residues is again, TMS.[14] In fact, TMS derivatization has also been proven useful in the discrimination between Pyrodex® and Triple Seven® residues by derivatizing fuels (*e.g.*, dicyandiamide, benzoic acid, and 3-nitrobenzoic acid) within each respective product.[15]

Finally, the analysis of fuel oil and other ignitable liquid products by GC–MS is routine in many forensic laboratories, of which most follow international standards for the isolation, analysis, and classification of these products.[16]

Some of these products have been utilised as fuels for commercial and improvised explosives. The classification and inter-comparison of such fuels can be extremely beneficial when determining the uniqueness of a liquid fuel among different samples. For instance, fuels such as diesel fuel and petrol (gasoline) exhibit considerable similarity among different refineries throughout the world but may show seasonal differences as well as differences arising from the crude material originally utilised or refinery processes (*e.g.*, some petrol formulations contain *n*-alkanes to a greater extent than others). However, some specialty products can be unique in their composition (to include additives for scents or distinct physical properties) and provide greater clarity on the type of product utilised.

1.3.2 Headspace and SPME Techniques

These techniques involve introducing the sample into the GC system by volatilising the material, or fractions of it, and either introducing this gas into the system or collecting it onto an amenable solid fibre which is introduced and heated again to release the captured compounds. While one would think that explosives would readily vaporise due to their chemical instability, this is often not the case. In fact, many explosives have a low vapour pressure (10^{-4} torr or lower at 27 °C (80 °F)). The common explosives that have a relatively higher vapour pressure include triacetone triperoxide (TATP), diacetone diperoxide (DADP), nitromethane (NM), nitroglycerine (NG) and ethylene glycol dinitrate (EGDN).[17] The low vapour pressure limits the usefulness of techniques like headspace and solid-phase micro-extraction (SPME). However, some components of commercially produced explosives can be targeted in addition to these select few volatile organic explosives. Some such targets include smokeless powder and plastic explosive components that have more volatile compounds. Common detection limits for these types of components using SPME or headspace techniques range from the low picogram (pg) range to the single nanogram (ng) range.[18,19] Several studies have explored the use of SPME or headspace for the analysis of organic peroxide explosives with detection limits in the single nanogram range.[20,21] A particular study utilised ambient laser desorption in conjunction with SPME for the analysis of some nitro-aromatic explosives. In this study, non-aromatic explosives were also examined (HMX and RDX), but were not detectable using this technique, perhaps showing a specificity of the SPME fibre for specific classes of materials.[22]

1.3.3 LC–MS and HRMS and MS^n

Liquid chromatography offers a wider ability to characterise organic components related to explosives since high temperatures are not involved in the separation of the chemical species. Because of this, LC–MS related techniques are perhaps the most used mass spectrometry class for the analysis of organic explosives. Depending on the sensitivity of the detector, these techniques are routinely capable of detecting explosive-related compounds at nanogram levels with some techniques capable of reaching detection limits in the femtogram (fg) region. The ability to ionise, the ionisation method, stability in solution, and selectivity of the detector are common factors that can affect the sensitivity.

Commonly, two types of ionisation methods are utilised in LC–MS analysis, electrospray ionisation (ESI) and atmospheric pressure chemical ionisation (APCI). To improve sensitivity, it often becomes advantageous to enhance ionisation through the use of adduct ions. Common ionisation enhancement solutions include acetate, formate, nitrate, or chloride salts in negative mode ionisation. While negative ion modes are most commonly used for explosives, certain targets such as picric acid, DPA, and the centralites are well suited for positive ionisation detection. In addition, because of its amine structure, HMTD is also readily observed in the positive ionisation mode with APCI.[23]

High-resolution mass spectrometry (HRMS) is becoming more commonplace with Orbitrap and time-of-flight (TOF) instruments increasing the selectivity of the mass detector. Some applications using these techniques achieve mass accuracy within a few part-per-million (ppm), narrowing the range of possible chemical molecules through mass spectrometry alone. As this technique becomes more routinely used, one may expect that the use of direct analysis through various ionisation techniques may increase in the forensic setting. Currently, most HRMS applications still involve the use of an LC component for separation. Case study #1 (Section 1.6.2) nicely illustrates an example of the application of HRMS.

A recent study utilised HRMS for the analysis of eleven sugar and sugar alcohol residues and validated the method with post-blast samples. This particular method involves a post-column addition of an ammonium nitrate solution to generate adduct ions, delivering detection levels from a high to low ng level when within a mixture.[24] Sugars and sugar alcohols can be utilised as fuels for explosive mixtures as well as precursors to produce nitrated sugar alcohols such as ETN and MHN.

Other than isotopic profiles, chemical impurity profiles obtained from LC–MS as a result of the synthesis of home-made explosives (HME) can also be used in forensic casework to differentiate or associate samples. This can become a challenge because impurity levels change due to acid catalysed degradation over time, affecting the certainty of the forensic intelligence obtained regarding the production and origin of the explosive material.[25] With the combination of LC–MS and IRMS, predictions of possible linkage between suspected precursor materials and HME samples (ETN, for example) in forensic explosive investigations has been shown to be feasible.[26] However, caution must still be taken when combining these techniques as the repeatability of the data obtained using different instruments will have an impact on the robustness of the dataset for forensic intelligence purposes.

Unfortunately, due to their structure, organic explosives are mostly not suitable for MS–MS applications. There are exceptions to this statement, but when these organic molecules fracture in the mass spectrometer, the negative charge typically stays with similar species revolving around the nitro-group, thus making non-characteristic ions. Notable exceptions to this are aromatic explosives that can easily stabilize a radical or negative charge as part of their structure (*e.g.*, TNT, dinitrotoluenes, and dinitro-napththalenes). Under these circumstances, MS–MS can help easily and quickly identify the nitro-aromatic explosive present. In addition, some related compounds such as the centralites and diphenylamine may prove to be useful analytes for this type of analysis.

1.3.4 Direct Mass Spectrometry

Direct analysis in real time (DART) and desorption ionisation (DI) have found some utility in the analysis of explosives under ambient conditions. Desorption techniques can encompass thermal and laser applications, and this type of analysis could become more routine as ionisation techniques advance with technology. Perhaps the most significant advantage to these types of techniques is the minimal sample preparation they require. The distinct disadvantage is the ability to create gas-phase molecules from less volatile components. As discussed earlier, however, specific classes of explosives and compounds of interest are suitable for analyses *via* these techniques such as those found in smokeless powders.[11,27] As technology continues to advance for ambient ionisation techniques, one could suspect that these ionisation techniques paired with HRMS could offer powerful screening techniques in the near future.

1.3.5 Isotope Ratio Mass Spectrometry

Isotope ratio mass spectrometry (IRMS) has garnered increasing interest in the field of explosive analysis, and forensics generally, over the past several years. Its current use in casework is still limited, but significant research has started to explore the utility of IRMS and some if its limitations. IRMS has shown some usefulness in being able to differentiate certain explosives from each other, when other conventional analyses could not. Most IRMS research has focused on the isotopes of carbon, nitrogen, and oxygen. Although oxygen is not a commonly studied isotope, it can prove to be useful in some research studies since not all analytes of interest for explosives need external oxygen for the combustion and ionisation process for IRMS. At the heart of IRMS lies the need to understand the range of isotope ratios observed within a particular compound or a particular set of material. Even further, an understanding of statistical models is needed to aid in drawing conclusion from such measurements.

Studies have been performed with some organic explosives including PETN, RDX, and Tetryl, showing the ability to possibly correlate between some samples and a greater ability to differentiate between samples whose isotope ratios are different.[28–30] In addition, other work has focused on inorganic analytes such as ammonium nitrate, black powder, and potassium perchlorate.[31,32] Again, some correlation can be found among manufacturers and possibly among areas of origin. However, the most beneficial aspect of IRMS is the ability to show differences among samples.

Some studies have suggested that isotope patterns may be able to link (or differentiate) original material and post-blast explosive residues.[32] This could also mean that for some substrates, a chemical precursor could be linked (or differentiated) to a synthesised explosive and/or its residues.[26] However, further work is needed in this area to draw any practical conclusions, including studies of variances/similarities in chemical precursors from different products, different combinations of fuels/oxidisers and how those ratios affect post-blast products, matrix effects, *etc.*

While the desire to correlate samples together is common across various chemical disciplines, caution is required when it comes to such statistical models needing a large (and continuing) dataset to determine the true "background" of isotopic patterns for a particular compound.[32] With the globalisation of goods and materials, it can be easily assumed that commercial and amateur manufacturers of explosives would likely opt for the easiest and least expensive sources of chemical precursors. Thus, a defensible statistical model (likely continually updated) is required before the sourcing of an explosive compound could truly occur.

It is for these reasons that IRMS is unlikely to find application in a traditional forensic case in a courtroom for some time in terms of "source identification". However, it should be expected that IRMS use will continue to grow for intelligence purposes and most definitely for differentiation purposes in a forensic setting. An example of the use of IRMS for forensic intelligence is outlined in case study #3 (Section 1.6.4).

1.4 Inorganic Explosives

Unlike organic explosives, inorganic explosives are not based on a hydrocarbon backbone. Inorganic explosives most commonly consist of a mixture of an oxidiser and a fuel. The oxidiser or both the oxidiser and fuel can be an inorganic compound. There are also mono-molecular inorganic explosives such as ammonium nitrate and ammonium perchlorate. Oxidisers, or oxidising agents, are usually oxygen-rich ionic solids that decompose at moderate to high temperatures, liberating radical oxygen to react with the fuel to sustain the combustion reaction that propagates through the mixture. The selection of the oxidiser used greatly influences the explosive properties of the mixture. The melting point, the heat of decomposition and the oxygen content of the inorganic oxidiser will affect the sensitivity to ignition as well as the explosive power of the mixture (Table 1.4 reports examples of solid oxidisers).

Table 1.4 Common oxidisers and their properties.

Compound	Formula	Molecular weight	Melting point (°C)	Heat of formation (kcal mol^{-1})
Ammonium nitrate	NH_4NO_3	80.0	170	−87.4
Ammonium perchlorate	NH_4ClO_4	117.5	Decomposes	−70.6
Barium chlorate	$Ba(ClO_3)H_2O$	322.3	414	−184.4
Barium nitrate	$Ba(NO_3)_2$	261.4	592	−237.1
Potassium chlorate	$KClO_3$	122.6	356	−95.1
Potassium nitrate	KNO_3	101.1	334	−118.2
Potassium perchlorate	$KClO_4$	138.6	610	−103.4
Sodium nitrate	$NaNO_3$	85.0	307	−111.8
Strontium nitrate	$Sr(NO_3)_2$	211.6	570	−233.8

Table 1.5 Common types of commercial products using various nitrate oxidisers.

Inorganic compound	Commercial explosives
Ammonium nitrate	Dynamite, emulsion, watergel, ANFO, ANAL
Potassium nitrate	Black powder, black powder substitutes, pyrotechnics
Barium nitrate	Pyrotechnics
Strontium nitrate	Pyrotechnics
Sodium nitrate	Dynamite, emulsion, watergels, black powder, pyrotechnics

1.4.1 Families

Inorganic explosives can be grouped into various "families" with similar explosive performance and forensic properties depending, predominantly, on the oxidiser utilised in the mixture.

1.4.1.1 Nitrate and Nitrate Salts

The most commonly available and used inorganic oxidisers and inorganic salts are nitrate-based. Potassium nitrate, for example, is a very commonly used oxidiser in explosive mixtures, has a low melting point and is therefore readily ignitable; however it is highly endothermic and will require a highly exothermic fuel to help achieve rapid burning rates (Table 1.5 covers examples of various nitrate salts used in commercial explosives).

1.4.1.2 Chlorate and Perchlorate Salts

One of the most reactive and common oxidisers is potassium chlorate, owing to its low melting point (356 °C)[38] and exothermic heat of decomposition. Potassium perchlorate is less sensitive to initiation because of its higher melting point (610 °C).[38] Chlorate and perchlorate salts are most commonly observed in pyrotechnic formulations and black powder substitutes.

1.4.1.3 Azides

Azides are substances that contain the N_3-group and can exist as inorganic salts. The most commonly encountered and discussed azides in the explosives industry, are lead, silver and mercury azide.

Certain azides are commonly used in primer compositions within firearm ammunition cartridges.

1.4.2 HME Mixtures

Homemade explosive (HME) mixtures are explosive products that can be easily produced by mixing commonly available materials and require only two basic constituents: a fuel and an oxidiser. Inorganic explosives are typically a blended mixture of an inorganic salt and a powdered metal, non-metallic elements or organic compounds. The most common oxidisers used are the inorganic salts, such as nitrates and nitrate salts, chlorate and perchlorate salts. HME mixtures may have different performance characteristics compared to their commercial or military explosives equivalents. However, the accessibility of precursor chemicals, the general simplicity of blending oxidiser and fuel selections, as well as the vast range of relative ratios at which the mixture will still produce a viable explosive, make these types of homemade mixtures attractive for illicit use despite their potential lower performance characteristics (Table 1.6 reports examples of uses of inorganic chemicals in terrorist attacks).

Table 1.6 Examples of combinations of oxidiser and fuel mixtures used in real incidents.

Oxidiser	Fuel	Common name (if any)	Remarks (*e.g.*, used in terror attacks)
Potassium nitrate	Sulfur and carbon	Black powder, black powder substitute, pyrotechnics	Often repurposed in pipe bombs
Ammonium nitrate	Fuel oil	ANFO	Commercial mining explosive, Used in Olso bombing (2011)[33] and University of Wisconsin bombing (1970)[34]
Ammonium nitrate	Nitromethane	AN-NM	Oklahoma city bombing (1995)[34,35]
Potassium chlorate	Nitrobenzene	Rack-a-rock	A mining explosive used in the 1880s[36]
Potassium chlorate	Aluminium and sulfur	N/A	Detonated with TNT in Bali bombings (2002)[3]
Ammonium nitrate	Aluminium	Exploding target, reactive target, emulsions, water gels	New York City Marine Corps 5 K Run, Chelsea Device (2016)[37]

1.4.3 Analysis

FTIR and Raman spectroscopic techniques are typically used to analyse inorganic explosives and precursor chemicals. However, when only trace amounts of material are present, these spectroscopic techniques may no longer be adequate for the sample.

Unlike organic explosives, the analysis of inorganic explosives and their post-blast residues can be particularly challenging. When considering the very low sensitivity of chromatography and mass spectrometry detection equipment and the prevalence of the inorganic salts and compounds in the environment, interpretation of the analytical data can be a complex endeavour. The forensic reporting scientists must interpret the data obtained from the analysis of solvent extracts and provide their expert opinion based on what is deemed significant or insignificant, considering the environmental background and its contribution to the matrix. Inorganic anions and cations commonly associated with inorganic explosives are present in the environment, either naturally abundant or through contamination from common consumer products. When considering the implications of a potential misinterpretation or false positive result, any exhibit analyses performed in a forensic laboratory should be accompanied by associated blanks of the sampling media (*e.g.*, swabs) and controls from the field where possible (*e.g.*, control soil samples). Having pre-determined limits of quantitation (LOQs) will assist in determining which results are significant and warrant interpretation (0.2 $mg\,L^{-1}$ to 0.5 $mg\,L^{-1}$ are some of the values observed in the literature for IC–MS).[39–41]

1.4.4 IC–MS

While other techniques, such as capillary electrophoresis (CE), can be used to analyse inorganic explosives and their residues, ion chromatography (IC) is utilised in many but not all forensic laboratories. IC is advantageous due to it being a highly sensitive and selective analytical technique for water soluble inorganic species. Sample preparation is easy involving a water extraction of a surface or grain, followed by filtration and dilution (if necessary). The IC lacks specificity, in that the detector alone is insufficient for determining a solute based on retention time alone; however when coupled with a mass spectrometer, IC–MS can conclusively identify inorganic explosive compounds. IC–MS detects ions (positive and negative) in solution and relies on electrospray ionisation, detecting a range of ionic species from low-mass ions such as chloride to higher mass ions such as perchlorate. Typically, the level of detection for both anions and cations is in the low $mg\,L^{-1}$ range (approximately 0.3 $mg\,L^{-1}$ or less).

IC–MS systems typically utilise electrospray ionisation (ESI) as the interface between the liquid-based IC system and the gas-based MS system. In ESI, the analyte, dissolved in a liquid phase, emerges from a metal capillary to which a high voltage is applied. During the ionisation process, the resulting field at the needle tip charges the surface of the emerging liquid, dispersing it into a fine spray of charged droplets. The droplets are evaporated using a nitrogen gas stream, decreasing their size and increasing their charge density. At a point where the charge density on the surface becomes too great, the droplet breaks into finer droplets. This process repeats until ions (cations and anions) are desorbed from the droplets into the ambient gas entering the mass spectrometer. ESI is a soft ionisation technique, where little fragmentation of ions occurs.

Mass spectral identification is based on comparison of the relative ratios of the ion isotopes to those of a standard reference material. Since ion ratios are sensitive to concentration in solution, a calibration curve should be generated using serial dilutions of a standard reference material as points of comparison for the data. Most IC–MS systems in use are low-resolution providing reasonably accurate masses to approximately ± 0.1 amu. There are scientific publications which discuss IC interfaced with a high resolution Orbitrap MS such as the Thermo Q-exactive (± 0.0002 amu). Table 1.7 summarises ions associated with various explosive mixtures and the expected masses these ions, including their isotopes.

One limitation of IC–MS is the inability to detect inorganic species that are not water soluble, such as iron oxide, lead chromate, lead tetroxide (red lead), lead dioxide and, importantly, aluminium. Another challenge utilising this technique is when the matrix is complicated by environmental contamination or by multiple species present in the sample. Inorganic salts dissociate in aqueous solutions into their respective anions and cations. Where multiple anions and cations are identified, it is not possible to confirm the originating inorganic species.

1.4.4.1 Inorganic Impurities

IC–MS may be able to provide forensic intelligence where impurities or unexpected ions are observed in the analytical sample. Through comparison with commercial products, the data could provide information which would suggest that the sample is homemade. Significant amount of research is focussed on optimising explosive mixtures and compounds for their intended use while ensuring the safety of personnel working with them.[42] Commercial manufacturers have strict quality control processes governing the materials that may be used in energetic materials. HME, on the other hand, is far less likely to have been manufactured to the same standards raising the possibility that the final product may contain residues that differ from what is expected in similar commercial products. Of these, cations tend to be most useful given they are often present as spectator ions in explosive synthesis reactions. For example, the identification of K^+ or Na^+ ions in a nitroaromatic or nitrate ester may indicate the employment of an improvised nitration process using potassium or sodium nitrate (KNO_3 or $NaNO_3$) rather than nitric acid.

Table 1.7 Inorganic explosives uninitiated and initiated detection.

Explosive	Uninitiated	Initiated/post-blast
Black powder	KNO_3, C, S ($NaNO_3$)	NO_3^-, SO_4^{2-}, NO_2^-, CO_3^{2-}, HCO_3^-, SCN^-, OCN^-, S^{2-}, K^+, $S_2O_3^{2-}$, (Na^+)
Pyrodex	KNO_3, C, S, $KClO_4$, sodium benzoate, dicyandiamide (DCDA)	NO_3^-, SO_4^{2-}, NO_2^-, CO_3^{2-}, HCO_3^-, SCN,OCN^-, S^{2-}, K^+, Na^+, $S_2O_3^{2-}$, benzoate, DCDA, ClO_4^-, Cl^-
Triple seven	KNO_3, C, $KClO_4$, sodium benzoate, DCDA, 3-nitrobenzoic acid (3-NBA)	NO_3^-, NO_2^-, CO_3^{2-}, HCO_3^-, OCN^-, Na^+, K^+, benzoate, 3-NBA, DCDA, ClO_4^-, Cl^-
Pyrotechnic formulations	$KClO_3$, $KClO_4$, $Ba(NO_3)_2$, $SrNO_3$	NO_3^-, NO_2^-, Sr^+, K^+, Ba^{2+}, ClO_4^- ClO_3^-, Cl^-
Dynamite	$NaNO_3$	NO_3^-, NO_2^-, Na^+
Emulsions/ water gels	NH_4NO_3	NO_3^-, NO_2^-, NH_4^+
AN mixtures	NH_4NO_3	NO_3^-, NO_2^-, NH_4^+

Anion isotopes (where ≥0.1%)

	Isotope	Mass (amu)	Abundance (%)
Cl	^{35}Cl	34.97	75.8
	^{37}Cl	36.97	24.2
S	^{32}S	31.97	95.0
	^{33}S	32.97	0.8
	^{34}S	33.97	4.2
C	^{12}C	12.00	98.9
	^{13}C	13.00	1.1
N	^{14}N	14.00	99.6
	^{15}N	15.00	0.4
O	^{16}O	15.99	99.8
	^{18}O	18.0	0.2
Br	^{79}Br	78.92	50.7
	^{81}Br	80.92	49.3

Cations isotopes (where ≥0.1%)

	Isotope	Mass (amu)	Abundance (%)
K	^{39}K	38.96	93.3
	^{41}K	40.96	6.7
Mg	^{24}Mg	23.99	79.0
	^{25}Mg	24.99	10.0
	^{26}Mg	25.98	11.0
Ca	^{40}Ca	39.96	96.9
	^{42}Ca	41.96	0.6
	^{43}Ca	42.96	0.1
	^{44}Ca	43.96	2.1
	^{48}Ca	47.95	0.2

Table 1.7 (*Continued*)

Cations isotopes (where ≥0.1%)			
	Isotope	Mass (amu)	Abundance (%)
Sr	^{84}Sr	83.91	0.6
	^{86}Sr	85.91	9.9
	^{87}Sr	86.91	7.0
	^{88}Sr	87.91	82.6
Ba	^{130}Ba	129.91	0.1
	^{132}Ba	131.91	0.1
	^{134}Ba	133.90	2.4
	^{135}Ba	134.91	6.6
	^{136}Ba	135.90	7.9
	^{137}Ba	136.91	11.2
	^{138}Ba	137.91	71.7

Common anionic compounds observed in Inorganic explosive residues			
Anions	Mass (amu)	Ions detected	Masses of ions
ClO_2 (chlorite)	67.45	ClO_2^-	67/69
ClO_3 (chlorate)	83.45	ClO_3^-	83/85
		ClO_2^-	67/69
ClO_4 (perchlorate)	99.45	ClO_4^-	99/101
		ClO_3^-	83/85
		ClO_2^-	67/69
NO_2 (nitrite)	46.01	NO_2^-	46
NO_3 (nitrate)	62.00	NO_3^-	62
C_6H_5COO (Benzoate)	121.12	$C_6H_5COO^-$	121/122
PO_4 (phosphate)	94.97	$H_2PO_4^-$	97
		PO_3^-	79
		PO_2^-	63
CNO (cyanate)	42.04	CNO^-	42
S_2O_3 (thiosulfate)	112.13	$HS_2O_3^-$	113/114/115
		$S_2O_3^-$	112/113/114
		SO_3^-	80/81/82
		SO_2^-	64/65/66
		SO^-	48/49/50
SCN (thiocyanate)	58.08	SCN^-	58/59/60

1.5 Metals and Their Use in Explosives

Metals, and their oxides, are incorporated into many explosive formulations as fuels with their addition also providing a number of different benefits.

Elements such as strontium and barium, as ionic salts, are commonly included in pyrotechnic mixtures as colourants, while sulfur (a non-metal) has been used in black powder formulations for hundreds of years.[43] Out of all the metals available, one of the most important and commonly used metals is aluminium. The addition of aluminium into a formulation increases the flame and temperature released by the explosive formulation as well as the duration of the pressure wave produced. Different grades of aluminium are used in a number of different explosive applications, ranging from high quality aluminium particles used in high performing military formulations, detonators and commercial blasting slurries[44] to improvised explosives utilised by terrorists incorporating lower quality particles sourced from readily available commercial or domestic products. An example of an improvised explosive incorporating a metal fuel, and widely recognised as a favorite among terrorists, is ammonium nitrate and aluminium (AN–Al). Metals are also a key component in the broader design of improvised explosive devices or munitions as circuitry, including switches and power sources, containers and significantly as fragmentation elements to increase the lethality, or as a liner in sophisticated shape-charge devices increasing their penetration. Of course, the analysis and exploitation of these components are important aspects of a forensic investigation and provision of intelligence to investigations. Although these components will not be included in the discussion below, given the focus of the chapter on explosives, their analyses can be conducted using the same techniques and this chapter can be used to inform their exploitation.

1.5.1 Forensic Analysis of Metals

Forensic investigations of metal constituents in explosives commonly utilise a spectroscopic technique, such as X-ray fluorescence (XRF) spectroscopy or scanning electron microscopy coupled with energy dispersive X-ray spectroscopy (SEM–EDS), for analysis. Both of these are non-destructive techniques, always a favourable attribute in forensic applications, and capable of accommodating both large (millimetre) and small (sub-millimetre) sample sizes. SEM–EDS also has the additional advantage providing high magnification morphological information. Indeed, research has shown that morphological data alone, from optical as well as electron microscopy, can provide important forensic intelligence.[45,46] While these techniques have significant advantages, XRF as a technique is unable to detect elements lighter than sodium and is less sensitive generally to light elements, while SEM–EDS is poor at detecting nitrogen and carbon (both non-metallic elements of interest to explosive analyses) and more susceptible to variable surface topographies.[47] Both techniques are also relatively insensitive in terms of detection, on the order of 0.1% w/w, which, if using elemental composition solely as an attribute, may provide limited value in forensic intelligence.

1.5.2 ICP–MS

With respect to high sensitivity across a wide range of elements, inductively coupled plasma–mass spectrometry (ICP–MS) is the gold standard technique. ICP–MS differs in the way that it produces ions for analysis in the mass spectrometer. As outlined in previous sections, most MS applications used in forensic science use chemistry to separate species before applying physical principles, in the form of electric charges or electron bombardment, in the MS source to produce ions for measuring. ICP, on the other hand, produces ions *via* an argon plasma (*i.e.*, ionised gas) created by loading a copper coil which is wound around a 'torch' containing argon with a high-frequency alternating current at relatively low power.[48–50] This current, together with a high-voltage discharge, starts an ionisation process in which a certain number of argon ions generate additional argon ions, which propagate further, resulting in the generation of a large amount of heat. The result is a stream of argon plasma at an extremely high temperature (approx. 10 000 K). Samples are introduced into the plasma stream *via* a nebuliser where they are aerosolised, with droplets that are fine enough being carried into the torch by the argon gas. Once it reaches the torch, the fine droplets evaporate leaving behind the dried analytes which are subsequently decomposed and dissociated into their individual atoms and finally ionised. The amount of ionisation that occurs differs for each element. One of the favourable conditions of using argon as the plasma source is its high ionisation potential. The amount of energy required to produce a singly charged argon ion (*i.e.*, its first ionisation potential, or IP) is higher than that of almost all other elements (F, Ne, and He are the exceptions), while lower than the ionisation potential for doubly charged ions (*i.e.*, second IP) for all but a small number of elements.[50] This property creates a stream of predominantly singly charged ions from the sample emanating from the torch for analysis using a mass spectrometer. Given the different energies required for ionisation by each element, none are ionised equally. Ions closest to argon in the periodic table have a first IP most similar to argon and therefore are ionised the least. These include fluorine, neon and helium which are not ionised at all, and nitrogen, oxygen, chlorine and bromine which experience a low degree of ionisation. While some of these, namely, nitrogen and oxygen, and to a lesser extent chlorine, are important elements, when it comes to explosives these are all easily measured using other techniques. Fortunately, most other elements are far more easily ionised with elements consisting of electron structures where only one electron is present in the outer shell exhibiting the highest proportion of ionisation. Argon's first IP is also lower than, or relatively similar to, the second IP of almost all other elements, meaning most elements will not form doubly charged ions and the few that do will only generate a very small proportion, simplifying the ultimate interpretation of the resulting mass spectra.

Once a sample is transformed into ionised atoms, they then travel through an interface into the vacuum of the mass spectrometer and are focussed through electrically modulated lenses into the inner components of the mass spectrometer. The interface acting as the bridge between the atmospheric pressure/high temperature plasma and the low vacuum/temperature spectrometer is composed of metal cones each with small holes at their apex, allowing ions from the plasma core to pass through while deflecting extraneous plasma flow. The holes must be large enough to limit blocking while being small enough to facilitate high analysis performance by maintaining optimal vacuum conditions. Lower vacuum pressures reduce the number of gaseous molecules present, limiting the number of interactions between ions and molecules, thereby increasing the transmission efficiency of ions into the mass spectrometer, which, in turn, minimises peak broadening and background artefacts, increasing resolution and sensitivity. Once past the vacuum interface, ions and neutral masses not yet deflected, enter the lens component. Here electric charges are used to repel and attract the ions in order to focus them in order to maximise their transmission efficiency and the instrument's performance. At the last stage of the lens assembly, the ions are electronically separated from any neutral masses and light photons by deflecting them from the axis along which they have travelled with the neutral species and photons to this point.

The next stage of an ion's path through an ICP–MS depends on what type of configuration the system has, most commonly categorized as a 'single-quad' or 'triple-quad' system. For forensic intelligence applications of explosives, high sensitivity is required with triple quad systems preferred over single quad systems to achieve this. Because of this, only triple-quad systems are considered in this chapter. The superior ion sensitivity from this configuration is produced by utilising an isolating-removing-measuring process across the three different parts of the system. In the first quadrupole (Q1), masses of interest (including on-mass interferences) are isolated from the stream of ions from the source. The next stage involves a collision/reaction cell where a reaction gas is added which inhibits interference ions and reacts with other ions, usually in the form of polyatomic ions changing their mass, and removing them from the ion stream. Finally, the now "purified" ion stream consisting only of the target analyte then moves to the last quadrupole (Q2), where it is measured.[50] Using this triple-quad configuration routinely provides sensitivity in the range of parts per trillion (ppt), with some approaching detection limits in the parts per quadrillion (ppq) range. ICP–MS also allows simultaneous measurement of multiple elements (typically 20–30) in a single analysis and offers a large linear range spanning eight orders of magnitude with fine enough resolution to conduct isotopic measurements.

Sample introduction is a significant issue with the technique requiring species to be dissolved in liquid. When focussing on the analysis of metals, this generally requires digestion of the sample, using either acid or alkali solutions often at increased temperatures, increasing the risk of contamination due to the increased handling and use of chemical reagents.[47] An alternative to this is the use of laser ablation (LA), which offers the ability to directly sample solid material by imparting enough energy onto the material so that it transforms into a gas and this technique is used across various other forensic evidence types including glass and tapes.[52,53] As part of an LA–ICP–MS arrangement, the gas is then transferred into the ICP–MS system and analysed. The combination of LA with ICP–MS provides a number of advantages, including being semi-destructive, only requiring very small areas (approximately 100 μm) of the material to be consumed as a sample; little to no sampling preparation; and allowing the sampling to be conducted across multiple axes (*i.e.*, 2D or 3D sampling). Results obtained from laser-ablated samples however tend to have greater variation in them given the chaotic nature of the ablative effect. Given this variation, it can be more advantageous to use more traditional techniques (*i.e.*, XRF) in a forensic analysis of intact explosives and LA–ICP–MS for post-blast debris.[54]

1.5.3 Applications of ICP–MS in Forensic Intelligence

The volume of effort required to exploit trace elemental analysis of explosives for forensic intelligence is unquestionably greater than that listed in research publications. Given the relationship between explosives and national security, most case details and innovations relating to their analyses are tightly held on secure systems.

The published research efforts have predominantly involved inorganic mixtures (including sparkler compositions and ammonium nitrate based fertilisers) with C4 explosive formulations and homemade urea nitrate also investigated.[54,55] Sparklers are readily available products often encountered in relatively small pipe-bomb devices, while AN/CAN/urea fertilisers are important industrial/agricultural products manufactured and stored in large quantities potentially posing a national security risk and with a well-known history of use in terrorist devices (*e.g.*, Oklahoma City bombing, Afghanistan/Iraq). Each study has attempted to create a database of explosive entities from different manufacturers/types/batches to provide a means to 'match' samples/sources; however, given the number of elements analysed, the resulting data are often complex.[54,56–59] To overcome this, the most common approach is to use statistical analyses with the goal of linking an unknown sample to one (or more) possible entities to establish a source. The statistical analyses utilised vary but generally include principle component analysis (PCA), linear discriminant analysis (LDA) or partial least squares-discriminant analysis (PLS-DA). Each of these computations is complex and automated with the resulting outcomes often needing to be accepted without the ability to verify the result by other means. This becomes challenging when being incorporated into intelligence where, ideally, the characteristics being used to establish a link are known and understood well enough to nominate the significance of the linkage. For this purpose, it seems that LDA, followed by a likelihood ratio/Bayesian analysis approach, is the most promising process moving forward as it is the most robust and utilises a process already widely accepted and used by the forensic community.

Data collected using laser ablation sampling become especially difficult to interpret. Coupled with the heterogeneous nature of compositions when sampled at a small scale, the variation in the plasma produced during the ablation process creates instances where trace elemental profiles of a sample as a whole cannot be easily established.[54] Taking this technique a step further and sampling post-blast debris *via* laser ablation is even more difficult. The chaotic nature of explosions can create a non-uniform spread of reaction products each with their own makeup. This makes forensic intelligence assessments almost impossible for such samples.

1.6 Practical Examples

Below are some specific examples of forensic intelligence research and applications from explosive samples. These provide an appreciation of the outcomes that are delivered in the in real world, which at times, can be limited.

1.6.1 Operationally Relevant Research

In an unpublished study, methyl ethyl ketone peroxide (MEKP) samples were synthesised under different reaction conditions, and purified and analysed using liquid chromatography mass spectrometry (LC–MS), an adaptation of published methodology.[60] The analyses were run on a Shimadzu LC–MS-8030, coupled with LC-30AD pumps, SIL-30AC autosampler and CTO-20AC column oven (see Table 1.8 for running conditions), delivering a limit of detection of 1 pg Reserpine.

MEKP is an organic peroxide formed by reaction of methyl ethyl ketone (MEK) with acidified hydrogen peroxide (Scheme 1.2 for possible reaction mechanisms).[61]

Table 1.8 LCMS conditions employed in the analyses of MEKP samples.

	LC conditions		MS conditions	
Pump A	10 mM ammonium formate + 0.1% formic acid		Interface	DUIS
Pump B	Methanol		Mode	Positive
Pump gradient	Time (min)	%B	Interface temp	350 °C
			DL temp	250 °C
	1.00	55		
	7.00	65		
	8.00	90		
	16.00	90		
	18.00	55		
	20.00	55		
Flowrate	0.2 ml min^{-1}		Nebulizing gas flow	1.50 L min^{-1}
Column temp	40 °C		Drying gas flow	10 L min^{-1}
Column	Shim-pack XR-ODS 2.0 mm id × 75 mm		Scan range	150–1000
Injection vol	10 μL		Data acquisition	20 min

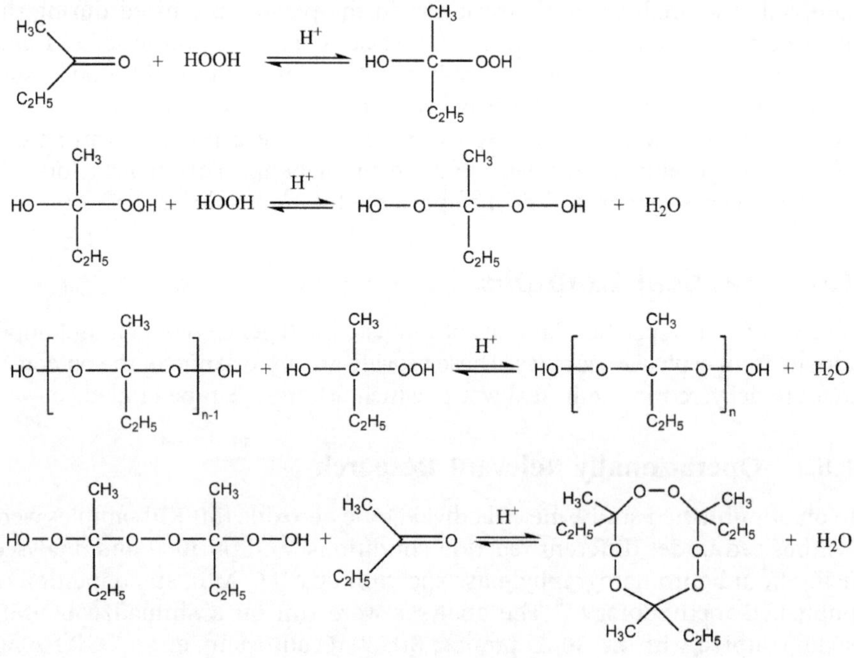

Scheme 1.2 Possible reaction mechanism of MEK.

Possible peroxides formed (Scheme 1.3) with compounds 1–7, which have been identified in the literature,[62] and compound **8**, which was deduced from the LC–MS analysis.

Species with *m*/*z* as high as 580 and 668 were detected (Figure 1.2).

This demonstrates the ability of laboratory analyses to detect and deduce precursors used in the explosive synthesis assisting investigations by opening up avenues of enquiry into suppliers/sources of chemicals and/or recipes and the level of knowledge of those who manufactured it.

1.6.2 Case Study #1

As part of an investigation into an attack on a building, swabs of the damaged building surfaces were collected for the analysis of explosive residues. These swabs were heavily sooted and contained potential chemicals used by first responders as part of the matrix to be deconvoluted. To this point, no forensic intelligence had been provided as the initial analysis did not conclusively indicate the presence of any energetic material due to peaks being obscured by potential interferences. Further analysis was conducted on acetone extracts of the swabs utilising a UPLC-HRMS (Orbitrap) technique, relying on the Orbitrap's higher selectivity. This method utilised nitrate adducts to enhance ionisation in the mass spectrometer (Figure 1.3) and required the nitrate adduct ion plus two isotopic ions for identification.

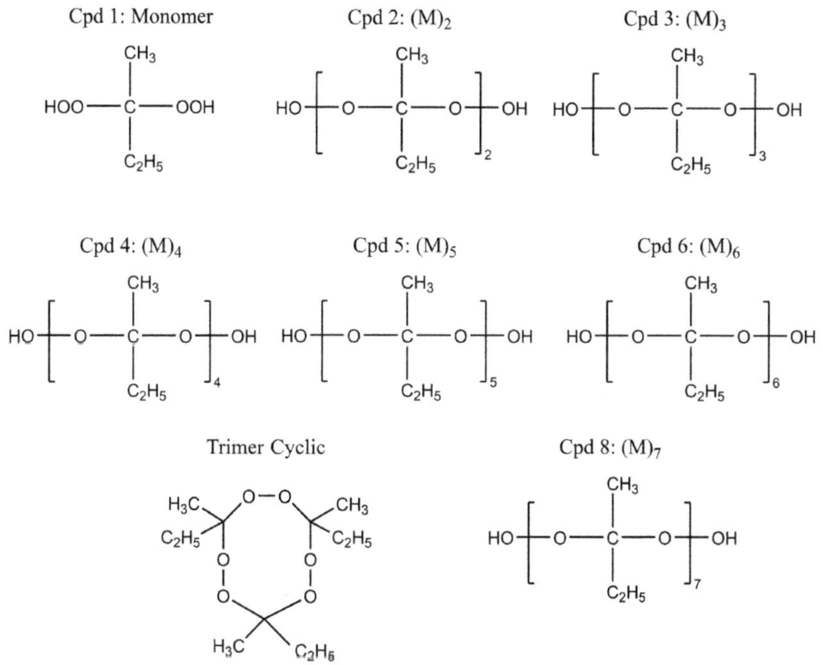

Scheme 1.3 Possible peroxides of MEKP.

Upon analysis using this method, the laboratory was able to identify between approximately 1–50 pg of HMX and PETN residues on the swab of the building materials. RDX could not be identified, at the approximate 100–500 fg level, as the threshold required to detect the second isotopic ion was not met (Figure 1.4). However, the selectivity of the HRMS showed utility in being able to detect accurate masses of single mass unit ions with an interfering matrix.

1.6.3 Case Study #2

A post-blast investigation of a vehicle was conducted in a residential area. The vehicle suffered minimal damage and it was clear that the device functioned only partially. Remnants from the explosive main charge were spread throughout the vehicle's interior and later determined to be a mixture of ammonium nitrate and a medium petroleum distillate. Visual examination of the material provided forensic intelligence that the ammonium nitrate had been harvested from cold packs. Precursor chemicals recovered from the scene provided additional forensic intelligence, suggesting the possible manufacture and involvement of HMTD, most likely as an improvised detonator and booster. The vehicle was left at the scene for two days before being moved to an indoor facility for exploitation. During the first day of the vehicle exploitation, the interior of the vehicle was examined utilising an

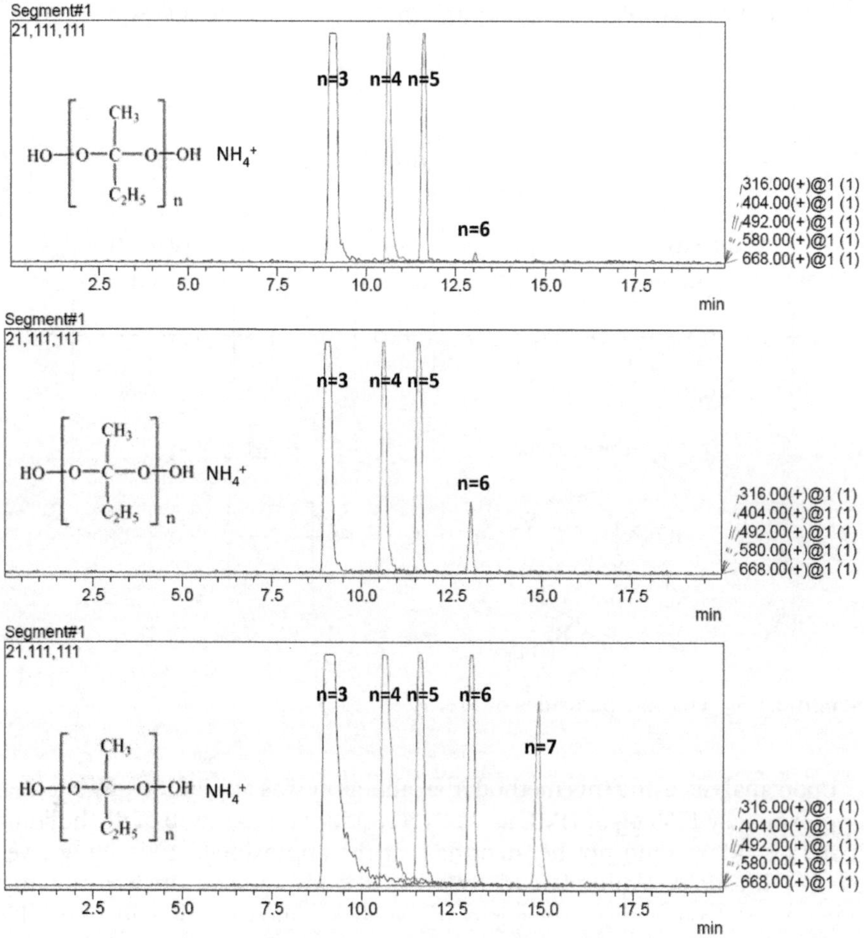

Figure 1.2 LCMS chromatograms of MEKP synthesised using MEK and agricultural grade hydrogen peroxide (17%) in the presence of cleaning agents containing 8% sulfuric acid (top) and 8% hydrochloric acid (middle), compared to MEKP synthesised using MEK and 30% hydrogen peroxide in the presence of 98% sulfuric acid (bottom) over 48 hours.

IMS system to determine if HMTD residues were present post blast, with one swab from the roof liner providing a positive HMTD response. The roof liner was subsequently cut out from the vehicle and packaged; however, the exhibits collected during this examination were not submitted to the laboratory for another three weeks following their collection.

Activated charcoal strips were used to absorb vapours from within the exhibit bags over an 18 hour period and desorbed according to a standard procedure. Portions of the roof liner, as well as other items recovered from within the vehicle, were also directly extracted with acetone and then filtered, concentrated and analysed using GC–MS (CI). None of the exhibits

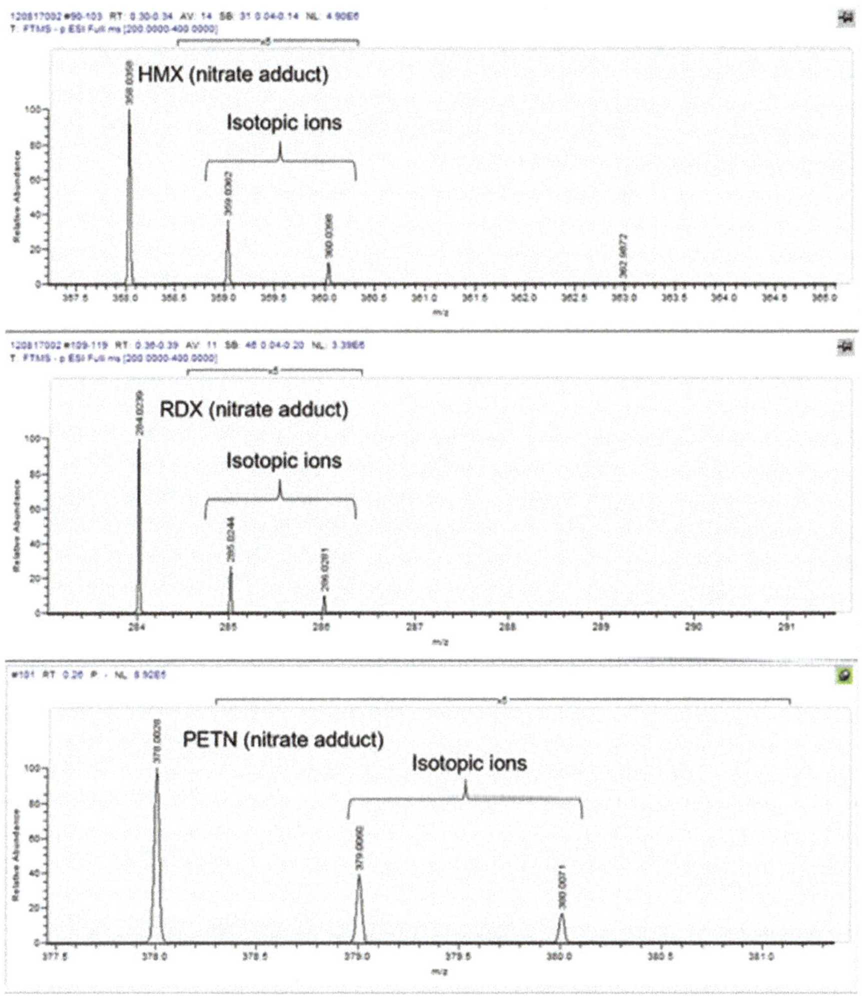

Figure 1.3 HRMS spectra of specific high explosives (1.6 pg) with nitrate adduct and isotopic ions.

analysed in the laboratory confirmed the previous forensic intelligence of the presence of HMTD.

This example demonstrates that, although multiple opportunities for forensic intelligence can exist in a single case, often they can also be fleeting. Volatile explosives, such as HMTD, require that laboratory analyses on samples suspected to contain them be conducted as soon as possible to prevent the loss of the explosive material. Not only does this provide the greatest opportunity to formulate good intelligence, it is also important for court proceedings as presumptive on-site testing results from techniques such as IMS are not considered confirmatory results for court, and are not included in forensic reports.

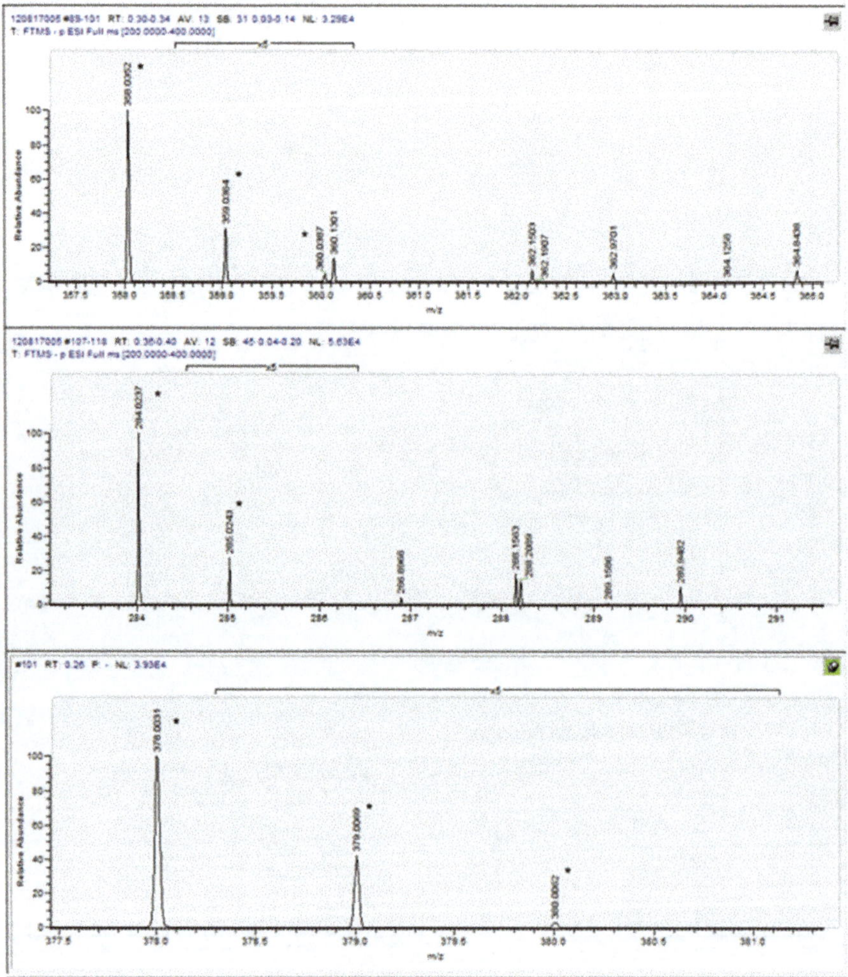

Figure 1.4 HRMS spectra of an acetone extract of the explosive residue swab with nitrate adduct and isotopic ions. The components with an asterisk relates to corresponding isotopic peaks in reference to compared standards in Figure 1.3.

1.6.4 Case Study #3

A device was seized as part of an investigation with the main charge identified as pentaerythritol tetranitrate (PETN) using spectroscopic techniques. Unlike the previous case studies involving post-blast residues, where the identity of the explosives was a significant forensic intelligence goal, much was already known about this sample leading to more involved questions being asked. Where could this material have been obtained? Has it been stolen from a mining site or from the military? Has it been homemade? Crucially, each of these questions help to frame the overarching question,

"How many more devices like this are out there, and what ongoing risk is posed to the community?"

PETN is typically used in commercial detonators, where it is present as a crystalline powder, as well as in detonating cords and boosters as part of a melt-cast formulation. Microscopic analysis was initially conducted to confirm the absence of fragments of plastic sheathing used to wrap PETN formulations into detonating cords. A combination of GC–MS and LC–MS, along with IC, was subsequently employed to explore different aspects of the sample. GC–MS was unable to confirm the presence of plasticisers. Analysis *via* LC–MS allowed the identification of pentaerythritol trinitrate, an intermediate product in the synthesis of PETN. While not utilising mass spectrometry, IC analysis also indicated the presence of ionic impurities, which may have been confirmed had an IC–MS been available in the laboratory. Holistically, the analyses provided the forensic intelligence that the PETN was homemade.

As the investigation progressed, a second PETN seizure from a known source was identified and it was hypothesized that the two shared the same source. A sample of the second PETN seizure was obtained and the carbon isotope ratios (δ13C) of each sample were measured. Published research exists on IRMS analysis of PETN; however, the value of the data is constrained by sample size limitations with most PETN production tightly secured as part of government apparatus and regulations. Additionally, PETN is produced with a very high (approx. 100%) purity through the nitration of pentaerythritol (PE) in a 1 : 1 PE : PETN ratio.[63–65] Carbon atoms within the PE precursor remain unaffected during this process; consequently, the corresponding isotope ratios of PETN directly relate to the ratios within the precursor. Although not available for this analysis, greater discrimination is provided by adding δ15N values to the dataset. The δ13C values for both seized samples with an approximately 1‰ difference were observed, three times greater than the measurement uncertainty, indicating different sources (see Figure 1.5) (the error bars represent expanded measurement uncertainty at 95% confidence intervals ($k = 2$)).

The analytical results, which allowed the differentiation between two samples, and the expert interpretation of the results in the context of the reaction process and wider research, offered two forensic intelligence outcomes:

(1) The samples come from the one production facility however pentaerythritol supplies had changed;
(2) The samples are unrelated and came from different production facilities, each with
    (a) their own supplier; or,
    (b) multiple suppliers.

The ambiguity of the intelligence provided is inherent to all real-world applications, given the reaction chemistry, and is compounded by the limitations of obtaining contemporary samples for analysis. Obtaining and

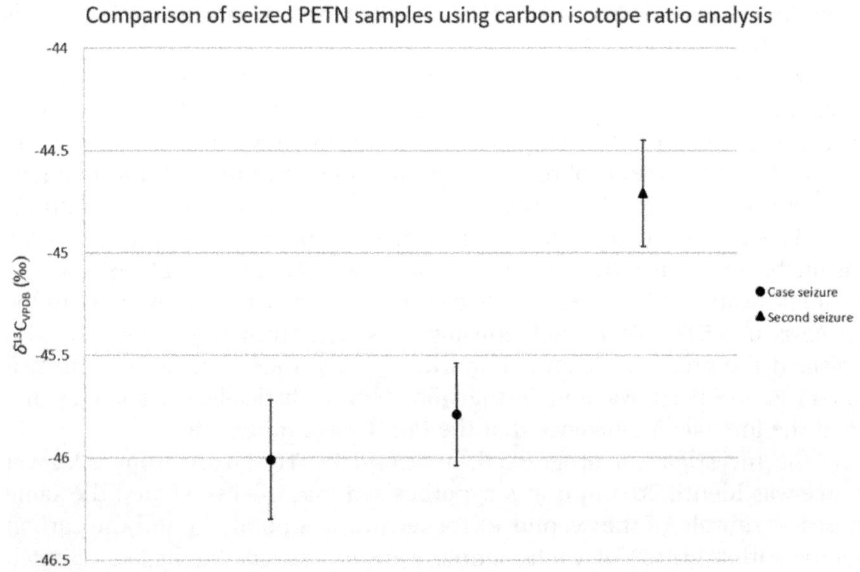

Figure 1.5 Comparison of seized PETN samples using carbon IRMS.

sharing this data is one of the major challenges of providing forensic intelligence in casework with calls for addressing this highlighted in research.[66,67]

1.7 Opportunities and Challenges

The exploitation of explosive samples using mass spectrometry to gather information for assessment and the provision of forensic intelligence is a field full of both numerous opportunities and challenges.

1.7.1 Databases – Future Perspectives

Generally, it is not feasible to build a single reference database for all explosives. Most commercially available explosives are controlled or prohibited items, which limits their availability. Large variabilities in the production of homemade explosives such as the type of precursors used, synthesis conditions, the stability of the explosives and its impurities, makes it extremely challenging to build a comprehensive database to provide forensic intelligence. Taking into consideration the challenges presented, the establishment of an explosive database to be used or shared internationally is a work in progress.

As with all science, forensic intelligence, evidence, or otherwise, the greater the number of similarities that are observed, the stronger the correlation between the samples being observed. In this respect, datasets from multiple MS analyses, including the combination of LA–ICP–MS and IRMS data, have also been combined in some studies, providing better results than single techniques only.[56] The establishment of databases where explosive samples are treated as entities within it, and the results of each technique listed as an attribute, could only lead to improvement. However, this will not overcome the challenges of open *vs* secure databases as discussed at the beginning of the chapter.

In an ideal world, law enforcement agencies across the globe, or subsets of them as formal trusted international and domestic partners, would share access to a secure database containing detailed chemical results. This would allow the comparison and potential linking of seized explosives, potentially identifying previously unknown terrorist or criminal networks and/or facilitating the monitoring of explosives trends over time and across regions. Of course, we do not live in an ideal world and chemical analysis capabilities vary broadly across countries, not to mention geopolitical barriers. This should not mean that such solutions cannot be endeavoured with like-minded and capable partners.

1.7.2 Sample Preparation

No studies yet identified, with respect to LA–ICP–MS analysis, have first separated samples into their individual explosive components prior to elemental analysis. Such an approach is often conducted in forensic laboratories for analyses using the technology currently used (as discussed previously). The inclusion of this step in forensic intelligence practices would improve the quality of the resulting assessments by providing a greater understanding of the origins of the correlations between samples. For example, from what explosive ingredient, say aluminium and ammonium nitrate, is a trace element profile observed? Analysing this formulation as a whole, or even as particles, could result in incorrect linkages if the presence of the trace element is due to the aluminium in one sample and the ammonium nitrate in the other.

Despite the challenges arising from variabilities in LA samples and trace, post-blast collections, there exists published literature on LA–ICP–MS analysis of organic gunshot residue (oGSR) particles.[68] This research included the collection of oGSR using traditional SEM stubs with a modified adhesive and subsequent analysis using separate techniques on each of the stub. One half of the sample underwent Raman spectroscopy analysis (necessitating the modified adhesive) while the other half was sampled and analysed using LA–ICP–MS (using optimised LA parameters to limit ablation of the aluminium stub). While demonstrating its success, the elements detected in the study were related to the ammunition itself (cartridge and/or projectile) or their primer compositions. While propellants can be encountered as charges in IEDs, the components detectable in this study are not. However, the study does highlight the opportunities to conduct LA–ICP–MS examinations on trace, post-blast samples. Previous studies have demonstrated that it is possible to capture explosive reaction products on SEM stubs metres away from detonations of multiple compositions including ammonium nitrate/aluminium and PE4 (RDX within a plasticiser matrix).[69] These were analysed using SEM–EDS (discussed previously), providing an ability to automatically image and analyse the sample. Particles of interest were highlighted based on their elemental composition. The superior sensitivity capability of ICP–MS in conjunction with LA sampling provides an avenue for trace elemental analyses for forensic intelligence studies on samples collected in a similar manner and it may be an avenue worth exploring.

1.7.3 Chemistry Limitations

Limitations too are also imposed by the explosive chemistry itself. Instances can occur at both ends of the spectrum, where either too little or much information is present for a meaningful determination. As demonstrated in case study #3, reaction mechanisms constrain the atoms are involved in explosive synthesis and can result in production of insufficient information. Transferring that to the final product will limit the determinations that can be made. Conversely, too much information can be gleaned from the analyses of some samples, for instance, from trace impurities from HMEs. Depending on the specifics of a reaction, the quality of its precursor chemicals, the steps taken to purify the final product, cleaning processes between batches, and so forth, a number of chemicals could be included in the final product at a range of concentrations. Analysis using highly sophisticated systems may become overwhelmed with data, particularly if the sample is not prepared properly. It can also become very difficult to link different batches of HME manufactured in the same clandestine laboratory if the trace chemical, or impurity, "profile" changes significantly. Often these types of laboratories will use whichever chemicals they can obtain to make their explosive of choice, which could even be dictated by the chemicals available.

1.7.4 Analysis of Wastewater

Wastewater is already analysed in sewerage systems around the world for drugs[70] and is discussed extensively in Chapter 10. With the ongoing and ever evolving threat posed by terrorist utilisation of IEDs across the globe, the analysis of wastewater could provide valuable forensic intelligence for the early disruption of potential security threats. PhD research at the University College, London, has demonstrated the it is possible to detect peroxide and organic explosives using LC–MS analysis from wastewater collections in London, UK.[71] The proposal within this study, suggesting the collaboration between forensic intelligence efforts and surveillance operations, which could include both physical and/or digital, to conduct targeted analyses of specific areas or addresses in real-world applications is a very valid one, especially when the specific case circumstances, for example, legislative restrictions or the operational security measures of targets, prevent covert activities to collect more direct samples from taking place. Achieving this is not without its challenges, including the complexity of the matrix, the dilution factor and the varying solubilities across the entire range of explosives and their precursors. Notwithstanding, opportunities exist to develop this methodology further and to deploy it into national security efforts.

1.8 Conclusion

The production of forensic intelligence from explosives samples has a history of over two decades, predominantly with the incorporation of portable systems into, or near, the crime scene to provide rapid presumptive results to feed into the investigation. Only in recent years however, has this been recognised as 'forensic intelligence', prompting more determined efforts to increase the scope of intelligence that forensic science can deliver. The improvement of portable systems, especially those using mass spectrometry techniques, will continue to be led by the private industry in the chemical detection market, in collaboration with law enforcement and military partners. The further development of these types of systems presents a significant opportunity to improve forensic intelligence produced during the earliest stages of an investigation from in-field analyses. Improvements to miniaturisation, sensitivity and specificity of current technologies are expected to continue as technology evolves, improving the accuracy and timeliness of forensic intelligence produced from the scene. Although technically challenging, this could be considered relatively easy given that each system can be improved upon in isolation. The most challenging aspect will be to incorporate multiple, highly sensitive, highly specific and complex analyses, including anticipated mass spectrometry, addressing current gaps such as inorganic explosives, from multiple samples into a coherent form allowing comparisons and assessments to be made to formulate meaningful intelligence. This cannot be done without sharing case data, most likely outside one's own agency to partners requiring security solutions. Additionally, whether this can be achieved without complex multivariate data treatment is unknown. Is it enough to rely on PCA results, or minor differences in first derivative data, and call it intelligence or does the person producing it need to have some understanding of what the data treatment has simply highlighted and be able to explain it? To some extent, this will depend on the outcome with significant or critical intelligence perhaps necessitating human understanding and explanation. Remembering that a key parameter of forensic intelligence is its utility, the longer this process takes, the more insightful it should be, increasing the likelihood of it providing significant or critical intelligence and enhancing human understanding and reasoning. To this end, should complex multivariate data be used at all? Will artificial intelligence be accepted instead?

The opportunities are great, but so are the challenges.

References

1. O. Ribaux, S. J. Walsh and P. Margot, The contribution of forensic science to crime analysis and investigation: forensic intelligence, *Forensic Sci. Int.*, 2006, **156**(2–3), 171–181.
2. E. Vermeij, P. Zoon, R. Gerretsen and V. Otieno-Alego, The outcome of the forensic triage preceding disaster victim identification in the downing of Malaysia Airlines flight 17, *Forensic Sci. Res.*, 2022, **7**(3), 566–575.
3. D. Royds, S. W. Lewis and A. M. Taylor, A case study in forensic chemistry: The Bali bombings, *Talanta*, 2005, **67**(2), 262–268.
4. online Smokeless Powder Database, 2023. Available from: http://www.ilrc.ucf.edu/powders/ [cited 7 February 2023].
5. Explosives Database 2014. Available from: http://expdb.chm.uri.edu/index.php, [2nd March 2023].
6. S. Benson, N. Speers and V. Otieno-Alego, Portable Explosive Detection Instruments, in *Forensic Investigation of Explosions*, ed. A. Beveridge, International Forensic Science and Investigation Series, CRC Press, 6000 Broken Sound Parkway NW, Suite 300 Boca Raton, FL, 2nd edn, 2012, pp. 691–723.
7. P. E. Leary, B. W. Kammrath and J. A. Reffner, Portable Gas Chromatography–Mass Spectrometry: Instrumentation and Applications, in *Portable Spectroscopy and Spectrometry. 1: Instrumentation and Applications*, ed. R. A. Crocombe, P. E. Leary and B. W. Kammrath, 1st edn, John Wiley & Sons, Ltd, 2021, pp. 367–389.
8. D. Burns, S. Mathias, B. J. McCullough, C. J. Hopley, D. Douce and N. Lumley, *et al.*, Ambient ionisation mass spectrometry for the trace detection of explosives using a portable mass spectrometer, *Int. J. Mass Spectrom.*, 2022, **471**, 116735.
9. ASTM, Standard Practice for Characterization and Classification of Smokeless Powder, ASTM International, 2016.
10. ASTM, Standard Test Method for Analysis of Organic Compounds in Smokeless Powder by Gas Chromatography-Mass Spectrometry and Fourier Transform Infrared Spectroscopy, ASTM International, 2017.
11. E. Lennert and C. Bridge, Analysis and classification of smokeless powders by GC–MS and DART-TOFMS, *Forensic Sci. Int.*, 2018, **292**, 11–22.
12. A. S. Chajistamatiou and E. B. Bakeas, A rapid method for the identification of nitrocellulose in high explosives and smokeless powders using GC–EI-MS, *Talanta*, 2016, **151**, 192–201.
13. S. Moore, M. Schantz and W. MacCrehan, Characterization of Three Types of Semtex (H, lA, and 10), *Propellants, Explos., Pyrotech.*, 2010, **35**, 540–549.
14. J. Nowicki and S. Pauling, Identification of Sugars in Explosives Residues by Gas Chromatography-Mass Spectrometry, *J. Forensic Sci.*, 1988, **33**(5), 1254–1261.
15. B. J. Routon, B. B. Kocher and J. V. Goodpaster, Discriminating Hodgdon Pyrodex((R)) and Triple Seven((R)) Using Gas Chromatography-Mass Spectrometry, *J. Forensic Sci.*, 2011, **56**(1), 194–199.

16. ASTM, Standard Test Method for Ignitable Liquid Residues in Extracts from Fire Debris Samples by Gas Chromatography-Mass Spectrometry, ASTM, 100 Barr Harbor Drive, PO Box C700, West Conshohocken, PA, 2019.
17. H. Östmark, S. Wallin and H. G. Ang, Vapor Pressure of Explosives: A Critical Review, *Propellants, Explos., Pyrotech.*, 2012, **37**(1), 12–23.
18. M. Joshi, K. Rigsby and J. R. Almirall, Analysis of the headspace composition of smokeless powders using GC–MS GC uECD and ion mobililty spectrometry, *Forensic Sci. Int.*, 2011, **208**(1–3), 29–36.
19. G. Sauzier, D. Bors, J. Ash, J. V. Goodpaster and S. W. Lewis, Optimisation of recovery protocols for double-base smokeless powder residues analysed by total vaporisation (TV) SPME/GC–MS, *Talanta*, 2016, **158**, 368–374.
20. D. Muller, A. Levy, R. Shelef, S. Abramovich-Bar, D. Sonenfeld and T. Tamiri, Improved Method for the Detection of TATP After Explosion, *J. Forensic Sci.*, 2004, **49**(5), 935–938.
21. Q. T. Gauthier and A. G. Simon, Storage and decomposition of pure and clandestine recipes of peroxide explosives, *Forensic Chem.*, 2022, **31**, 100458.
22. C. A. Seneviratne, S. Ghorai and K. K. Murray, Laser desorption sample transfer for gas chromatography/mass spectrometry, *Rapid Commun. Mass Spectrom.*, 2016, **30**(1), 89–94.
23. C. M. Marsh, R. F. Mothershead 2nd and M. L. Miller, Post-Blast Analysis of Hexamethylene Triperoxide Diamine using Liquid Chromatography-Atmospheric Pressure Chemical Ionization-Mass Spectrometry, *Sci. Justice*, 2015, **55**(5), 299–306.
24. Y.-H. Tsai, C.-W. Tsai and C. A. Tipple, A validated method for the analysis of sugars and sugar alcohols related to explosives via liquid chromatography mass spectrometry (LC–MS) with post-column addition, *Forensic Chem.*, 2022, **28**, 100404.
25. K. Bezemer, L. McLennan, L. van Duin, C-J Kuijpers, M. Koeberg and J. van den Elshout, *et al.*, Chemical attribution of the home-made explosive ETN – Part I: Liquid chromatography-mass spectrometry analysis of partially nitrated erythritol impurities, *Forensic Sci. Int.*, 2020, **307**, 110102.
26. K. Bezemer, L. McLennan, R. Hessels, J. Schoorl, J. van den Elshout and A. van der Heijden, *et al.*, Chemical attribution of the homemade explosive ETN - Part II: Isotope ratio mass spectrometry analysis of ETN and its precursors, *Forensic Sci. Int.*, 2020, **313**, 110344.
27. E. Lennert and C. M. Bridge, Rapid screening for smokeless powders using DART-HRMS and thermal desorption DART-HRMS, *Forensic Chem.*, 2019, **13**, 100148.
28. J. D. Howa, M. J. Lott, L. A. Chesson and J. R. Ehleringer, Carbon and nitrogen isotope ratios of factory-produced RDX and HMX, *Forensic Sci. Int.*, 2014, **240**, 80–87.
29. J. D. Howa, M. J. Lott and J. R. Ehleringer, Observations and sources of carbon and nitrogen isotope ratio variation of pentaerythritol tetranitrate (PETN), *Forensic Sci. Int.*, 2014, **244**, 152–157.

30. A. Chajistamatiou, Y. Angelis, P. Kiousi, M. Tsivou and E. Bakeas, Discrimination of tetryl samples by gas chromatography – Isotope ratio mass spectrometry, *Forensic Chem.*, 2019, **12**, 42–45.
31. H. Brust, M. Koeberg, A. van der Heijden, W. Wiarda, I. Mugler and M. Schrader, *et al.*, Isotopic and elemental profiling of ammonium nitrate in forensic explosives investigations, *Forensic Sci. Int.*, 2015, **248**, 101–112.
32. N. Gentile, R. T. Siegwolf, P. Esseiva, S. Doyle, K. Zollinger and O. Delemont, Isotope ratio mass spectrometry as a tool for source inference in forensic science: A critical review, *Forensic Sci. Int.*, 2015, **251**, 139–158.
33. S. J. Sollid, R. Rimstad, M. Rehn, A. R. Nakstad, A.-E. Tomlinson and T. Strand, *et al.*, Oslo government district bombing and Utoya island shooting July 22, 2011: The immediate prehospital emergency medical service response, *Scand. J. Trauma, Resusc. Emerg. Med.*, 2012, **20**, 3.
34. J. T. Thurman, Identification and Recognition of Commercial, Improvised, and Military Explosives, in *Practical Bomb Scene Investigation*, CRC Press, Boca Raton, 3rd edn, 2021, p. 84.
35. M. Dzikansky, G. Kleiman and R. Slater, Evaluating Suicide Bombing, in *Terrorist Suicide Bombings: Attack Interdiction, Mitigation, and Response*, CRC Press, 2012, p. 139.
36. I. D. Rae and J. H. Whitehead, Rackarock: On the Path from Black Powder to ANFO, in *Gunpowder, Explosives and the State: A Technical History*, ed. B. J. Buchanan, Routledge, 1st edn, 2006, pp. 367–382.
37. R. Cross, Explosives used in New York bombing, *C&EN Global Enterp.*, 2016, 7.
38. C. Mocella and J. A. Conkling, *Chemistry of Pyrotechnics: Basic Principles and Theory*, CRC Press, Boca Raton, 2nd edn, 2010.
39. K. Szomborg, F. Jongekrijg, E. Gilchrist, T. Webb, D. Wood and L. Barron, Residues from low-order energetic materials: the comparative performance of a range of sampling approaches prior to analysis by ion chromatography, *Forensic Sci. Int.*, 2013, **233**(1–3), 55–62.
40. F. G. M. Mauricio, V. R. M. Abritta, R. de Lacerda Aquino, J. C. L. Ambrosio, L. P. L. Logrado and I. T. Weber, Evaluation of interferers in sampling materials used in explosive residue analysis by ion chromatography, *Forensic Sci. Int.*, 2020, **307**, 109908.
41. L. Barron and E. Gilchrist, Ion chromatography-mass spectrometry: A review of recent technologies and applications in forensic and environmental explosives analysis, *Anal. Chim. Acta*, 2014, **806**, 27–54.
42. K. Tomasello, J. Adams and M. Sharp, *Origin of Test Requirements and Passing Criteria for the Qualification of Primary and Booster High Explosives*, presented in part at PARARI 2011, Brisbane, Australia, November 2011.
43. M. S. Russell, *The Chemistry of Fireworks*, Royal Society of Chemistry, Thomas Graham House, Science Park, Milton Road, Cambridge, UK, 2000.

44. J. Akhavan, *The Chemistry of Explosives*, Royal Society of Chemistry, Thomas Graham House, Science Park, Milton Road, Cambridge, UK, 3rd edn, 2011.
45. D. Ommen, J. Baldaino, C. Saunders, J. Hietpas and J. Buscaglia, Characterization and differentiation of aluminum powders used in improvised explosive devices. Part 2: Micromorphometric method refinement and preliminary statistical analysis, *J. Forensic Sci.*, 2022, **67**(2), 505–515.
46. K. Pun and A. Gallusser, Macroscopic observation of the morphological characteristics of the ammunition gunpowder, *Forensic Sci. Int.*, 2008, **175**(2–3), 179–185.
47. ASTM, Standard Test Method for Forensic Comparison of Glass Using Micro X-Ray Fluorescence (u-XRF) Spectrometry, ASTM International, 2017.
48. S. C. Wilschefski and M. R. Baxter, Inductively Coupled Plasma Mass Spectrometry: Introduction to Analytical Aspects, *Clin. Biochem.*, 2019, **40**(iii), 115–133.
49. R. S. Houk, V. A. Fassel, G. D. Flesch, H. J. Svec, A. L. Gray and C. E. Taylor, Inductively Coupled Argon Plasma as an Ion Source for Mass Spectrometric Determination of Trace Elements, *Anal. Chem.*, 1980, **52**(14), 2283–2289.
50. R. Thomas, *Practical Guide to ICP–MS: A Tutorial for Beginners*, CRC Press, 6000 Broken Sound Parkway NW, Suite 300, Boca Raton, FL, 3rd edn, 2013.
51. D. R. Lide, *Handbook of Chemistry and Physics*, CRC Press LLC, 84th edn, 2004, pp. 2003–2004.
52. C. Martinez-Lopez, T. Trejos, A. Mehltretter and J. Almirall, Elemental analysis and characterization of electrical tape backings by LA–ICP–MS, *Forensic Chem.*, 2017, **4**, 96107.
53. T. Hoffman, R. Corzo, P. Weis, E. Pollock, A. van Es and W. Wiarda, *et al.*, An inter-laboratory evaluation of LA–ICP–MS analysis of glass and the use of a database for the interpretation of glass evidence, *Forensic Chem.*, 2018, **11**, 65–76.
54. K. Bezemer, R. Woortmeijer, M. Koeberg, W. Wiarda, P. Schoenmakers and A. van Asten, Multicomponent characterization and differentiation of flash bangers - Part II: Elemental profiling of plastic caps, *Forensic Sci. Int.*, 2018, **290**, 336–348.
55. C. Mahoney, R. Kelly, L. Alexander, M. Newburn, S. Bader and R. Ewing, *et al.*, Bayesian Integration and Classification of Composition C-4 Plastic Explosives Based on Time-of-Flight-Secondary Ion Mass Spectrometry and Laser Ablation-Inductively Coupled Plasma Mass Spectrometry, *Anal. Chem.*, 2016, **88**(7), 3598–3607.
56. H. Brust, M. Koeberg, A. van der Heijden, W. Wiarda, I. Mugler and M. Schrader, *et al.*, Isotopic and elemental profiling of ammonium nitrate in forensic explosives investigations, *Forensic Sci. Int.*, 2015, **248**, 101–112.

57. J. A. D'Uva, D. DeTata, C. D. May and S. W. Lewis, Investigations into the source attribution of party sparklers using trace elemental analysis and chemometrics, *Anal. Methods*, 2020, **12**(41), 4939–4948.
58. C. G. Fraga, A. V. Mitroshkov, N. Mirjankar, B. P. Dockendorff and A. M. Melville, Elemental source attribution signatures for calcium ammonium nitrate (CAN) fertilizers used in homemade explosives, *Talanta*, 2017, **174**, 131–138.
59. J. D'Uva, D. DeTata, R. Fillingham, R. Dunsmore and S. Lewis, Synthesis and characterisation of homemade urea nitrate explosive from commercial sources of urea, *Forensic Chem.*, 2021, **26**, 100369.
60. D. DeTata, P. Collins and A. McKinley, A fast liquid chromatography quadrupole time-of-flight mass spectrometry (LC–QtoF–MS) method for the identification of organic explosives and propellants, *Forensic Sci. Int.*, 2013, **233**(1–3), 63–74.
61. J. Zhang, W. Wu, G. Qian and X.-G. Zhou, Continuous synthesis of methyl ethyl ketone peroxide in a microreaction system with concentrated hydrogen peroxide, *J. Hazard. Mater.*, 2010, **181**(1–3), 1024–1030.
62. N. A. Milas and A. Glubovic, Studies in Organic Peroxides. XXV. Preparation, Separation and Identification of Peroxides Derived from Methyl Ethyl Ketone and Hydrogen Peroxide, *J. Am. Chem. Soc.*, 1959, **81**(21), 5824–5826.
63. J. Yinon, *Forensic and Environmental Detection of Explosives*, John Wiley & Sons Ltd, West Sussex PO19 IUD, England, 1999.
64. R. Meyer, J. Kohler and A. Homburg, *Explosives*, Wiley-VCH, Verlag GmbH, Weinheim, 5th edn, 2002.
65. J. Howa, M. Lott and J. Ehleringer, Observations and sources of carbon and nitrogen isotope ratio variation of pentaerythritol tetranitrate (PETN), *Forensic Sci. Int.*, 2014, **244**, 152–157.
66. G. Pierrini, S. Doyle, C. Champod, F. Taroni, D. Wakelin and C. Lock, Evaluation of preliminary isotopic analysis (^{13}C and ^{15}N) of explosives A likelihood ratio approach to assess the links between semtex samples, *Forensic Sci. Int.*, 2007, **167**(1), 43–48.
67. J. D. Howa, M. J. Lott, L. A. Chesson and J. R. Ehleringer, Carbon and nitrogen isotope ratios of factory-produced RDX and HMX, *Forensic Sci. Int.*, 2014, **240**, 80–87.
68. Z. Abrego, N. Grijalba, N. Unceta, M. Maguregui, A. Sanchez and A. Fernandez-Isla, *et al.*, A novel method for the identification of inorganic and organic gunshot residue particles of lead-free ammunitions from the hands of shooters using scanning laser ablation-ICPMS and Raman micro-spectroscopy, *Analyst*, 2014, **139**(23), 6232–6241.
69. N. Abdul-Karim, C. Blackman, P. P. Gill, R. Morgan, L. Matjacic and R. Webb, *et al.*, Morphological Variations of Explosive Residue Particles and Implications for Understanding Detonation Mechanisms, *Anal. Chem.*, 2016, **88**(7), 3899–3908.

70. European Drug Report 2022: Trends and Developments, Publications Office of the European Union, Luxembourg: European Monitoring Centre for Drugs and Drug Addiction, 2022.
71. S. C. Gamble, Forensic detection of explosives in wastewater system: Implications for intelligence gathering, PhD thesis, University College London, 2016.

CHAPTER 2

The Application of Isotope Ratio Mass Spectrometry to Forensic Casework

K. JONES* AND J. MEIKLE

Forensics, Australian Federal Police, GPO Box 401, Canberra, ACT 2601, Australia
*Email: Kylie.Jones@afp.gov.au

Abstract

The purpose of this chapter is to discuss the use of isotope ratio mass spectrometry (IRMS) and isotopic abundances in their application to forensic science and forensic intelligence. Key considerations for developing models and databases for this purpose are discussed, and key thematic examples are used to show the diversity, depth and impact that these types of highly discriminating measurements can have. The chapter ends with a number of case examples, which further support the impact of the use of IRMS in the realm of forensic intelligence.

2.1 Introduction and Background

There has been a significant shift across law enforcement to utilise forensically derived information in support of a wider range of activities. Although focus needs to remain on supporting the triers of fact and the judiciary, other activities such as intelligence gathering to support investigative lead generation and disruption activities to prevent criminal activities from

New Developments in Mass Spectrometry No. 14
Applications of Mass Spectrometry for the Provision of Forensic Intelligence: State-of-the-art and Perspectives
Edited by Simona Francese and Stephen Bleay
© The Royal Society of Chemistry 2024
Published by the Royal Society of Chemistry, www.rsc.org

occurring, also benefit from the expertise that forensic science can provide. At its core, forensic science is fundamentally concerned with answering questions and reconstructing events, and as such, examiners are now more commonly being asked to trace the origin of materials, the methods employed in their construction or the ways in which they have been used or misused depending on their application. Isotope ratio mass spectrometry (IRMS) has the potential to contribute significantly to answering these questions, advancing the understanding of the source or method of production, or to demonstrate authenticity. Alongside other forms of intelligence or forensic information, isotopic comparisons can be used to deduce trends or convergences across instances of criminal activity, or to reveal relevant contextual investigative information.

There are a large number of reference materials available that explain the technique and underlying theory of IRMS in detail.[1–4] The purpose of this chapter is to present the basic knowledge required to understand the use of stable isotopes in forensic science, and to demonstrate the effectiveness of using this highly discriminating technique in intelligence and casework scenarios.

The most effective applications of IRMS seek to add value where other techniques may have reached their limit, whether through the identification of an investigative gap, shift in crime methodology or the identification of a new organisational problem. When this has been demonstrated, IRMS can provide discrimination or comparability between materials where chemical identification has its limits. Alongside other information, a more holistic picture can then be overlaid or added to other analytical results to strengthen the evidence given back to the investigative team.

IRMS started gaining attention in forensic science in the late 1990s. The technique itself had been used for many years prior, but it became more accessible with the introduction of continuous flow methods of sample combustion that allowed higher throughput and automation of sample measurement. These instruments established relative rather than absolute measurements, and methods have since continued to evolve, with instrumentation becoming simpler to operate, cheaper to maintain and therefore easier to produce high quality results. Groups such as the Forensic IRMS (FIRMS) Network have focussed on ensuring that forensic scientists have access to good practice guidelines that assist in producing high quality measurements and robust interpretation of IRMS results.[5,6]

Within forensic science, the use of IRMS has evolved over time, and focus has shifted across a range of domains and evidence classes. Large-scale efforts have focussed on the use of IRMS for the comparison of explosives, sourcing of drugs (including precursors), food provenance and tracing human movement or origins. Impact has also been observed in the measurement and comparison of physical materials such as office papers, polymers, cable ties and wood, of which papers and polymers will be discussed here as examples to demonstrate their utility.

The primary function of IRMS is to measure and quantify the abundance of the stable isotopes within samples. Stable isotopes are different forms of an element that vary in the number of neutrons present in the nucleus, but do not undergo radioactive decay. This leads to mass number differences within atoms of the same element – for example, carbon has six protons and six electrons, but can vary in the number of neutrons present between six (^{12}C) or seven (^{13}C). When carbon has eight (^{14}C) neutrons, a radioactive isotope is formed. IRMS measures and then calculates a ratio between the number of light and heavy stable isotopes within a sample. For carbon, this means the ratio of $^{12}C/^{13}C$, which is expressed in permil (‰).

As the proportion of the heavy isotope within a sample is generally very low and it elutes later than the lighter isotope,[7] multiple detectors with fixed molecular weight targets are used to measure the compound of interest. Light stable isotopes measured include carbon, oxygen, hydrogen, nitrogen and sulphur, while heavier stable isotopic elements including lead, strontium and zinc are measured using inductively coupled plasma mass spectrometry methods. An example of carbon and hydrogen spectra obtained from a typical bulk measurement of a material using an IRMS instrument is shown in Figure 2.1.

Stable isotopes can be described as both fixed and transient – while the total proportion of heavy isotopic species is stable, their proportion within a material changes through uptake, chemical reaction or mixing, *e.g.*, in cellulosic materials, the hydrogen and oxygen abundances are formed based on water uptake and mixing with water that is already present within the plant, prior to being chemically fixed as cellulose molecules are formed. Processes such as evaporation also have significant effects in warmer climates, leading to the fractionation of water sources, and because heavier isotopes react at a slower rate, this creates isotopic differences between reactant and product pools. In general, the net effect of these processes is predictability in the isotope ratios observed at a single location, and variability between locations for naturally occurring materials. Although there is nuance in this statement (*e.g.*, due to seasonal shifts in precipitation), what is most important to understand is that stable isotopic abundances vary, and it is this variability which can be used to compare one or more samples to determine whether the samples have originated from the same source or not. More information on whole earth isotopic cycling can be found in Dawson and Siegwolf.[8]

For industrially produced materials, the isotopic variability of a particular material is further complicated by the processing used to transform one or more precursor materials into a final product. Using paper as an example, the final sheet of paper that is measured in forensic casework is the product of at least three source materials ($2 \times$ cellulose sources $+ 1\times$ bulk additive such as calcium carbonate) and is subjected to a process that involves harsh chemicals, whitening agents, significant volumes of water, heat, and pressure. Previous work aimed at understanding the impact of this process,

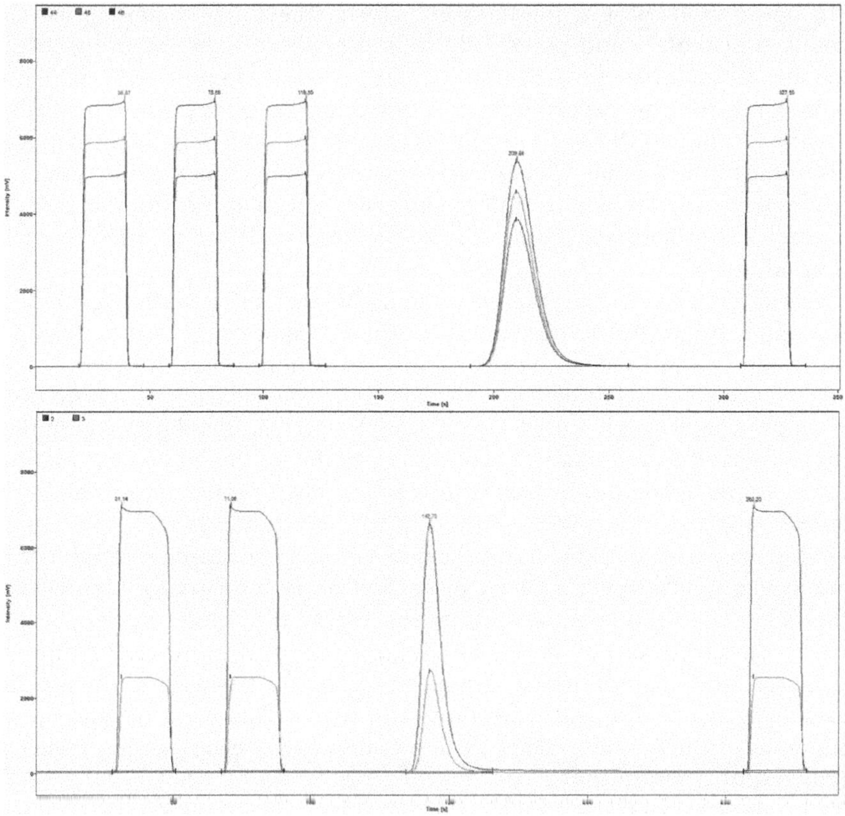

Figure 2.1 Typical spectra measuring carbon (top) and hydrogen (bottom) using isotope ratio mass spectrometry. Sample peaks are at the centre (rounded top), with peaks from reference gas on either side, which are used for peak scaling. A ratio for the sample is determined by calculating the peak area difference between the two molecular weight peaks.

for example, demonstrated that the main isotopic difference was induced through mixing, but isotopic differences were observed throughout the process in response to the different stages of production.[9,10]

Given the production-based effects and mixing in industrially produced materials, it cannot be assumed that end products purchased, used and recovered in the forensic context retain the isotopic abundance values imparted by the location from where the source materials were derived. What is measured instead is a mixed signal that contains contributions from the production method and the facility from which the product was made. For products that are produced at scale, this could also include differences created through different grades or brands of the same product where the proportion or type of precursor materials change, or batch differences, wherein the source materials vary over time.

In addition to production-induced effects, precursor materials can be produced at scale – and potentially shared between manufacturing plants globally. An example of this is the woodchip industry in Australia – which is responsible for the exportation of woodchips produced from the same growing location as the wood produced for paper production and distributed within Australia. The net effect of this was that papers produced in Australia were isotopically similar to office papers produced in facilities in China, although most samples could still be discriminated due to differences in other additives.[11]

Isotopic databases are used to understand variability within a material class and serve as the foundation for making forensic comparisons. They are constructed to test the assumptions and constraints that may be present within a set of materials and that would impact the veracity of an opinion if not appropriately understood – for example, are the similarities or differences observed between one or more samples due to the inherent variability of the sample or are they because the samples have originated from different sources?

Using databases to understand the natural range or expected variability of isotope ratios within an industry or sub-set of an industry for a particular material is the fundamental information required to guide interpretation. Efforts must be directed towards ensuring that databases are well designed, measured and maintained to ensure appropriate representation for the market for which the case questions will arise. This can and should be different for different situations, and the following questions can be used as a guide during planning:

- Where is the casework likely to originate from?
- Are there sub-segments within the industry from which the material originates, or can it be expected to be internationally consistent?
- How many brands and varieties of material are available within the industry?
- What is the most common material that could be used as a base?

Databases can be large, general and pre-prepared, or smaller and built around specific case circumstances. Large, general sample databases are best supplemented at the time of case comparison with small contemporaneous collections that are matched in terms of matrix and time to the samples of interest. Testing can also be conducted in advance, based on the expected casework sample type, to assess the difference between a general database and a specific subset of that material – an example of this would be a paper database populated with plain copy paper compared to the expectation of receiving lined notepaper within casework. When similarities are observed, or when the casework question pertains to the source, in some cases, the database could also be used to indicate where a material has (or has not) been produced. Likewise, when two samples cannot be discriminated, using a representative database of materials may enable the formulation of probability statements indicating whether the material in the case is relatively common or rare.

The construction of a database of samples is never complete. Change is a constant factor in manufactured materials in particular and these changes can result in changes in the measured isotopic composition. For example, changes have been observed over the last decade in the origin of papers available for purchase in Australia due to changes in trade and importation legislation.[12] The recommended approach for database construction is summarised in Figure 2.2.

The nature of forensic casework samples once again adds further complexity and variability which must be accounted for during examination and interpretation. For a material to be detected as part of an incident or investigation, it cannot be in the same 'fresh from the packet' condition in which it was sold. The effects of normal usage are numerous and can include contact with other materials, exposure to weather or water, physical handling or further processing such as printing for papers, or heat sealing for polymers. Methods used for forensic examination can also influence a material, *e.g.*, fingermark enhancement using wet chemicals, or tape lifting methods for the collection of trace DNA. Understanding the impacts of normal and forensic handling is a critical requirement during interpretation as, in many cases, the questioned materials are being compared to database or actual comparison materials that are clean or unused. This may lead to the creation of false negatives during interpretation.

The unique range of factors at play in the isotopic comparison of manufactured materials is shown in Figure 2.3, where each point along the line is a point where fractionation or mixing may occur, potentially altering the isotopic abundance value of the final measured material.

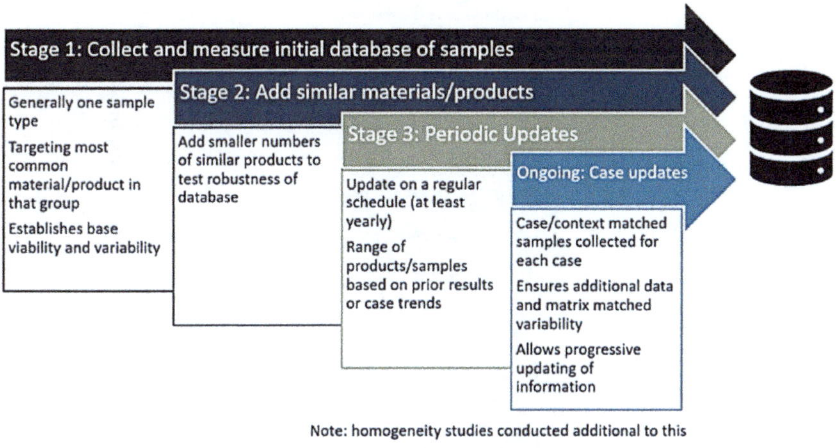

Figure 2.2 Isotopic database construction and maintenance approach recommended for the materials to be used in forensic casework.

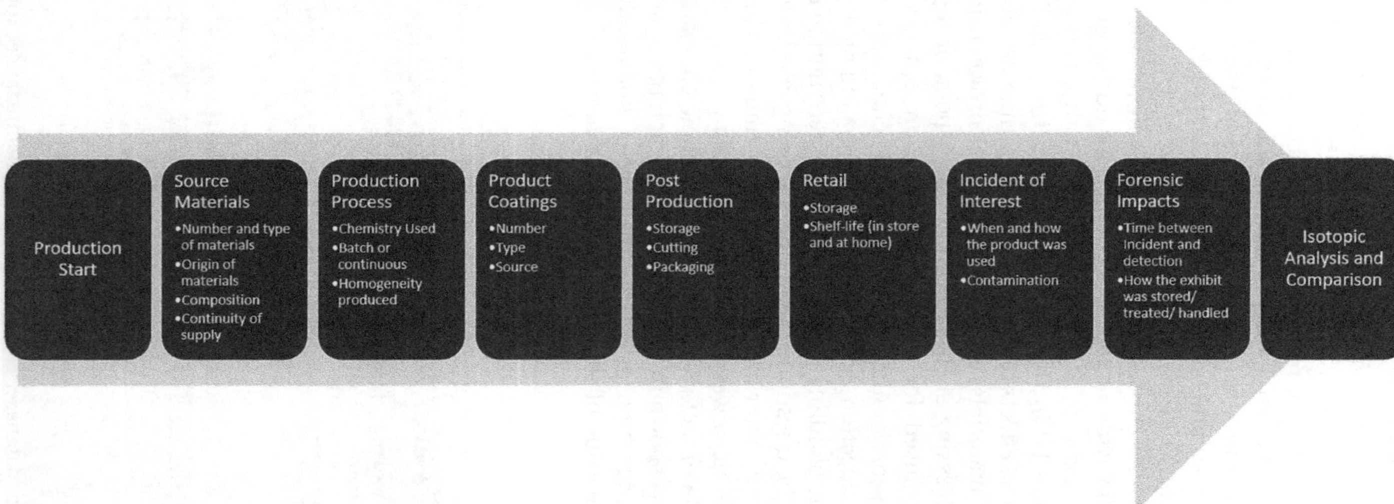

Figure 2.3 Possible sources of fractionation or mixing experienced across the lifecycle of an industrially produced material, before it arrives for forensic examination.

As for any forensic technique, the parameters used for quality control and assurance of the values produced are fundamental to avoid additional variability. A significant body of work already exists to define the best practice in the laboratory methodology used in measuring and correcting results for IRMS.[5,13] Validation of methods is crucial to ensure that the technique and the results it produces are robust and can be used within a court setting. Databases are also generally collated and measured over time, and samples will be measured across analytical sequences. The veracity of the database then relies on having a robust and repeatable method that is traceable to international standards to ensure that variability between samples is due to the material itself rather than the effects of poor instrument control. For example, soft polymer materials are homogenous and hence can be compared using their measurement uncertainty as an indicator of whether they have been produced from the same source.[14] If this measurement uncertainty was unnecessarily increased due to poorly controlled instrument operating conditions, this could result in larger measurement uncertainty values and hence lead to an increased risk of false associations between samples that could otherwise be discriminated.

IRMS can be used in a range of contexts to elucidate aspects of criminal activity that may not be immediately clear (*e.g.*, drug synthesis pathways), or to link separate offences together through the examination of methodology and/or products (*e.g.*, polymer and paper comparisons). Although a proportion of crime is moving to the digital space, there will always be a place for the forensic examination of physical items – and ensuring that highly discriminating techniques such as IRMS have the methodology, databases and interpretation frameworks in place to meet the needs of law enforcement is essential. Forensic intelligence inherently involves a multi-disciplinary and layered approach for providing value to investigations, and isotopic abundance measurements must be placed within the context of the case examination – be that alongside results from other techniques or within the context of the evidence item itself.

2.2 Examples of Use of IRMS in Forensic Science

The following sections demonstrate the wide range and effectiveness of the use of IRMS within forensic science including intelligence applications. The examples selected are diverse – from demonstrating the provenance and authenticity of food products, to providing information on the production methodology used to manufacture illicit drugs, to the chemical comparison of materials used in incidences of offending such as document fraud.

2.2.1 Food and Wine

The focus on food testing has traditionally been to ensure that the products are safe for consumption. While counterfeiting of commodities is not new, as a consequence of globalisation, authentication of food and beverages has become increasingly important in protecting the brand and trade value across borders. Counterfeit and fraudulent products aim to redirect the value of high priced or exclusive items into illicit markets[15] and while the initial aim may be to redirect money through sales, the damage can be far higher when the brand value drops as a consequence.[16]

Internationally, wine is a high value product and premium wines command higher prices which are often based on the location of growth and production. Wine isotopic composition databases are often used when the geographical origin of grapes used for the production of wines is in question, or to routinely track and measure authentic wines to allow subsequent detection of suspected fraudulent wines in future investigations.

A number of researchers have used a combination of $\delta^{13}C$ of ethanol in wine and $\delta^{2}H$ and $\delta^{18}O$ in the 'wine water' to narrow down the geolocation of growth. The differences observed in relation to the $\delta^{2}H$ and $\delta^{18}O$ values of wine are derived from five differences – latitude effects, continental effects, altitude effects, temperature effects, and rainfall amounts.[17] Success has been achieved in determining the locations using databases from South Brazil,[18] Argentina,[19] China,[20] and European countries with focus on Italy and France.[17,19–24] In one study, an Italian researcher analysed 41 samples from three different grapes grown in two different regions in Italy, with the results showing that the vintage, grape variety, and region all had an influence on the isotopic compositions.[21] This has also been shown in South Brazil where two grape varieties from two geographical locations were differentiated using an isotope ratio database of 58 samples.[23] In one of the largest studies using isotopic and elemental analyses,[20] 600 verified wine samples imported into China from France, Spain, Italy, Australia, USA, South Africa, and Chile as well as samples purchased within China were analysed. Isotopically, French wines could be discriminated from others with an accuracy of 85%; however the accuracy of discrimination for wines produced in other countries was lower, between 74.4% and 18.2%, with an overall average of 53.4% correctly identified.

Economically motivated adulteration occurs when an intentional substitution, alteration or addition is made to a food or beverage product to reduce costs.[25] An example of this may be the spiking of honey or maple syrup with corn syrup to lower the cost of production. For plant material, the isotopic abundance values are influenced by the CO_2 present in the atmosphere and the photosynthetic mechanism used to assimilate the carbon into plant tissue. The basic principles of isotopic fractionation revolve around the reaction rates of the isotopes of the element of interest, *i.e.*, the reaction rate of the light isotope C^{12} is faster and it is therefore used in preference to the heavier isotope C^{13}. The most common type of photosynthesis (C3 photosynthesis or the Calvin cycle) uses a three-carbon enzyme to convert CO_2 *via* intermediates, to sucrose. This process is used by temperate climate plants that have continuous and unrestricted access to the atmosphere, meaning that the sucrose production process favours the freely available C^{12}.

This is in contrast to higher temperature C4 plants, which use a four-carbon enzyme and typically have an atmospherically closed system for carbon metabolism. In these systems, C^{12} is expended and the plant must use what is available (the C^{13}), resulting in a higher proportion of C^{13} being assimilated into sucrose. C3 plants include cereals, barley, oats, rice, soybeans and wheat, and have a $\delta^{13}C$ isotopic composition between −35 and −18‰,[26] while C4 plants include corn, maize, sugarcane and millet have a $\delta^{13}C$ isotopic composition between −15 and −8‰.[26] For food authentication, this means that C3 products (*e.g.*, honey) that are spiked with sugarcane or high fructose corn syrup, for example, will have carbon isotopic values that have shifted outside the expected range.[27–31] In a similar way, the relationship between $\delta^{2}H$ and $\delta^{18}O$ can be used to identify if local water is added to the wine water.[21,28]

The use of isotope ratios in food forensic authenticity has been extremely valuable and remains a prominent monitoring tool. What assists in promoting, growing and underpinning the models in wine authenticity, particularly, is government regulation, investment and collective standards that describe the technique, measurement factors and frequency of measurement required for growers to submit sample data to demonstrate authenticity.[32,33] For example, the European Commission requires wine producers of countries in Europe to measure isotopic compositions of wine produced directly from grown grapes and submit the results to the EU-wine database.[17,24] Since 1992, over 1400 samples from 18 European countries have been added to the database annually.[17] The European Commission also holds a smaller database of wines from non-European countries. Approaching the databasing problem holistically allows the industry to address and mitigate authenticity concerns, while also allowing the monitoring of both the tracking and patterns of criminal activities.

2.2.2 Illicit Drugs

Internationally, policing agencies invest in illicit drug monitoring programs to link and monitor drug seizures. Programs like the Australian Illicit Drug Intelligence Program,[34] the European Monitoring Centre for Drugs and Drug Addiction EU Drug Markets research publications[35] and the long-standing operations run by the US Drug Enforcement Agency such as the Heroin Signature Program[36] aim to piece together the local and international illicit drug market. As part of these programs, it is essential for law enforcement organisations to link seizures with their location of origin or production, providing the opportunity to understand drug markets including international movement of illicit material to enable the disruption of future importations. Chemical profiling is one aspect of these programs, and when it includes the use of stable isotope measurements, it is possible to elucidate the location of growth from plant-based drugs such as cocaine and heroin. For chemically manufactured substances such as methamphetamine, isotopic drug profiling can provide information on the synthesis pathway used to produce the drug.

Derived from the Coca plant, seized cocaine samples produced in 19 known growing regions within South America, specifically Colombia, Bolivia, and Peru have been measured.[37–39] In a comprehensive study involving 500 samples, δ^2H, $\delta^{18}O$, $\delta^{13}C$, and $\delta^{15}N$ results were used in combination with alkaloid profiling to produce isotopic landscapes (isoscapes) of cocaine.[39] The accuracy of predicting the geographic location of origin of samples using these methods was shown to be greater than 96%.[38]

Similar results were observed in heroin seizures where known growing regions include Southeast and Southwest Asia, South America, and Mexico,[40,41] with the $\delta^{13}C$ of heroin being shown to be reflective of the characteristics of both morphine and acetic anhydride used in the production. More recent studies have suggested that the poppy growing location has a greater influence on the carbon and nitrogen isotope ratio values than post-growing heroin processing.[41]

An example of the importance of a database of known samples for casework comparisons was presented by Casale *et al.*[40] In this case, the isotopic compositions from heroin samples seized from a merchant ship off the coast of Australia were compared to a database consisting of 221 heroin samples of known origin. The results showed that these samples could be discriminated isotopically and using trace alkaloid measurements from all other samples within the database. The opinion given was that this evidence demonstrated that the samples' growing conditions were different from those of the samples in the database. In this case, the results corroborated other evidence that suggested that the origin of heroin was isotopically different from the origin of the samples contained in the database, but more holistically, results of this kind inform investigators that there may be a different growing region or new syndicate that has not yet been identified.

IRMS analysis of chemically synthesised methamphetamine is effective in indicating the synthesis pathway. The carbon and nitrogen isotopic compositions can give indications of whether the pseudoephedrine precursors are sourced from ephedra plants indicating natural production, or if they are produced from propiophenone, therefore suggesting a synthetic origin, or fermented from benzaldehyde, which is considered semisynthetic.[42–45] The carbon results will also indicate when samples are made from the less common phenylpropan-2-one precursor.[46] The synthetic route between ephedrine/pseudoephedrine and methamphetamine has been shown to be free of fractionation and therefore isotopic abundance values do not indicate the synthetic route used in production, instead reflecting the isotopic abundance of the precursor material directly.[45,47,48] Similar intelligence programs have begun to profile other synthetic drugs including methylenedioxymethamphetamine (MDMA)[49,50] and fentanyl[51,52,53] as cases involving these drugs are on the rise internationally.

Routine screening and monitoring of the isotopic abundances of drug seizures, alongside other chemical and physical morphology traits, enable tracing and linking of seizures over time. The intelligence gained from these types of programs, usually undertaken at the national level, allows law enforcement agencies to understand the scale of importations originating from particular locations, and to direct resources at disrupting future importations either through overt and direct action, or through diplomatic liaison with other nations.

2.2.3 Humans

The pinnacle of forensic isotopic applications would be taking a non-invasive sample from a person (*e.g.*, hair and nails) and definitively provide intelligence information to link that person to the locations they have spent time in. Having utility in a range of applications including asylum claims, missing persons or unidentified remains cases where electronic data or documentation may be questionable or missing, isotopic studies in this context are based on a foundation of "you are what you eat," *i.e.*, that the isotopic abundances of the food and drink consumed by a person are metabolised and fixed into human tissues and are indicative of the geographical origin of those products. Although human based geo-location is the focus here, it should be noted that animal and archaeological studies have also added significant value to the underlying knowledge within this field.

On a global scale, researchers have measured the isotopic compositions within populations in an attempt to provide a correlation between locations and isotopic abundances measured from human tissue. In one of the first studies to be conducted that compared values from individuals residing in different countries, analysis of 60 people from nine countries demonstrated relatively small but distinct isotopic differences.[54] The same study showed intra-person stability of stable isotopes when an individual did not travel beyond a fixed location. This finding was mirrored by Kusaka *et al.*[55] who measured the isotope ratio values of 1305 Japanese people and suggested that the source of their diet was homogeneous due to the low overall variation of $\delta^{13}C$ and $\delta^{15}N$. When looking between locations, other research measuring hydrogen, carbon, nitrogen, oxygen and sulphur isotopes demonstrated clear differences for individuals that lived in Australia using $\delta^{34}S$ and $\delta^{2}H$, and Europeans by $\delta^{13}C$ when compared to Central and South American populations.[56]

Looking specifically at $\delta^{13}C$ values, human tissues have been shown to be correlated to atmospheric CO_2 fixed in organic molecules during plant photosynthesis. The isotopic values of the plants consumed are transferred to higher trophic levels in the food chain through direct ingestion or *via* animal products that have been fed with the same plant materials. Figure 2.4 shows the expected carbon and nitrogen isotopic values of a range of dietary inputs and how those values change through trophic level shifts. While the isotopic values of most modern humans who are omnivorous and reside in the same location have been shown to be largely heterogeneous, more broadly, some populations are distinct from others. The key example here is the US population – whose $\delta^{13}C$ values consistently differ from those of other regions internationally (*e.g.*, Europe) due to the high proportion of corn in the US diet.[57]

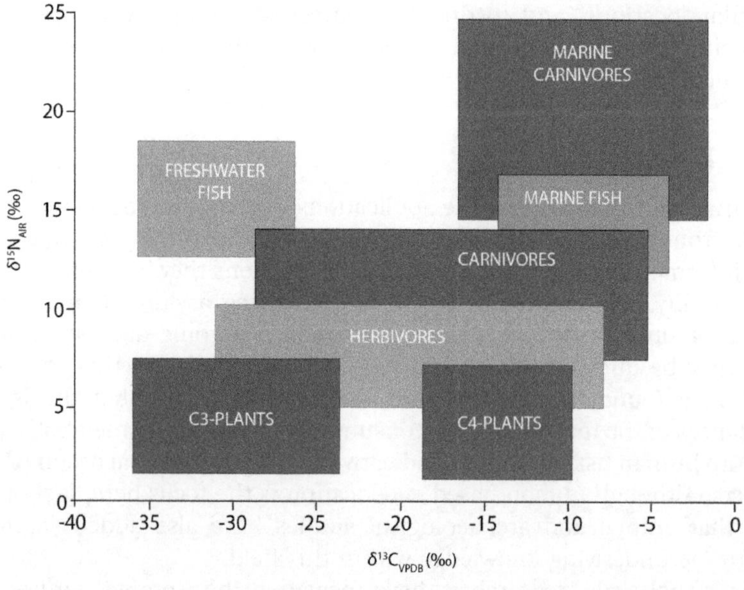

Figure 2.4 Plot of typical carbon and nitrogen isotopic abundances showing differences in the photosynthetic pathway (C3 *vs.* C4 plants) and trophic level shifts. Reproduced from ref. 58, https://doi.org/10.3390/nu14173594, under the terms of the CC BY 4.0 license https://creativecommons.org/licenses/by/4.0/.

While carbon isotopic abundances are indicative of the plant material type and origin, nitrogen isotope values are indicative more broadly of dietary preference. The critical factor here is the proportion of plant to animal protein – with studies demonstrating the clear difference in the isotopic values observed in hair samples of omnivores, vegetarians who eat dairy and eggs, and vegans.[58,59] Other researchers have reported that certain diets have effects on the isotopic composition of human hair. People who eat a lot of seafood, for example, have altered sulphur and nitrogen isotope ratio values.[60,61] These types of results can also be used to reverse-engineer the diet of archaeological samples – analysing the hair of ancient Egyptian remains, for example, has demonstrated a diet with a high intake of C3 plants, little seafood consumption and high portion of meats.[62]

In the context of forensic intelligence, examining carbon and nitrogen isotopic values can help paint a more comprehensive picture of a person's lifestyle and movement. Information such as whether a person is vegetarian or vegan can provide contextual information, for example, in cases involving unidentified human remains. Nitrogen isotopic abundances have also been used to indicate the periods of starvation in individuals with values changing due to catabolism of muscle proteins due to dietary deficiency.[63,64] Newer methods that specifically target the stable isotopes of amino acids in hair[65] have been developed to investigate correlations between isotopic abundance and behavioural traits such as alcohol consumption. When considering all factors relating to humans and human behaviour, however, complexity exists and the individual context needs to be considered to ensure that correct interpretation occurs – *e.g.*, studies have shown that lifestyle factors such as pregnancy and stress impact the $\delta^{15}N$ of hair[66,67] which could contraindicate other factors if not accounted for.

Large studies have been successful in showing correlation between tap water $\delta^{2}H$ and $\delta^{18}O$ isotopic values and location averaged isotopic values of hair.[68,69] Isoscapes, or isotopic landscapes, are models developed to express the differences in isotopic values across a geographical area. These models can then be used as tools to estimate the probability of a sample's measured isotopic values belonging to different regions of interest.[70] The assumption in these models, with the largest models to date focussing on North America, is that the isotopic values measured in tap water are regionally representative of the beverages (drinking water and others) and food water consumed by individuals at that particular location. While this assumption may hold true in the context of North America, where for the most part, food and beverages are consumed within the locality in which they are produced, this may not be the case for other populations.[71] It has been demonstrated is that case or context specificity is crucial, and that isoscapes or models (while useful) cannot be extrapolated outside of the context for which they were built.

Where models do exist, researchers have used $\delta^{13}C$, $\delta^{15}N$ and $\delta^{18}O$ data alongside heavy isotopes to assist in the identification of unknown remains. In one example, researchers created an isotopic profile for a male individual who was found drowned in Amsterdam, with results suggesting that the individual had originated from Poland, Slovakia or the Ukraine–Romania–Bulgaria region. Through investigation and eventual identification of the man, this consensus was supported as the individual was confirmed to be from South-West Poland.[72]

While pinpointing or providing an estimate of the probability of an individual being in a particular location requires modelling on a large scale, what can be particularly useful in a forensic intelligence context is the measurement of hair along its length to infer movement of an individual over time. A number of studies have demonstrated the effect of change of location on the stable isotopes of hair,[73–75] and in a number of examples, measurement of $\delta^{18}O$ of hair has assisted with the identification of unidentified human remains, with individuals shown to have moved between geographic locations in the recent periods before death.[76,77]

Human tissue types other than hair are also used to provide comparative location information. While hair and nail samples provide records of recent movements, long bones such as the femur provide an average isotopic record of the location/s of the last 20 years due to bone remodelling and turnover.[78] Teeth, which include the measurement of carbonate and phosphate samples, can be used to measure geographic history that dates back to childhood or adolescence.[79] Isotopic measurements of these types of samples are useful in cases involving unidentified human remains such as those previously cited, comingled remains[79,80] and in repatriation of remains as a result of conflict.[81,82]

2.2.4 Paper and Other Cellulosic Materials

Paper and packaging remain ubiquitous in modern society, even taking into account a shift towards the use of digital technologies. The use of paper examinations within forensic science, however, is under-utilised or not considered in comparison to ink analysis, mainly due to the lack of accessibility to a technique that can easily measure and compare papers chemically. Paper is by design, inhomogeneous and other techniques that attempt to measure the surface of the sheet are affected by those inconsistencies or are overwhelmed by cellulose signals, for example, when trying to measure filler components. IRMS solves this issue by taking a sample size large enough to account for the general bulk signal of the paper, and taking a useful measurement of the cellulose signal for comparison with other samples.

IRMS has demonstrated the ability to discriminate 89% of papers in a database of 125 samples of office papers collected from across Australia.[11] While this database continues to be updated annually, effects from changes in importation legislation[12] are contracting and lowering the discrimination factors due to loss of variability in the production locations of the paper. This, however, continues to change as further legislative or economic factors shape the industry in Australia. While discrimination factors are affected, these changes do introduce an interesting potential for this technique to provide the ability to date a particular paper – which is useful in a range of circumstances but most useful in cases where a document purports to be older than it actually is. For papers that have been produced within Australia where there is only one paper production mill, changes in formulation or source materials may additionally provide the opportunity to date papers.

In terms of comparisons, the inhomogeneity of paper sheets was observed to have an effect when two or more sheets were being directly compared for the purposes of determining whether they may have originated from the same source, and this relatively high intra-sample variability requires careful consideration during interpretation.[11]

Other applications that utilise IRMS for analysis of paper and cellulosic materials include the measurement and linking of packaging from pharmaceutical counterfeiting, alongside extensions of the work already undertaken in the environmental and production space to track and trace exported and imported wood-based materials in legal and illegal logging.

2.2.5 Polymers and Plastics

Plastics are often found at scenes of crime, where the information that comes from comparisons can be used for further investigations. Traditional forensic analysis includes physical features such as size, shape and striations as well as chemical analysis such as infrared methods (such as Fourier transform infrared spectroscopy, FTIR) to identify the polymer type. IRMS analysis adds a highly discriminating comparison to the analysis and is able to discriminate between samples that are otherwise indistinguishable using traditional examination methodologies.

A range of different types of polymers have been investigated and used within forensic casework with high levels of discrimination observed across all reported studies. These studies have included rigid polymers (typically nylon-6 and polypropylene) such as cable ties, which are often used in criminal activities to bind or restrain, with 100% discrimination achieved using $\delta^{15}N$ and $\delta^{13}C$ to compare 19 samples of different brands.[83] Likewise soft polymers (polyethylene), which are commonly used to produce consumer materials such as cling films and clip seal type bags, demonstrated very low intra-sample variability and high discriminability.[84–87] The most extensive study calculated a discrimination power of 95% using $\delta^{13}C$ and $\delta^{2}H$ within a database of 74 cling film samples.[85] Clip seal bags are still under investigation in the authors' laboratory but they have demonstrated promise for high discriminability in a smaller pilot study.[84] In particular, a further point of comparison to support an interpretation that demonstrates that samples have originated from the same production source would be to add a comparison of the polymer zip component (typically polyethylene) of clip seal type bags, which is added to the bag polymer during the construction stage, which is subsequent to base polymer manufacture.

Other polymeric sample types such as tapes, ropes, heat sealer and freezer bags are also commonly encountered in forensic casework. Adhesive tapes, which are usually constructed using polypropylene for packing tapes, or polyvinylchloride for electrical tapes have demonstrated similar isotopic variability and discriminability once adhesive layers are removed.[88,89] In other studies, researchers have first removed and then measured the isotopic values of the adhesive layer as a separate point of comparison;[90] however, in practice in forensic casework, the authors of this chapter have attempted this and found that due to use of the tape in the conduct of the offence the evidence is associated with, the adhesive is often highly contaminated with extraneous material and therefore cannot be easily compared with a clean comparison/known sample roll.

Additive manufacturing, commonly known as three-dimensional printing (3D printing), is an emerging problem for law enforcement internationally, with traditional barriers to manufacturing complex items such as cost and accessibility being removed. Additive manufacturing most often uses polymer substrates and home-printing of items such as firearms or device components is a tangible threat, particularly in countries where items may be more limited in availability. When devices are sold or trafficked, and hence are seized in places that are unrelated to the suspect production location, linking these devices to the printer that produced them will be of key concern in demonstrating the production source and in supporting prosecution activities.

Many of the printing filament materials used in additive manufacture are organic polymers such as poly-lactic acid (PLA), acrylonitrile butadiene styrene (ABS), and polyethylene terephthalate glycol (PETG). PLA is normally produced from corn or sugarcane starch and hence produces a signal consistent with other C4 plant derived samples, which results in $\delta^{13}C$ values that are more enriched than those of petrochemically derived polymers.[91] As PLA is one of the most commonly used polymers, research and assessment of the

PLA polymer will continue to assess the intra- and inter-sample variability and to explore any usage effects that may cause fractionation and prevent the direct comparison of printed devices with the unprinted/stock filament.

2.3 Using Isotopes Within Forensic Casework and Intelligence

As forensic intelligence processes aim to gather layering information to provide context and to support the identification of converging factors, measuring and monitoring isotopic abundances holds value within a particular context to provide the opportunity to establish connections over time. One such example may be the measurement of drug-related wrappings and packages that are detected either through seizure, or through programs that identify packages in the international mail stream. Overlaying similarities or differences in the isotopic abundances of polymers alongside morphological features and biometric linkages such as fingerprint or handwriting comparisons, would allow the use of a broader range of features to identify patterns and to track changes or discrete linkages over time.

Within individual forensic cases, the improvement in results achieved using traditional methods through the use of IRMS is primarily derived from the ability to give opinions at the source level. A secondary improvement is also derived from using the isotopic abundance values alongside the detailed knowledge of the associated industry or application of the materials, allowing an examiner to address the case specific set of propositions that require clarification or elucidation – which are often at the activity level. An example of this type of proposition may be a question of support for the proposition that a particular location was used for the packaging of an item, given the materials collected from the scene.

The following case studies, which are summaries of real cases examined and reported in the Australian Federal Police Forensics Command, focus on the use of the isotopic comparisons of papers and polymers to demonstrate how isotopic abundance results can be used within a forensic context to provide information to the investigator which was not otherwise available.

2.3.1 Forensic Case Studies

2.3.1.1 Case 1 – Taxation Fraud

An individual was suspected of operating a corporation for the purposes of evading taxation. A number of documents were submitted for examination, for the purposes of a signature comparison, but there were no specimen signatures available for comparison. The documents included:

- five single page documents with printed entries that indicated that they were minutes of an annual general meeting held for the corporation, dated across a period of 7 years, with two calendar years missing and

- a number of multi-page stapled documents with printed entries, indicating that they were contracts. The documents were copies of the same contract but there were some dissimilarities where it appeared that the content may have been altered between the different copies. The main contract of interest was the one that contained an original ink signature and it was apparent that this document had been stapled and re-stapled at least 3 times, due to the presence of additional staple holes in the top left corner.

A physical and chemical examination was undertaken which included IRMS of the paper of each of the pages.

For the single page documents, the propositions that were set were as follows:

H1: The pages were produced on the date printed on each document.
H2: The pages were produced on a different date to that printed on each document.

Using IRMS, the pages exhibited similar carbon and oxygen isotopic abundance values that could not be discriminated from each other. Within the context of the documents, which were being purported to have been produced over a 7-year period, this evidence supported H2, in that the documents were produced at the same time rather than over the seven years indicated by their dates. On comparing the isotopic abundances with the paper database at the time, the following observations were made:

- The abundance values aligned with those samples in the database that had been produced within Australia;
- A closer analysis of the Australian papers within the database (remembering that there is only one production facility in Australia) demonstrated their variability, which meant that even papers from the same production source isotopically varied over time;
- The mill within Australia did produce different grades of paper, which were expected and observed to contain differences in formulation which affected the isotopic values. However, even within the most common brand for this mill, variability was observed across years – so unless the exact same package of paper was kept and used over seven years, it was unlikely that being loyal to a particular brand would have produced the results observed; and
- Given the office/business context within which the documents were produced, it was again deemed unlikely that the same ream (a ream contains 500 sheets) was used over the seven-year period.

Additional evidence observed through document examination results showed that indentations (impressions of previous writing produced on top of the page being examined) were present on one of the documents that was produced 3 years prior to the other documents allegedly being printed.

In relation to the single pages, the finding presented to the court was that the evidence provided strong support for the proposition that the documents were produced on a date different from that printed on the documents, rather than over a period of 7 years. Additionally, comments were made that there was evidence that the documents were produced at the same time.

In examining the main contract of interest, the propositions that were set were as follows:

H1: The contract of interest has been altered.
H2: The contract of interest has not been altered.

The IRMS results of the pages in the contract showed that page 5 of the contract was isotopically different using both carbon and oxygen isotopes. Alongside differences in the alignment observed in the formatting of the printed entries on this page, the evidence provided strong support for the proposition that the contract had been altered.

2.3.1.2 Case 2 – Counterfeit Currency

A series of counterfeit \$50 notes were detected in circulation nationally across Australia, representing the first instance of such counterfeits mimicking the polymer substrate used by the Reserve Bank of Australia to produce genuine Australian currency. Both the scale and complexity of offending were of concern, particularly given the investment undertaken to move to a polymer-based currency. While key security features were absent in the counterfeit notes, the look and feel of the notes were sufficiently consistent with genuine currency for them to evade detection in general circulation.

In this case, investigators had identified a potential source for the notes, and conducted a search warrant, locating 2 × rolls and offcuts of a clear, unprinted polymer for comparison with the suspect notes. The polymer was identified as polypropylene using FTIR, and the samples were also measured for carbon and hydrogen isotope ratios. A contemporary database was constructed using commercially available polypropylene, mostly in the form of overhead projector sheets. An existing polyethylene database was used as a further reference for the comparison.

Isotope ratio comparisons were performed by sampling from the clear window of the \$50 notes to avoid any potential contamination from the constituents of the printed ink. The propositions that were set were as follows:

H1: The counterfeit \$50 notes were produced using the polymer samples seized.
H2: The counterfeit \$50 notes were produced using a different polymer.

Measurement and comparison of the $\delta^{13}C$ and $\delta^{2}H$ abundances of the 2 × rolls and the offcuts showed differences between the each of the rolls and the offcuts. The duplicate $50 notes, which all contained the same serial number and were physically consistent, were isotopically consistent. Comparing the rolls, offcuts and the $50 notes isotopically, one of the rolls could not be discriminated from the $50 notes, while the other roll and the offcuts were significantly different. This provided support for the proposition that this roll could have been used to produce the $50 notes.

On comparing the result to the database of materials collected and measured, it was observed that while polypropylene samples were collected and demonstrated variability, it was unclear how variable printable rolls of clear polypropylene were. Through market research, it was also clear that these types of materials are considered quite rare and, as such, there would likely only be a limited number of sources from which to purchase this material.

Overall, this meant that while similarities in isotopic abundance were observed between the $50 notes and the roll seized, little evidential strength could be attached to the finding. While a request was made to the investigative team for more time and resources to conduct further research on this type of material, this was not supported due to other evidence in the case.

2.3.1.3 Case 3 – Black Plastic Shrink Wrap

Drug samples are often wrapped in multiple soft polymers such as cling films or black plastic packaging including garbage bags. When two or more seizures are compared, the comparison can be performed on the drug samples themselves or the plastics in which they are wrapped. When wrappings are found within the residence of a person of interest (POI), these can be compared to the wrappings of the drugs.

In 2021, police seized eight packages containing a crystalline substance wrapped using black plastic shrink wrap. Further intelligence identified a location of interest, and a search warrant was executed at this location where, alongside other forensic evidence, two small rolls of black plastic shrink wrap were located, each with a single cut end. The plastics from both the packages and the rolls were similar in overall appearance but there were some differences in the width of the plastic when unrolled from the packages.

The propositions that were set were as follows:

- H1: The plastic from the packages originated from the rolls seized from the location of interest.
- H2: The plastic from the packages originated from a different roll.

The polymer was identified as polyethylene using FTIR. IRMS was used to measure the $\delta^{13}C$ and $\delta^{2}H$ values of each of the samples including the plastic from the packages, and from each of the two rolls. A contemporary database of black polyethylene was constructed and added to an existing database of 120 clear polyethylene samples. The black polyethylene samples were easily

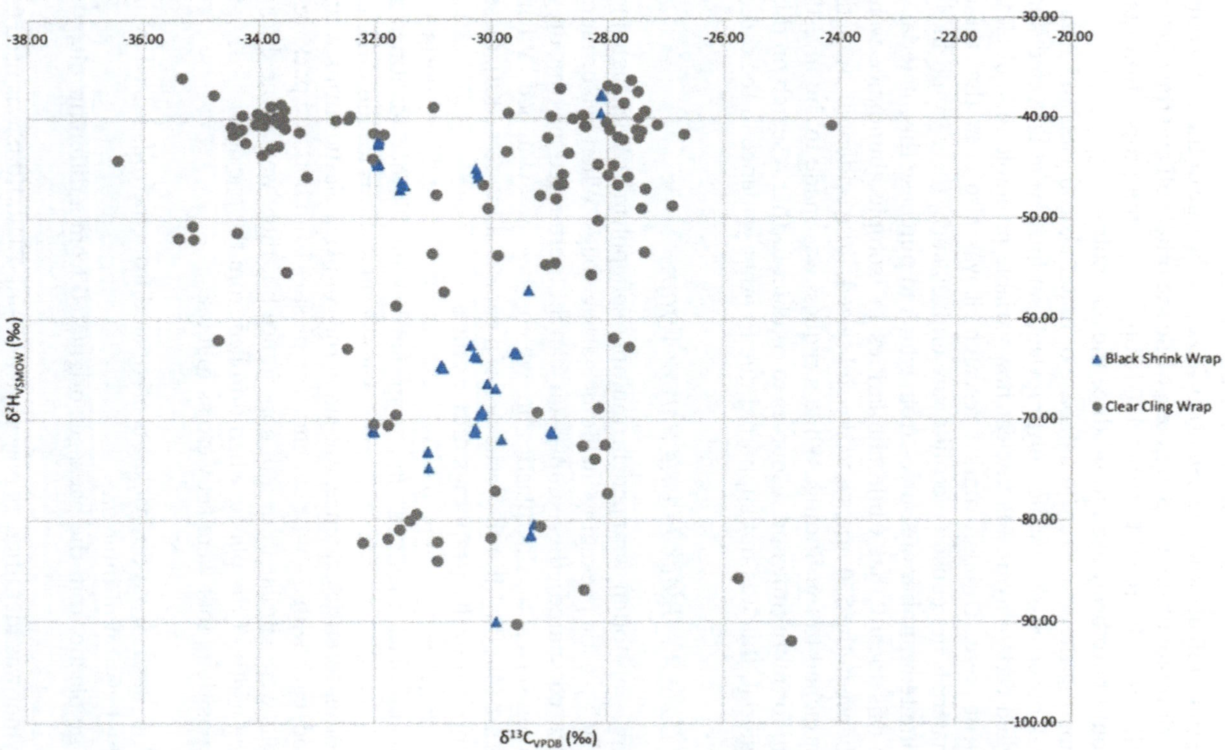

Figure 2.5 Plot of hydrogen and carbon isotopic abundances in polyethylene samples showing consensus between the general polyethylene database (clear cling wrap samples) and the contemporary, matrix and case matched samples (black shrink wrap).

and freely sourced (generally marketed as "pallet wrap") and demonstrated variability consistent with the broader clear polyethylene database. This is demonstrated in Figure 2.5.

The isotope ratio results showed that the polyethylene from the packages and from either of the 2 × rolls could not be discriminated, but physically neither of the rolls could have been the source of the black plastic on the packages – one of the rolls has fused edges, rendering it unsuitable for use, while the other was approximately two centimetres smaller in width compared to the plastic used on the packages. This meant that while there was isotopic evidence to support the proposition that the plastic from the packages originated from the rolls, there was also evidence for H2 – that the plastic from the packages originated from a different roll. Given the cut edges on the seized rolls, which did not physically fit together, this may have meant that a third cut roll had been used which was missing (possibly consumed in wrapping the packages) or that an additional roll from the same brand/source was used. Overall, this advice was given in the context of a finding of equal support for H1 and H2.

2.4 Conclusions

The use of isotope ratio mass spectrometry to measure the isotopic abundances of light stable isotopes has and will continue to positively contribute to forensic investigations across a range of evidence types, in both traditional and non-traditional settings. IRMS also holds the potential and ability to be further applied across a wide range of other applications where specific problems or questions need to be answered.

The power and utility of any application using IRMS, however, comes with a requirement to understand the industry or context in which the material exists and has been used, and to question, research and fully explore the underlying assumptions and/or limitations that may be present. This is achieved through the construction and upkeep of representative material specific databases, but this requires foresight, investment and capacity to ensure that databases remain up-to-date. Databases, therefore, must be matched to the operational need and application requirements of the law enforcement organisation using them and represent an investment in potential future casework.

While it is not possible to provide statistics on the operational use of IRMS (due to lack of intelligence based monitoring software and information not kept or in public domain), within every context to which IRMS is applied, the benefit of using this technique has been shown to have a high impact – from applications across illicit drug manufacturing, human movement and dietary patterns to the ability to link or disassociate physical evidence such as paper and polymers in a way that was previously not possible. Operationally, the ability to conduct comparative based casework is key, and while definitive information can be provided which indicates different sources for materials within the same chemical class (*e.g.*, difference between physically consistent polyethylene bags), IRMS is also able to add value when materials cannot be isotopically discriminated, suggesting a common origin. Added to this are the applications of IRMS in forensic intelligence, which will continue to grow in scale and scope in the future – both to monitor and identify convergences over time and across offences.

References

1. S. Benson and K. Jones, *Anal. Tech. Forensic Sci.*, 2021, 267–293.
2. B. Fry, *Stable isotope ecology*, Springer, 2006.
3. W. Meier-Augenstein, *Stable isotope forensics: an introduction to the forensic application of stable isotope analysis*, John Wiley & Sons, 2011.
4. Z. Sharp, *Principles of stable isotope Geochemistry*, 2nd edn, 2017.
5. P. J. Dunn and J. F. Carter, *Good Practice Guide for Isotope Ratio Mass Spectrometry*, 2nd edn, 2018.
6. S. P. G. Doyle and G. van der Peji, *FIRMS guidance for the forensic interpretation of isotope ratio data*, 1st edn, FIRMS, 2020.
7. R. E. Criss, *Principles of stable isotope distribution*, Oxford University Press on Demand, 1999.
8. T. E. Dawson and R. Siegwolf, *Stable isotopes as indicators of ecological change*, Academic Press, 2011.
9. K. Jones, S. Benson and C. Roux, *Forensic Sci. Int.*, 2013, **233**, 355–364.
10. K. Jones, S. Benson and C. Roux, *Forensic Sci. Int.*, 2016, **268**, 151–158.
11. K. M. Jones, PhD thesis, University of Technology, Sydney, 2016.
12. W. Zhou and D. Peng, *American Journal of International Law*, 2021, **115**, 94–101.
13. N. Gentile, L. Besson, D. Pazos, O. Delemont and P. Esseiva, *Forensic Sci. Int.*, 2011, **212**, 260–271.
14. J. Meikle, K. Jones, S. L. Cresswell, S. E. Boyd and J. F. Carter, *Forensic Chem.*, 2021, **23**, 100320.
15. J. F. Carter and L. A. Chesson, *Food forensics: stable isotopes as a guide to authenticity and origin*, CRC Press, 2017.
16. B. Lecat, J. Brouard and C. Chapuis, *Br. Food J.*, 2017, **119**, 84–104.
17. N. Christoph, A. Hermann and H. Wachter, *BIO Web of Conferences*, 2015, **5**, 20150502020
18. S. V. Dutra, L. Adami, A. R. Marcon, G. J. Carnieli, C. A. Roani, F. R. Spinelli, S. Leonardelli, C. Ducatti, M. Z. Moreira and R. Vanderlinde, *Anal. Bioanal. Chem.*, 2011, **401**, 1571–1576.
19. M. Horacek, N. Ogrinc, D. A. Magdas, D. Wunderlin, S. Sucur, V. Maras, A. Misurovic, R. Eder, F. Čuš, S. Wyhlidal and W. Papesch, *Front. Sustainable Food Syst.*, 2021, **5**, 638941.
20. H. Wu, L. Tian, B. Chen, B. Jin, B. Tian, L. Xie, K. M. Rogers and G. Lin, *Food Chem.*, 2019, **301**, 125137.
21. B. Raco, E. Dotsika, D. Poutoukis, R. Battaglini and P. Chantzi, *Food Chem.*, 2015, **168**, 588–594.

22. F. Camin, N. Dordevic, R. Wehrens, M. Neteler, L. Delucchi, G. Postma and L. Buydens, *Anal. Chim. Acta*, 2015, **853**, 384–390.
23. S. V. Dutra, L. Adami, A. R. Marcon, G. J. Carnieli, C. A. Roani, F. R. Spinelli, S. Leonardelli and R. Vanderlinde, *Food Chem.*, 2013, **141**, 2148–2153.
24. European Commission: Joint Research Centre, JRC activities on European Wine databank, DPR-EC-02706.4, 2022.
25. U.S. Food and Drug Administration, Economically Motivated Adulteration (Food Fraud) 2023, https://www.fda.gov/food/compliance-enforcement-food/economically-motivated-adulteration-food-fraud.
26. J. R. Ehleringer and T. E. Cerling, *Encyclopedia of global environmental change*, 2002, vol. 2, pp. 186–190.
27. G. Versini, F. Camin, M. Ramponi and E. Dellacassa, *Anal. Chim. Acta*, 2006, **563**, 325–330.
28. G. J. Martin and M. L. Martin, *Phytochem. Rev.*, 2003, **2**, 179–190.
29. C. T. Chen, B. Y. Chen, Y. S. Nai, Y. M. Chang, K. H. Chen and Y. W. Chen, *J. Food Drug Anal.*, 2019, **27**, 175–183.
30. A. Schellenberg, S. Chmielus, C. Schlicht, F. Camin, M. Perini, L. Bontempo, K. Heinrich, S. D. Kelly, A. Rossmann, F. Thomas, E. Jamin and M. Horacek, *Food Chem.*, 2010, **121**, 770–777.
31. X. Zhou, M. P. Taylor, H. Salouros and S. Prasad, *Sci. Rep.*, 2018, **8**, 14639.
32. Resolution Ocno, Determination by isotope ratio mass spectometry $^{13}C/^{12}C$ of wine ethanol or that obtained through the fermentation of musts, concentrated musts or grape sugar, 2001.
33. European Commission, European Wine DataBank, 2020, https://knowledge4policy.ec.europa.eu/food-fraud-quality/eu-wine-databank_en.
34. M. Collins, *Aust. J. Forensic Sci.*, 2017, **49**, 591–604.
35. The European Monitoring Centre for Drugs and Drug Addiction (EMCDDA), EU Drug Markets: In-depth analysis 2023, https://www.emcdda.europa.eu/publications/eu-drug-markets_en.
36. Drug Enforcement Administration (DEA), The 2018 Heroin Signature Program, 2020.
37. J. R. Ehleringer, J. F. Casale, M. J. Lott and V. L. Ford, *Nature*, 2000, **408**, 311–312.
38. J. R. Ehleringer, D. A. Cooper, M. J. Lott and C. S. Cook, *Forensic Sci. Int.*, 1999, **106**, 27–35.
39. J. R. Mallette, J. F. Casale, J. Jordan, D. R. Morello and P. M. Beyer, *Sci. Rep.*, 2016, **6**, 23520.
40. J. Casale, E. Casale, M. Collins, D. Morello, S. Cathapermal and S. Panicker, *J. Forensic Sci.*, 2006, **51**, 603–606.
41. R. A. Thompson, D. R. Morello, S. Panicker, S. G. Toske and L. Li, *Drug Test. Anal.*, 2022, **14**, 505–513.
42. Y. Makino, Y. Urano and T. Nagano, *Bull. Narc.*, 2005, **57**, 63–78.
43. J. Meikle, S. L. Cresswell, S. E. Boyd, J. F. Carter and S. P. Collett, *Forensic Chem.*, 2019, **16**, 100174.
44. H. Salouros, G. J. Sutton, J. Howes, D. B. Hibbert and M. Collins, *Anal. Chem.*, 2013, **85**, 9400–9408.

45. M. Collins, A. T. Cawley, A. C. Heagney, L. Kissane, J. Robertson and H. Salouros, *Rapid Commun. Mass Spectrom.*, 2009, **23**, 2003–2010.
46. M. Collins and H. Salouros, *Sci. Justice*, 2015, **55**, 2–9.
47. H. Salouros, M. Collins, A. Cawley and M. Longworth, *Drug Test. Anal.*, 2012, **4**, 330–336.
48. G. E. David, A. Coxon, R. D. Frew and A. R. Hayman, *Forensic Sci. Int.*, 2010, **200**, 123–129.
49. J. Cormick, J. F. Carter, T. Currie, C. Matheson and S. L. Cresswell, *Rapid Commun. Mass Spectrom.*, 2023, **37**, e9446.
50. J. Miller, R. Puch-Solis, H. A. S. Buchanan and N. Nic Daeid, *Data Brief*, 2023, **47**, 108931.
51. J. F. Casale, M. Lott and J. R. Mallette, Office of Justice Programs' National Criminal Justice Reference Service, 2017.
52. Drug Enforcement Administration (DEA), Fentanyl Profiling Program Report, 2021.
53. J. F. Casale, M. Lott and J. R. Mallette, *Investigation of the Use of Deuterium and Oxygen in Illicit Fentanyl Analysis*, U.S. Department of Justice, USA, 2020.
54. I. Fraser, W. Meier-Augenstein and R. M. Kalin, *Rapid Commun. Mass Spectrom.*, 2006, **20**, 1109–1116.
55. S. Kusaka, E. Ishimaru, F. Hyodo, T. Gakuhari, M. Yoneda, T. Yumoto and I. Tayasu, *Sci. Rep.*, 2016, **6**, 33122.
56. E. Mutzel Rauch, C. Lehn, O. Peschel, S. Holzl and A. Rossmann, *Int. J. Legal Med.*, 2009, **123**, 35–40.
57. L. O. Valenzuela, L. A. Chesson, G. J. Bowen, T. E. Cerling and J. R. Ehleringer, *PLoS One*, 2012, 7, e34234.
58. K. W. Alt, A. Al-Ahmad and J. P. Woelber, *Nutrients*, 2022, **14**, 14173594.
59. K. J. Petzke, H. Boeing and C. C. Metges, *Rapid Commun. Mass Spectrom.*, 2005, **19**, 1392–1400.
60. C. P. Bataille, M. M. G. Chartrand, F. Raposo and G. St-Jean, *PLoS One*, 2020, **15**, e0237105.
61. C. Lehn, A. Hameder and M. Graw, *Int. J. Legal Med.*, 2023, **137**, 251–258.
62. A. Touzeau, R. Amiot, J. Blichert-Toft, J.-P. Flandrois, F. Fourel, V. Grossi, F. Martineau, P. Richardin and C. Lécuyer, *J. Archaeol. Sci.*, 2014, **46**, 114–124.
63. F. M. Neuberger, E. Jopp, M. Graw, K. Puschel and G. Grupe, *Forensic Sci. Int.*, 2013, **226**, 22–32.
64. L. J. Reitsema, *Am. J. Hum. Biol.*, 2013, **25**, 445–456.
65. M. P. V. Matos and G. P. Jackson, *Anal. Chem.*, 2020, **92**, 3014–3022.
66. B. T. Fuller, J. L. Fuller, N. E. Sage, D. A. Harris, T. C. O'Connell and R. E. Hedges, *Rapid Commun. Mass Spectrom.*, 2004, **18**, 2889–2896.
67. B. T. Fuller, J. L. Fuller, N. E. Sage, D. A. Harris, T. C. O'Connell and R. E. Hedges, *Rapid Commun. Mass Spectrom.*, 2005, **19**, 2497–2506.
68. G. J. Bowen, J. R. Ehleringer, L. A. Chesson, E. Stange and T. E. Cerling, *Water Resour. Res.*, 2007, **43**, 2788–2793.

69. L. A. Chesson, B. J. Tipple, J. R. Ehleringer, T. Park and E. J. Bartelink, *Forensic anthropology: theoretical framework and scientific basis*, 2018, pp. 127–148.
70. S. T. M. Ammer, E. J. Bartelink, J. M. Vollner, B. E. Anderson and E. M. Cunha, *J. Forensic Sci.*, 2020, **65**, 1049–1055.
71. G. J. Bowen, *Annu. Rev. Earth Planet. Sci.*, 2010, **38**, 161–187.
72. L. Font, G. van der Peijl, C. van Leuwen, I. van Wetten and G. R. Davies, *Sci. Justice*, 2015, **55**, 34–42.
73. L. Hu, M. M. Chartrand, G. St-Jean, M. Lopes and C. P. Bataille, *Front. Ecol. Evol.*, 2020, **8**, 568943.
74. C. Lehn, E. M. Kalbhenn, A. Rossmann and M. Graw, *Int. J. Legal Med.*, 2019, **133**, 935–947.
75. C. Lehn, C. Lihl and A. Rossmann, *Isot. Environ. Health Stud.*, 2015, **51**, 68–79.
76. C. P. Bataille, S. T. M. Ammer, S. Bhuiyan, M. M. G. Chartrand, G. St-Jean and G. J. Bowen, *PLoS One*, 2022, **17**, e0275902.
77. L. A. Chesson, B. J. Tipple, L. V. Youmans, M. A. O'Brien and M. M. Harmon, in *New perspectives in forensic human skeletal identification*, Elsevier, 2018, pp. 157–173.
78. W. Meier-Augenstein and I. Fraser, *Sci. Justice*, 2008, **48**, 153–159.
79. L. A. Chesson, W. Meier-Augenstein, G. E. Berg, C. P. Bataille, E. J. Bartelink and M. P. Richards, *Forensic science and humanitarian action: Interacting with the dead and the living*, 2020, pp. 285–310.
80. G. E. Berg, L. A. Chesson, J. Yuryang, S. Youngsoon and E. J. Bartelink, *Forensic Sci. Int.*, 2022, **336**, 111319.
81. W. R. Belcher, C. Y. Shiroma, L. A. Chesson, G. E. Berg and M. Jans, *WIREs Forensic Sci.*, 2022, **4**, e1446.
82. L. A. Chesson and G. E. Berg, *WIREs Forensic Sci.*, 2022, **4**, e1439.
83. L. M. Nienaber, S. L. Cresswell, J. F. Carter and T. Peter, *Sci. Justice*, 2018, **58**, 67–75.
84. K. Jones, F. Koens and T. Simpson, *Sci. Justice*, 2018, **58**, 276–281.
85. J. Meikle, K. Jones, S. L. Cresswell, C. Matheson and J. F. Carter, *Forensic Chem.*, 2023, **33**, 100486.
86. E. Taylor, J. F. Carter, J. C. Hill, C. Morton, N. N. Daeid and R. Sleeman, *Forensic Sci. Int.*, 2008, **177**, 214–220.
87. F. A. Idoine, J. F. Carter and R. Sleeman, *Rapid Commun. Mass Spectrom.*, 2005, **19**, 3207–3215.
88. J. F. Carter, P. L. Grundy, J. C. Hill, N. C. Ronan, E. L. Titterton and R. Sleeman, *Analyst*, 2004, **129**, 1206–1210.
89. M. E. Dietz, L. A. Stern, A. H. Mehltretter, A. Parish, V. McLasky and R. T. Aranda, *Sci. Justice*, 2012, **52**, 25–32.
90. M. Horacek, J. S. Min, S. Heo, J. Park and W. Papesch, *Rapid Commun. Mass Spectrom.*, 2008, **22**, 1763–1766.
91. Y. Suzuki, F. Akamatsu, R. Nakashita and T. Korenaga, *Chem. Lett.*, 2010, **39**, 998–999.

CHAPTER 3

ICPMS and IRMS for Elemental and Isotopic Forensic Investigations

G. S. WALKER

Forensic, Environmental and Analytical Chemistry, College of Science and Engineering, Flinders University, GPO Box 2100, Adelaide 5001, South Australia, Australia
Email: stewart.walker@flinders.edu.au

Abstract

Inductively coupled plasma mass spectrometry (ICPMS), for determining the elemental composition, and isotope ratio mass spectrometry (IRMS), for determining the isotopic ratio of light stable isotopes, are very powerful techniques for providing forensic intelligence, either individually or in combination, and can be applied to a large range of materials of forensic interest. Elements and isotopes vary from location to location worldwide and so the measurement of their presence, their concentration and their relative amounts can provide useful intelligence to a number of forensic investigations. Not only do the hydrogen : deuterium and $^{18}O:^{16}O$ isotopic ratios provide information about the relative sources of water but the concentration of other elements is also important; chlorine content helps in distinguishing between saline and fresh water, and strontium isotope ratios provide useful information for comparing different sources of water. As a consequence, the combination of IRMS and ICPMS is a very powerful approach for the provision of intelligence in forensic environmental investigations.

New Developments in Mass Spectrometry No. 14
Applications of Mass Spectrometry for the Provision of Forensic Intelligence: State-of-the-art and Perspectives
Edited by Simona Francese and Stephen Bleay
© The Royal Society of Chemistry 2024
Published by the Royal Society of Chemistry, www.rsc.org

3.1 Background

3.1.1 Fundamentals

Developed in the 1990s, inductively coupled plasma mass spectrometry (ICPMS) is used for the determination of most of the elements that may be present in a material. Samples are introduced into an argon plasma at ~10 000 K which causes breakdown into (mostly) elements together with the formation of some charged species in the plasma, which can then be directed through skimmer cones into a mass spectrometer. This enables the detection and identification of almost the whole periodic table, from ^{6}Li, ^{7}Li up to ^{238}U. Samples for ICPMS analysis can be in the gaseous (*e.g.*, mercury vapour), liquid (*e.g.*, aqueous solutions with 2% nitric acid to keep ions in solution) or solid form (*e.g.*, released from a solid by laser ablation).

Knowledge of the existence of isotopes is as old as the principle of mass spectrometry itself. Isotopes were discovered by J. J. Thompson (1856–1940), who is credited with inventing the first prototype mass spectrometer in 1913 when he conducted experiments by sending a beam of charged neon ions through electric and magnetic fields under reduced pressure to study 'canal rays' and noticed that two areas developed on the photographic paper which corresponded to two isotopes of neon – ^{20}Ne and ^{22}Ne. Subsequently, William Ramsay (1852–1916) used mass spectrometry to identify 212 of the 287 naturally occurring isotopes. These discoveries were followed by the discovery of deuterium by Harold Urey (1893–1981), who pioneered the use of oxygen isotopes in paleoclimate research.

The basic principles behind the application of stable isotopes ratios in analysis are as follows:

(1) These isotopes are stable and neither decrease nor increase in abundance over time (except the formation of ^{14}N from the decay of ^{14}C).
(2) The ratio of the heavier to lighter isotope provides an indication of the sources of the elements. If they vary from source to source, or place to place, or during a natural process such as evaporation or a manufacturing process such as synthesis of illicit drugs – then the ratios can be used to link (or discriminate between) materials containing similar or dissimilar ratios of isotopes.
(3) Changes can occur in the ratio due to natural or anthropogenic processes such as evaporation and condensation – for example, water vapour above a body of water will contain more of the lighter isotopes than the heavier ones because water containing one or more of the heavier isotopes ($^{1}H_2{}^{18}O = 20$ amu and $^{2}D^{1}H^{16}O = 19$ amu, for example) requires more energy to evaporate than the lighter water $^{1}H_2{}^{16}O = 18$ amu. If the atmospheric water vapour now moves over land and is precipitated as rain, the opposite process occurs and the rain will contain more of the heavier isotopes than the lighter isotopes. So as the clouds move from open sea to land, the water in the

clouds become more depleted in the heavy isotopes. This process is called fractionation and can occur whenever there are two phases with an energy difference so that, if the process does not go to completion, there will be a difference between the final isotopic abundance in both phases.

(4) One consequence of fractionation is that for an accurate measurement of isotopic ratios, care must be taken to ensure that ALL the materials in a sample are transferred and analysed. An example is when water of varied salinity is analysed for H and O isotopes. Analysis of saline solutions requires the water to be evaporated from the salt; however, in this process, if any water is trapped in the residual salt then it will be enriched in heavier isotopes whereas the evaporated water will be depleted in heavy isotopes.

Currently IRMS has been developed for a number of geological, biological, environmental, zoological and cosmo-chemical applications where there are differences in the isotopic abundance between samples. Most applications are for the light isotopes – hydrogen ($^{2}D:^{1}H$), carbon ($^{13}C:^{12}C$), nitrogen ($^{15}N:^{14}N$), oxygen ($^{18}O:^{16}O$) and sulphur ($^{34}S:^{32}S$) – however other isotopes are also used including chlorine ($^{37}Cl:^{35}Cl$), which occurs in a characteristic 3:1 ratio, other metal elements including strontium, which has four stable isotopes ^{84}Sr, ^{86}Sr, ^{87}Sr and ^{88}Sr, and lead which also has four stable isotopes ^{204}Pb, ^{206}Pb, ^{207}Pb, and ^{208}Pb. Table 3.1 presents elements, isotopes, nominal masses and accurate masses for major elements[1] and examples of isobaric interferences.[2]

As hydrogen, carbon, nitrogen, oxygen and sulphur are major building blocks of the organic materials forming living organisms, their analysis has many applications related to the environment and living things.

Although they are now widely used and accepted in mainstream scientific areas and are gaining acceptance as critical evidence in court cases, the enormous potential of utilising these mass spectrometry techniques in forensic investigations is yet to be realised. There are several possible reasons for IRMS and ICPMS not being used to their full potential, including cost, time, availability and specialisation of instruments, as well as the complexity of sample preparation and presenting results to non-scientific audiences including judges and juries. However, both techniques share the potential to provide essential intelligence that can assist forensic investigations without being used as evidence in the court.

3.1.2 IRMS and ICPMS and HR-MC-ICPMS Instruments

3.1.2.1 IRMS Instruments

Classic magnetic sector mass spectrometry instruments form the basis for light stable isotope ratio measurements. The magnetic sector can separate ions containing the isotopes into different radii and place each one in a

Table 3.1 Elements, isotopes, nominal mass, exact mass, abundance and difference for isobaric separation. Reproduced from ref. 1 and 2 with permission from IUPAC, Copyright 2009, 2016.

Element	Isotope	Nominal mass	Exact mass	Abundance %
Hydrogen	H	1	1.00783	99.98
Deuterium	D or ^{2}H	2	2.01410	0.016
Carbon	^{12}C	12	12.0000	98.89
	^{13}C	13	13.0034	1.11
Nitrogen	^{14}N	14	14.0031	99.63
	^{15}N	15	15.0001	0.37
Oxygen	^{16}O	16	15.9949	99.76
	^{18}O	18	17.9992	0.204 ($^{17}O = 0.037$)
Sulphur	^{32}S	32	31.9721	95.04
	^{33}S	33	32.9715	0.75
	^{34}S	34	33.9679	4.22
Chlorine	^{35}Cl	35	34.9689	75.76
	^{37}Cl	37	36.9659	24.23
Iron	^{56}Fe	56	55.939	
Argon oxide	$^{40}Ar^{16}O$	56	55.9475	Difference between ^{56}Fe and $^{40}Ar^{16}O = +0.0085$
Iron	^{58}Fe	58	57.9333	
Argon oxide	$^{40}Ar^{16}O$	58	57.9332	Difference between ^{58}Fe and $^{40}Ar^{18}O = -0.0001$
Arsenic	^{75}As	75	74.922	
Argon chloride	$^{40}Ar^{35}Cl$	75	74.9313	Difference between ^{75}As and $^{40}Ar^{35}Cl = 0.0093$

specifically placed Faraday cup. This setup offers several advantages for measuring isotope ratios, including

(1) the Faraday cups can be set-up for measuring specific isotopes – with higher amplification for the lower abundance isotopes than the major isotope – for example, the natural abundance[2] of ^{13}C ranges from 1.11% to 98.89% of ^{12}C and for oxygen isotopes, ^{18}O is 0.204%, ^{17}O is 0.037% and ^{16}O is 99.76%, so the measurement of CO_2 would require three Faraday cups with appropriate amplification, set up for collecting and detecting $^{12}C^{16}O_2$ at *m*/*z* 44, $^{13}C^{16}O_2$ (and $^{12}C^{17}O_2$) at *m*/*z* 45, and $^{12}C^{16}O^{18}O$ (and $^{13}C^{17}O_2$) at *m*/*z* 46 to determine $\delta^{13}C$ and $\delta^{18}O$,

(2) each ion that enters the Faraday cups is counted so that the absolute ratio of each isotope is determined – in contrast to other ion detectors such as electron multipliers that rely on the ion having sufficient energy to start a cascade of electrons, and

(3) the use of multiple Faraday cups enables the simultaneous measurement of the isotopes rather than sequentially (as with a quadrupole mass spectrometer), giving a more accurate comparison of the isotope ratio from the same sample flow.

Attempts have been made to use other types of mass spectrometers for isotope ratio analysis. However, the ability of magnetic sectors with multiple dedicated Faraday cup collectors to separate and detect multiple ions simultaneously, results in more accurate values; this is because the ions are detected at the same time and hence are not subject to fluctuations arising from sequential separation – as is the case with a quadrupole – and analysis with a single detector at different times.

Having detected the ions with different isotopes, a comparison is made with the ratio of a standard material so that all IRMS values are related and relatable.

The difference between the ratio of the heavier to lighter isotope in a sample is measured against the ratio of the heavier to lighter isotope in a reference material. For example, the ratio of $^{15}N:^{14}N$ (eqn (3.1[1]) is expressed as $\delta^{15}N$ in parts per thousand ‰.

$$\delta^{15}\text{N sample, standard} = \left(\frac{\left(\frac{^{15}\text{N}}{^{14}\text{N}}\right)\text{sample} - \left(\frac{^{15}\text{N}}{^{14}\text{N}}\right)\text{standard}}{\left(\frac{^{15}\text{N}}{^{14}\text{N}}\right)\text{standard}} \right) \times 1000‰ \tag{3.1}$$

Parts per thousand, or one tenth of a percent, is used as the values tend to be small and so reporting as a percentage would involve a lot of unnecessary zeros. As an example, nitrogen is measured as gaseous N_2 and cups detect ions of *m/z* 28 due to $^{14}N_2$, *m/z* 29 due to $^{14}N^{15}N$ and *m/z* 30 due to $^{15}N_2$ and the ratio is compared to the ratio of ^{15}N to ^{14}N in air which is defined as zero. A negative value indicates depletion, meaning that the proportion of the heavier isotope to lighter isotope, ^{15}N to ^{14}N in this case, is less than that for the global standard reference material. A positive value indicates enrichment of the material, signifying that there is a higher proportion of the heavier isotope to light isotope compared to the reference standard.

Developments over time have altered the size of IRMS's and also the amount of off-line sample preparation.

3.1.2.2 Sample Preparation for IRMS

The element to be analysed must first be converted into a form that is amenable to introduction into the mass spectrometer so that (i) it can be readily ionised to a single charge and (ii) can be separated based on the different *m/z* and detected simultaneously using an array of detectors.

Examples include nitrogen converted to and presented as N_2 gas, carbon as CO_2, oxygen as CO_2, and sulphur as SF_6 or SO_2.

Historically, sample preparation was performed off-line, and once in the appropriate form, the gas was introduced into the IRMS. For carbon-containing compounds, this involved combustion to CO_2 and cryogenic separation to separate CO_2 from water and nitrogen. For water, the separation could be achieved by either shaking with carbon dioxide of a known isotopic ratio to reach equilibrium and then analysing the CO_2 or by reacting with palladium to produce H_2.

The other main development in light isotope ratio determination that has been instrumental in increasing its use and application to forensic investigations has been the introduction of on-line sample preparation, enabling interconnection between sample preparation and analysis in one unit. Prior to this, each sample had to be processed off-line to produce a gas in a form that could be loaded up to a sampling device on the IRMS. This was time consuming and required a range of off-line gas lines to convert, separate and isolate the gaseous product without any fractionation.

Two examples of current on-line processing are elemental analysis coupled with IRMS and gas chromatography combustion coupled with IRMS.

Coupling elemental analysis with IRMS enabled the combination of carbon and nitrogen isotope analysis from the same sample. A sample would be combusted to fully oxidise the carbon content into CO_2, for example, in a furnace containing Cr_2O_3 and Ag-coated CoO at 1020 °C. These metals and temperatures are required in order to ensure complete combustion of carbon to carbon dioxide but since the combustion might also produce some nitrogenous oxides, a copper-based reduction furnace is employed afterward that does not affect the CO_2 but reduces any nitrogenous oxides formed to N_2. Trapping of water followed by separation with a chromatographic column enables separate introduction of the carbon and nitrogen containing compounds in a mixture into a magnetic sector mass spectrometer.

One problem that may be encountered with elemental analysis (EA)-IRMS is the amount of non-combustible material in the sample that can lead to accumulation of 'ash' or un-combusted carbonate in the furnace, leading to reduction in efficiency with each sample as the next sample is further away from the hottest part of the furnace. This limits the number of samples that can be run between inlet clear-outs.

The ability to separate different organic compounds by gas chromatography and then combust each compound in turn, followed by feeding the generated CO_2 and N_2 generated into IRMS, opened up a route for determining not only the bulk isotopic ratio but also the ratio of each individual component in a mixture, which increases the amount of information that could be obtained for an environmental sample. This is also important when trying to apportion contribution of pollutants to several sources.

3.1.2.3 Developments of IRMS

Many developments have improved the samples, preparation and scope of investigations that IRMS can address. The best example of changes that have occurred in the development of IRMS are those related to the processes used for analysis of sulphur isotopes. Initially, sulphur isotopes were determined by conversion of sulphur containing samples to SF_6 by reaction with BF_5, which produced characteristic peaks at 146 *m*/*z* for molecule ions containing ^{32}S and 148 *m*/*z* for ^{34}S or related peaks two mass units apart due to SF_5^+ *etc.* This process had some advantages and some disadvantages.

An advantage is that the molecule produced, SF_6, was a stable, non-toxic gas with the only other component being mono-isotopic fluorine. The disadvantages were that SF_6 had molecular ions of 146 *m*/*z*, for ^{32}S, and 148 *m*/*z*, for ^{34}S, hence requiring a large magnet for sufficient separation. BF_5 was also a dangerous compound that required safeguards to be put in place to prevent explosions.

The use of IRMS for discrimination of sulphur isotopes in materials, that had previously been mainly used for mineralogical applications, increased and expanded into more environmental studies with the acceptance of analysis of SO_2 as an alternative to SF_6. This had the advantage that it required a smaller magnet to separate *m*/*z* 64 and *m*/*z* 66, but did pose an issue that the molecule contained oxygen (which itself has isotopes 2 mass units apart, ^{16}O and ^{18}O), and the combined presence of SO_2 and any moisture in the system could lead to the formation of corrosive sulphurous acid, H_2SO_3. With the subsequent acceptance of SO, $m/z = 48$ and 50, as an alternative ion for sulphur isotopic determination, the size of the magnet has been further reduced. As a result, the application of sulphur isotope ratio analysis, although not as prevalent as other stable isotopic analyses, has increased and therefore its use in providing intelligence in forensic investigations has seen a corresponding increase.

3.1.2.4 Multiple Isotopic Ratio Determination

The development of continuous flow sample inlets enables the introduction of standard gases in-between the sample gases for calibration and has enabled the development of multi-isotopic analysis beyond C and N to C, N and H, O and S isotopes from the same sample. A typical outcome for combined carbon and nitrogen analysis with two reference samples followed by a nitrogen sample and a carbon sample and then two more reference samples is shown in Figure 3.1.

A development from the combined analysis of $\delta^{13}C$ and $\delta^{15}N$ from the same sample through combustion/reduction using an elemental analyser and separation of CO_2 and N_2 using a GC column was the introduction of a second GC column for the separation of SO_2 and the removal of excess oxygen and water. Fry[3] claimed replicated values of less than 0.1‰ for $\delta^{13}C$, $\delta^{15}N$ and $\delta^{34}S$. The coupling of commercially available EA with removal

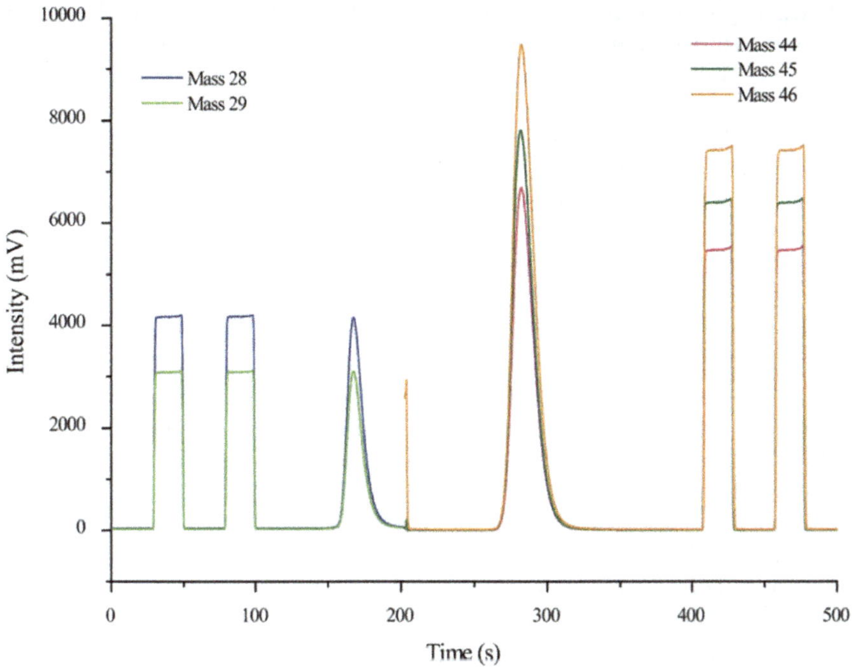

Figure 3.1 Two reference gas peaks for N_2 followed by a sample peak for N_2 ($^{14}N_2$ $m/z = 28$ and $^{14}N^{15}N$ $m/z = 29$), a sample peak for CO_2 followed by two reference gas samples for CO_2 ($^{12}C^{16}O_2$ $m/z = 44$, $^{13}C^{16}O_2$ and $^{12}C^{17}O^{16}O$ $m/z = 45$ and $^{12}C^{18}O^{16}O$ and $^{13}C^{17}O^{16}O$ $m/z = 46$) for IRMS. Reproduced from ref. 7 with permission from the Royal Society of Chemistry.

of water *via* a P_2O_5 trap (to prevent interference of $^{32}S^{18}O^+$ with $^{34}S^{16}O^+$ as both have $m/z = 50$) serves to prevent the formation of sulphurous acid by the reaction of SO_2 and water.[4] Limitations include coping with the different range of concentrations of carbon, nitrogen and sulphur in materials[4] and the optimum requirements for sample preparation for combined $\delta^{13}C$, $\delta^{15}N$, δ^2H and $\delta^{18}O$ determination – for example, use of tin cups for $\delta^{13}C$ and $\delta^{15}N$ and silver cups for $\delta^{18}O$ and δ^2H.[5] Despite advances in this area, some applications dictate the analysis of multi-isotope data from separate analysis runs to optimise results for each element rather than making compromises.

3.1.3 ICPMS

Inductively coupled plasma mass spectrometers come in a variety of equipment types including magnetic, quadrupole and time-of-flight based instruments.

The means of presenting a sample to ICPMS also varies, with the capability to handle gas, liquid or solid samples, all being available. The basic principle of sample preparation is to introduce an aerosol of material into the argon plasma that reaches up to 10 000 degrees Kelvin. At these temperatures, (most) elemental bonds are broken and some of the ions formed in the gaseous plasma are introduced through a sample cone and skimmer cone, with an orifice diameter of around 0.75 mm into the mass spectrometer. This makes ICPMS an ideal technique for determining nearly all the elements and isotopes in the periodic table except (i) argon, as it is in the plasma, and (ii) some elements that do not form strong positive ions including those elements that are prone to isobaric interference; examples of isobaric interferences are: (i) elements with no isotopes such as arsenic $m/z = 75$ that is isobaric with argon and chloride $m/z = 75$ and (ii) lighter isotopes that suffer from polyatomic interference with oxides MX^+ and chlorides MCl^+ such as chromium, zinc and iron. Examples of the difference between isobaric interferences for arsenic and iron are presented in Table 3.1. Calcium is another typical example as its most abundant isotope $m/z = 40$ coincides with the large peak due to argon $m/z = 40$ and its other isotopes have polyatomic interferences – ^{42}Ca and $^{40}Ar^1H_2$, ^{43}Ca and $^{27}Al^{16}O$, ^{44}Ca and $^{12}C^{16}O_2$, $^{14}N^{16}O$, and $^{28}Si^{16}O$, ^{46}Ca and $^{14}N^{16}O_2$ and $^{32}S^{14}N$, and ^{48}Ca and $^{33}S^{15}N$, $^{34}S^{14}N$, and $^{32}S^{16}O$.

Magnetic sector instruments offer the best separation and ability to detect simultaneously which is advantageous for the accuracy and precision of isotopic ratios over quadrupole instruments where analysis is conducted sequentially or ToF-MS where ions hit the same detector in a short time range. Conventional quadrupole ICPMS has been used for some isotopic ratio investigations – including sulphur in environmental samples;[6] however, for the most accurate measurements, simultaneous values from magnetic sectors are required as this removes potential for variation in signals between sequential measurements. Isobaric interferences can be removed, to a certain extent using collision or reaction chambers, or totally by the application of high-resolution multi-collector ICPMS. Concentration of elements is measured in counts per second and depends not only on the initial concentration and the efficiency of ionisation of the particular element, but also on the length of time for which the mass spectrometer is set to focus on counting a particular ion. Analysis of the whole periodic table comes at the expense of amount of time spent in counting each ion; for this reason, selecting the most discriminatory elements reduces the time spent on other ions and increases the time spent on those yielding evidence. This also improves the quality of the analysis when only a small amount of sample is available but can raise questions from the defence as to why only some elements were selected. Detection limits can span from parts per hundred (percent), parts per thousand, down to parts per billion or even parts per quadrillion. This shows the power of ICPMS. While the defence may claim that parts per billion are very low concentrations and should not be considered as evidence (for example, as a result of transfer), there are samples that show differences not in the major elements but in the rare-earth and lanthanoid elements that can be detected at low levels through ICPMS.

3.1.4 HR-MC-ICPMS Instruments

Magnetic sector mass spectrometers have been developed to operate at high resolution in order to cope with isobaric interferences that impact the absolute isotope ratio measurement. Examples include separation of arsenic, which has a nominal monoisotopic mass-to-charge ratio (*m*/*z*) of 75 amu; this is usually a large peak due to the formation of argon chloride ions which also have a nominal mass of 75 amu. Users of non-high-resolution instruments either (1) apply correction factors to account for $^{40}As^{35}Cl^{+}$ by calculating 'peak intensity for $^{40}As^{35}Cl^{+}$ based on the peak at *m*/*z* 77 from $^{40}As^{37}Cl^{+}$ taking into account any contribution at $m/z = 77$ from selenium *etc.* or (2) use 'collision' cells to break up most of the polyatomic interferences into component mono-atomic elements.

Using high resolution instruments with high powered magnets, larger separation of *m*/*z* and narrow Faraday cups that can be precisely positioned, the arsenic peak at $m/z = 74.922$ can be separately detected from the $^{40}As^{35}Cl^{+}$ interference at $m/z = 74.9313$ – (Table 3.1).

3.2 Examples of Application of IRMS and ICPMS for Forensic Investigations

3.2.1 Metals and Minerals – Gold, Diamonds and Other Conflict Minerals

3.2.1.1 Diamonds

'Conflict minerals' include gold, diamond and other valuable commodities, and are defined as minerals where profits from their trafficking and sale may be used to fund violence and armed conflict.[9] Diamond mining has come under increased scrutiny as the potential for diamond smuggling is rife. Much of the initial research into analysis of conflict minerals for forensic intelligence was undertaken in Australia and South Africa as these are the countries with advanced analytical facilities and also have large gold and diamond industries. The Kimberley Process and Certification Scheme (KPCS), since its ratification in 2003, worked to remove 'conflict diamonds' from the global supply.[9] This requires an undertaking by diamond dealers to certify shipments as conflict-free and 'provide supporting certification'. Although mobile and non-destructive techniques would seem to be ideally suited to provide forensic intelligence in the field, applications have shown that IRMS and ICPMS can be useful in confirming origin and LA-ICPMS is required for discrimination as these materials can be quite similar. Moreover, uncut diamonds have 'excess' surrounding material that can be analysed without impacting the value of the gemstone. Diamonds, being mostly carbon but containing appreciable amounts of nitrogen, are well-suited for analysis using stable isotopes of these two elements. However, data reported by Stachel *et al.* in 2022[10] show that despite some narrow ranges in some diamond fields, other areas show, not only a difference between periodontic and eclogitic diamonds from the same field, but also a wide range of $\delta^{13}C$ values in some areas. Estimates of the market for smuggled diamonds comes from Brill *et al.*[11] from a study of one area in Africa, Central African Republic, that was subject to unstable governance. Despite a ban on Central African Republic (CAR) diamonds, Brill[11] claimed in 2014 that 140 000 carats worth US$24 million had been smuggled out of the country and passed off as having come from somewhere else. Comparison of different techniques for discriminating between sources of diamonds from different areas within the Central African Republic (Western and Eastern) and adjacent countries, Southern Democratic Republic of the Congo, Ghana, and South Africa showed that morphology and FTIR could provide only 'supplementary information' but LA-ICPMS was required to discriminate between diamonds from each area. In a case study, LA-ICPMS was also able to determine that of the three parcels of questioned diamonds, A, B and C, suspected parcels A and B were 'near-identical' to diamonds originating from CAR West, whereas parcel C was determined to be of mixed origin. These conclusions were based on comparison against a database of 1000 gem-quality diamonds from six different African countries. Nitrogen concentrations were inconclusive as were morphology and FTIR comparisons. LA-ICPMS could help determine similarities in the concentration trend of the light trace metal elements (from Ga to Eu, except for Y, Zr and Nb) for some of the CAR C diamonds and diamonds known to have come from the CAR East region. Other diamonds from the CAR C package showed a similar trend and concentration for the elements from Gd to U (except Th) to diamonds known to have come from the CAR West region. With the reopening of selected CAR mining areas, the ability to analyse and distinguish between diamonds from areas within CAR that have adhered to, and are compliant with the Kimberly Process, and those diamonds from rebel held areas, show a real-world application of these mass spectrometric techniques.

However, Smith, Smit and Shirey[12] concluded their assessment of the methods and challenges of establishing the geographic origin of diamonds with three limitations that are paraphrased below:

(1) Because comparable geological processes have produced diamonds, there are overlapping physical and chemical characteristics.
(2) Subsequently trace elemental data have not shown clear evidence of distinct geographic signatures.
(3) The only proven method – LA-ICPMS – is prohibitively slow, expensive, 'high-tech' and destructive, precluding large global-scale investigations into origin discrimination.

Their conclusion supports the use of LA-ICPMS as the only proven technique capable of discriminating at the level required for diamond provenancing but highlights the need of enhancing its accessibility so that it can become the mainstream technique used for providing forensic intelligence locating the place of origin of diamonds or providing indisputable proof that a diamond does not come from where it is claimed to have originated.

3.2.1.2 Gold Fingers

Similar to diamonds containing characteristic trace elements, gold extracted from specific sites will contain a range of trace elements in various proportions that reflect their place of origin. Native gold can be analysed for the full periodic table by ICPMS and this analysis can assist in assigning the possible place of origin. Being able to detect down to parts per billion, trillion or even quadrillion enables ICPMS to detect traces of elements in very low concentrations in otherwise highly purified elements. Even after gold has been purified to levels of 99.999%, ICPMS can still detect the trace elements present and compare them to enable discrimination between different samples of gold.

A laser ablation ICPMS (LA-ICPMS) method[13] was developed to analyse the gold standard GSB 04-3312-2016 (from Shandong Mokingran Jewelery Co., Ltd) and standard samples of 99.9909%, 99.9801%, 99.9488%, 99.892% and 99.8001% purity. The authors were able to determine 20 impurities at detection limits of down to 10 ppb for ruthenium, rhodium, palladium, silver, cadmium, antimony, iridium, platinum, lead and bismuth and up to 140 ppb for arsenic with R^2 of between 0.9913 to 0.9991. These trace elements could help distinguish between the origin of the different samples by comparing their trace element compositions.

Although, LA-ICPMS is classified as 'minimally destructive', as it leaves only a small crater, there will still be reticence from owners and museums to have the integrity of valuable artifacts compromised by even a small amount.

3.2.1.3 Egyptian Silver Jewellery from 2700 BC

A recent paper[14] has illustrated the vital use of multi collector high resolution ICPMS to determine isotopic ratios of lead isotopes to trace the most likely origins of silver jewellery belonging to Queen Hetepheres I, the mother of pyramid builder King Khufu (date of reign *ca.* 2589–2566 BC). Egypt does not have any indigenous silver deposits and so silver does not occur in archaeological finds prior to the mid-Bronze Age. Analysis was undertaken using modern techniques to try to identify the origin of the silver in this jewellery collection. X-ray fluorescence could detect lead at 0.1% of the amount of silver with an uncertainty of ±1%. Energy dispersive X-ray fluorescence had better detection limits of 0.001% lead in silver but larger uncertainties of ±10% of the measured value. Lead isotopes, present in the silver as impurities, were analysed using a multi collector inductively coupled plasma mass spectrometer which was incredibly sensitive, being able to detect background levels down to 20 picogram (2×10^{-14} g) and having reproducibility of at least 0.005% between replicate analysis of the same sample. Comparison of the lead $^{208}Pb:^{204}Pb$, $^{207}Pb:^{204}Pb$, and $^{206}Pb:^{204}Pb$ isotope ratios with silver of known areas in the region suggested that the silver was most likely to originate from the Cyclades – Aegean islands in Greece which were at a reasonable distance from Egypt and not the obvious source.

3.2.2 Environmental Forensic Investigations

Because of the ubiquitousness of compounds containing some of the key elements that lend themselves to isotopic ratio analysis by IRMS (C, H, N, O, and S), and other elements that are amenable to analysis by ICPMS, these two techniques have been used in a wide range of environmental forensic investigations.

The tenet of forensic toxicology was expressed in a French court in 1840 by Orfila who said:[15]

> I shall prove that there is arsenic in the body of Lafarge second that this arsenic comes neither from the reagents with which we worked nor from the earth that surrounded the coffin: also that the arsenic we found is not the arsenic component which is naturally found in every human body.

This was paraphrased for forensic environmental science by Walker in 2001 (G. S. Walker, unpublished material):

> Forensic Environmental Science is the collection and analysis of samples to prove that there is some **chemical (X)** in the environment second that this **chemical (X)** comes neither from the reagents with which we worked nor from the atmosphere or other contamination: also that the **chemical (X)** we found is not the **chemical (X)** component which is naturally found in the environment.

Further to identify the 'criminal' in environmental pollution, we have to also prove that the **chemical (X)** came from a particular source – or what proportion of **chemical (X)** came from that source. In order to achieve this, we should be able to distinguish between natural and anthropogenic chemical signals.

There are three options for detecting environmental anomalies:

(1) Measuring the concentration of a particular element and comparing it to the 'normal' background level to establish whether or not it is above or below the expected background. This requires a knowledge of the normal concentration levels from analysis of samples from previous times. For this approach to be effective, the areas must be sampled and analysed prior to any pollution event. An example is the breaching of a tailings dam in a nickel refinery in Northern Queensland that was used to store the contaminated water from the processing plant; the

refinery burst due to a cyclone which resulted in not only elevated levels of zinc in adjacent sediment and coral but also exposed the coral to below normal concentrations of chloride as they experienced a flush of non-saline water (G. S. Walker, unpublished material). Cubes of the coral that were cut in annual growth bands from an adjacent reef were analysed by ICPMS and elevated levels of zinc, compared to other terrestrial sourced metals, were detected in the segment of the coral corresponding to growth in the year in which the tailings dam burst.

(2) Determining that the concentration of an element or compound is above that stipulated in an environmental guideline. An example of this was a mining tailing dam in Western Australia that was used to store water and suspended solids from a mine that was gradually leaking, so the levels of arsenic in seafood in adjacent sound became elevated above the guidelines.

(3) The isotopic ratio of an element (or more than one element) differs from what is expected to be at a particular location. Examples of this would be a change in the carbon isotopic signal in an ecosystem due to change between the C4 and C3 balance of plants in the catchment or the detection of changes in the sulphur isotope ratio in air-borne plumes emanating from smelting operations. Plants that are designated as C3 produce a three-carbon compound in photosynthesis whereas C4 plants produce a four carbon compounds as the first step in photosynthesis. Consequently, the stable isotopic ratios for carbon differ with C3 being between −20 and −37‰ whereas C4 plants have a carbon isotope ratio range of −12 to −16‰.

3.2.2.1 *Natural and Spiked Strontium Studies in Environmental Forensic Investigations*

Strontium isotopes have been extensively used for comparing different water sources. Because strontium in drinking water comes from the underlying rocks and is then incorporated into humans and their hair, the occurrence of strontium isotopes has been mapped. For example, the strontium values for water across continental USA have been determined from analysis of bedrock, water, bones, teeth and antlers and these values can be used for tracing the origins of plants, aquatic-life and people.[16]

Strontium in fish otoliths has been analysed in cross-sections to determine changes that have taken place during the life of the fish. Strontium signals have either been shown to be constant – indicating that the fish remained in the same water – or have changed indicating a shift into different waters. Studies have been conducted that utilise the natural variations of strontium isotopes in the growing otolith[17] to introduce trace amounts of isotopes. This method involves 'spiking' the otoliths of juvenile fish fry with at least 0.45 mg of ^{84}Sr and 2.28 mg of ^{86}Sr per kilogram of body weight[18] that can then be released into the wild. When the fish are recaptured, analysis of the strontium $^{88}Sr/^{86}Sr$ and $^{88}Sr/^{84}Sr$ patterns can be used to indicate their movements. This showed that changes in habitat between locations with different strontium ratios can be manifested in fish, enabling discrimination between those that stay in the same water and those that move into different waters.[18]

Other studies have used well-controlled conditions[19] to expose fish fry to different strontium signals in both water and their feed, aiming to determine the relative importance of each source. They did this by allowing the fish to grow in water with a specific strontium ratio. Initially, fish were exposed to salt solution of between 3 and 5% for between 3.5 and 5 minutes (depending on batch) and then freshwater with 1% calcein, a fluorescent marker, that enabled differentiation of pre- and post-experimental growth. They were then kept in water with a different strontium 86 (0.14 $mg\,l^{-1}$) and barium 137 (0.17 $mg\,l^{-1}$) level and fed spiked food at 0.26 and 0.58 $mg\,l^{-1}$ over a short time. The water in tanks was then changed to prevent interaction between residual food and surrounding water and replenished with water with the original strontium ratio. After 31 to 39 days, depending on the specific experiment, subsequent analysis of sections of otoliths for calcium, barium and strontium by ICPMS showed that water was the dominant source of strontium and barium.

Other studies that have coupled isotopic and elemental analysis were applied to determine the flight paths and migratory routes of a variety of birds, fish and animals. Monti[20] and co-authors coupled multi-isotopic analysis of carbon, nitrogen and sulphur from feathers from 93 Palearctic ospreys with GPS tracking devices to 'ground-truth' the migratory patterns of these birds. These studies may become more important with reported climate changes altering the migratory patterns of birds and fish.

3.2.3 Drugs

Most drugs are organic in nature making them suitable for light stable isotopic analysis to provide intelligence on their potential source, routes of manufacture, and adulterating agents and allowing batch-to-batch or source-to-source comparisons to be made. Whereas in the past, specific compounds and DNA could be used to determine place of origin of cultivated drugs, the global scale of distribution of seeds means that 'Thai cannabis plants' with characteristic compounds could now be grown anywhere. Compound-specific analysis by separation of mixtures through GC followed by combustion and analysis of stable isotope ratios is a powerful technique for distinguishing between the isotopic signal of components in the mixture and allows the consideration of individual isotope ratios of the drugs and the cutting agents. Many cutting agents contain an organic component – starch, sugars, powdered milk and caffeine, for example – so these must be taken into account for bulk isotope analysis. Many cutting agents also contain a metal – magnesium silicate hydrated (talcum powder) or calcium carbonate (ground up chalk), sodium hydrogen carbonate (sold as baking soda) are just three examples of readily available and cheap materials that have been used as cutting agents. As a consequence, the analysis of metal content by ICPMS is also useful in a holistic study of components of a drug mixture. ICPMS has found application in determining metal compositions for forensic investigations which encompasses natural drugs – for example, linking trace elements in drugs grown in soil or hydroponically and also determining the synthesis route used for drug manufacture. In the latter case, ICPMS is used for detecting concentrations of metals used in a particular synthesis route – for example, lithium and/or mercury in the synthesis of amphetamine type substances (ATS). Consequently, there are numerous studies of different drugs in the literature of which the following are selected examples.

3.2.3.1 Example of Synthetic Drugs – Amphetamine-type Stimulants

Amphetamine and methamphetamine (MA) have seen increased usage and have been investigated using organic by-products to discriminate synthesis methods from phenyl-2-propanone, P2P, or from ephedrine/pseudoephedrine and also to discriminate between the large number of reported synthesis methods including the Leuckart, Nagai, Emde, Hypo, Moscow, Rosemund and Nazi/Birch methods.[21] Isotope ratios were described as 'breaking the limits of conventional impurity profiling by linking the seized MA and their batches to their synthetic origin'.[21] David *et al.*[22] considered the determinants of isotopic composition during HI/Pred (HI and red phosphorus method) synthesis of methamphetamine to determine which steps of the synthesis process resulted in observable differentiation of isotopes in the product drug. They found that the addition of hydrogen at the benzylic position, either to replace iodine, or to reduce unsaturation as the last step in the synthesis, led to significant fractionation with hydrogen preferred over deuterium. This was due to the kinetic effect where differences between the energy required for the reaction of hydrogen and deuterium led to preferential incorporation of the lighter hydrogen over the heavier deuterium. In their experiments, large positive δ^2H values in methamphetamine were taken as indicators of semi-synthetic origin. However, as the carbons in the framework of the initial reagent were retained in the product, this did not lead to fractionation in synthesis; as a result, no changes in $^{13}C:^{12}C$ were observed and hence carbon isotope ratios did not mirror the synthetic process but accurately reflected the carbon isotopes present in the original material. As such, carbon isotope ratios could be used for tracing the original starting materials. However, the nitrogen and hydrogen values from the different starting reagents used in this investigation were too 'similar' to show up differences due to the synthesis method. Consequentially, carbon isotope ratio comparisons could be used to determine a link between drugs and suspected precursors – or to indicate that the drugs were not synthesised from that starting material.

With regard to ICPMS analysis, synthetic routes to methamphetamine can use a variety of elements including mercury or lithium. Iodine was found to be an indicator of use of the Nagai route, and palladium and barium were found to be indicative of the Emde route.[23] Therefore, a combination of IRMS, for light isotope ratios and ICPMS, for metals, is a very powerful approach for providing including or excluding starting materials, synthesis routes and also batch-to-batch comparisons of seized drugs.

3.2.3.2 Example of Semi-synthetic Drugs from Natural Material – Heroin

Much research has been conducted on heroin profiling in the countries where heroin is produced or where use is endemic.[24] Lui *et al.*[25] attempted to discriminate between heroin from the "Golden Triangle" and the "Golden Crescent" by inorganic fingerprinting using multi-elemental ICPMS analysis of 417 authentic heroin samples. This method was then applied to 907 unknown samples and was proven to be reliable when compared with the results of the traditional method consisting of major alkaloid, acidic and neutral component profiling. The authors claim that inorganic fingerprinting not only produces a similar result but is also quicker and easier than determining the organic profiles.

3.2.4 Food and Beverages

3.2.4.1 Fundamentals

Similar to minerals and gems having an elemental fingerprint that can be linked back to their origin, there is a link between the concentration of elements in soil, water and fertiliser that can be transferred to the food or drink. An underlying understanding of the original composition of the soils, water, fertiliser and growing conditions coupled with an understanding of the selective process governing the incorporation of these signals into the food and drink product can be used to trace back and confirm (or dispute) the origin of the material. Many studies have reported investigations of such links – including on rice, coffee, tea and wine,[26] outcomes from IAEA research projects,[27] and those included in a 2020 review of IRMS and ICPMS for authentication of agri-food products.[28]

The provision of sufficient nutritious food and beverages, including water, is enshrined in Article 25(1) of the United Nations Universal Declaration of Human Rights[29] and also features in many of the UN's 17 Sustainable Development Goals[30] including those relating to the provision of food and water: 1. No Poverty, 2. Zero Hunger, 3. Good Health and Well-Being, 6. Clean Water and Sanitation, 12. Responsible Consumption and Production, those relating to sustainable living conditions, 8. Decent Work and Economic Growth, 10. Reduced Inequalities, 11. Sustainable Cities and Communities, and those relate to action for living with climate change and on land and water, 13. Climate Action, 14. Life Below Water, and 15. Life on Land. IRMS and ICPMS analysis has played a key role in providing evidence or intelligence to enable the realization of these goals. Examples where IRMS and/or ICPMS have been applied in investigations are presented below.

In addition to the use of strontium isotopes for determining the water source, other elements have been used for determining dietary food webs. Jaouen *et al.*[31] used zinc isotope ratios of bones and teeth as indicators of new food webs arising from modern changes due to climate changes and due to introduction of different plants (C3 and C4) or animals (pigs, buffalos, cows and other grazing animals). They used the most abundant isotopes of zinc (64 and 66) to track isotopic ratios in various herbivores (grazers, browsers and those with a mixed diet) and carnivores (hyenas and other carnivores – including crocodiles) in three different areas; Koobi Fora, Western Cape and Kruger Park. Herbivores could be distinguished because leaf eaters had lower isotopic values than browsers who obtain their carbon from leaves and grazers who predominantly consume grasses. However, as crocodiles can regrow teeth, they were found to have poor correlation between the isotopic value of the teeth and the current diet as the isotopic value may be influenced not only by the seasonal diet but also by the stage of growth of the teeth. Additional research on crocodiles[32] conducted in the Northern Territory of Australia showed that their diet, and therefore their carbon and nitrogen isotopic signals have changed due to the introduction of pigs and buffalo in this region leading to a change from aquatic to terrestrial food sources. The authors also considered the downstream impact on the addition of terrestrially sourced isotopes into the aquatic ecosystem. Crocodiles who have moved from a diet dominated by aquatic life to a diet with more introduced terrestrial species will be defecating waste with a different signal that will change the downstream signals as well.

Human diet has also been studied using isotopes. IAEA has published an assessment of zinc metabolites in humans using stable zinc isotope techniques.[33]

Archaeological research includes the analysis of human and animal remains from different epochs in attempts to determine the diets and changes in diets over time. This study also indicates the extent of veganism in different parts of the world at different times.[34] Analysis of well-preserved material, bone collagen and hair, from three distinct groups of population described as modern urban ($n=5610$), modern subsistence ($n=1167$) and those from people who were pre-AD-1910 Haber–Bosch process for production of ammonia from nitrogen in the air ($n=6879$), showed that the isotopic signal in the modern subsistence population overlapped with the pre-1910 people ($\delta^{13}C$ −30.1 to −11.7‰ and $\delta^{15}N$ of −1.8 to +15.1‰ compared with −31.4 to −10.4‰ and −2.3 to +22.3‰). However, the modern urban diet had much tighter ranges and also varied depending on the climatic location and diet.

The uptake of isotopes and metals into human and animal bodies has been used to identify the likely geographical origin and movement of human bodies. Adam – a headless limbless torso – was dragged from the river Thames in London in September 2001. He had no fingers for fingerprinting, no face for reconstruction or teeth or hair for identification of diet or elemental intake. Analysis of trace metals and isotopes in his bones was compared with known geological profiles and ultimately compared with bones from recently deceased people from locations in Nigeria to determine the most likely village or origin.[35]

Two unidentified corpses found in separate locations in Utah have been identified by comparison of oxygen and strontium isotopes in their hair. The isotopic ratio of both of these elements can be compared to those in drinking water in order to determine the potential places of origin. A woman's body was discovered rolled in a carpet near Salt Lake City in 1978. The isotope ratio indicated that she was originally from the Texas area and had recently resided in the south region of Utah. With this information, a search through missing person reports indicated that she was Patricia Barrett, who was 34-year old. "Saltair Sally", as she was called, was found in October 2000 in desolate brushland near Saltair on the banks of the Great Salt Lake, Utah. Investigation of oxygen and strontium isotopes indicated that she had moved in and out of the Salt Lake City water region a couple of times in the last 22 months and that she had gone to somewhere with similar water isotopes to the Northwest area of USA. Once this information was obtained, further investigations focused on people whose movements aligned with identified candidates from the reported missing people and identified the unknown woman, with DNA confirmation, as 20-year old Nikole Bakoles from Tacoma. Washington.[36] Although neither of these homicides have been solved, isotope ratios have provided forensic intelligence that led to their identification and highlighted potential suspects.

3.2.4.2 Examples of Food and Beverage Investigations

IRMS and ICPMS analytical processing has been applied to food and beverage investigations in a range of products. Selected, recent articles on using IRMS and/or ICPMS for food and beverage provenancing are presented to indicate the growing use of these techniques in this area.

These studies can be divided into three groups: those investigating techniques for general applications, those investigating specific products from specific neighbouring regions, and those investigating comparison of worldwide products.

Examples of investigations into products from specific regions include rice from the Basmati Himalayan geographic region[37] and strawberries from Slovenia.[38] Matos *et al.* combined whole-body compound specific isotope ratios of ^{13}C, ^{15}N, and cadmium and lead to discriminate Eastern oysters (*Crassostrea virginica*) from five different harvest areas in Gulf of Mexico, Florida, Louisiana and Texas.[39] Others have investigated if the product was as claimed – for example, a study[40] discriminated between Iberian pork that was 'acorn-fed, 'feed-raised' and had 'mixed-feed'. Watling *et al.* developed combined isotope ratios through IRMS and selected metal analysis through ICPMS to investigate wine, tea, wine, *etc.*[26] The ability to discriminate between teas from India, Sri Lanka, Pakistan and Taiwan using these techniques may provide evidence but being a supervised principle components discrimination process, the examiner selects which elements contribute to the discrimination process. The need to select which elements are included in the chemometric processing of the data to separate origins and the fact that the selected elements produce different conclusions means that this process may provide intelligence but could be challenged if presented in a court of law as principal evidence.

3.2.4.3 Isotopic and Metal Changes Due to Processing of Food

Having access to certified samples from the proposed locations of origin assists in the identification of links between questioned material and origins. Ideally, access to known materials – food, water and soil – from known growing areas would enable ground truthing of isotopic and metallic fingerprints. However, any food or drink material that undergoes aging or processing may have undergone changes in elemental or compound composition. For example, to investigate changes that may occur during roasting of coffee beans, Tan and Walker[41] undertook sequential roasting of green coffee beans to mild, medium and dark roast and then compared isotopic and multiple elemental compositions during the process. The roasting process produces browning compounds through the Maillard reaction. This condensation of an aldehyde and amine group with the loss of water, followed by loss of carbon dioxide from the carboxylic acid group, gives the 'roasted' flavour. This process would be expected to reduce the amount of water and the amount and isotopic distribution of oxygen, carbon and nitrogen. Results[41] indicated that there were some elements that could be considered as 'conservative' as they exhibited behavior suggesting preservation within the bean during the roasting process – for example, potassium and phosphorous increased slightly in percentage. Whereas some elements decreased indicating loss from the bean while others increased indicating less loss than other molecules. For example, due to the decrease in other components – especially water – the percentage carbon content increased despite some being lost in the form of carbon dioxide in off-gassing after roasting. The isotope ratio of carbon, however, did not alter dramatically – but, as carbon was initially around 50% of the coffee bean, any loss may have been small in comparison to the total amount. The nitrogen isotopic signal did change from green to mild, medium and dark roasted beans. As the shift in nitrogen isotope ratio is always in the enrichment direction (that is the nitrogens remaining in the bean tend to be more of the heavier isotopes) and these shifts are consistent in magnitude, there can be some linking of roasted beans back to potential green beans. Like other research carried out on drugs and explosives, a regular isotopic shift can enable the exclusion of some sources. If a roasted bean has a particular signal, then all green beans having an initial isotopic value that is more enriched than the roasted beans could not have been the original source of the green beans.

3.2.5 Wildlife Forensics – Not Just Ivory and Sharks' Fins

The use of isotope ratios and/or elemental fingerprinting to detect the real place of origin of animals, birds, eggs, reptiles, coral, fish, plants, trees, wood, and parts of animals or plants is increasing in applications. Animals with high profiles include elephants and rhinos, pangolins, tigers, shark fins and other endangered animals but there are also concerns relating to smuggling of other creatures including reptiles, birds and coral products. DNA may be used to link ivory tusks to specific de-tusked animals but can also provide insights into the actual origin of a sample.

In addition to deuterium: hydrogen and strontium isotopes as indicators of water consumed,[42,43] other isotopes have also been applied. Sulphur in bird feathers has been used for determining changes in the marine diet of birds and other studies have combined sulphur stable isotopes of bird feathers with GPS tracking[20] to determine their migratory patterns. A combination of microscopy, DNA analysis and isotope assessment through IRMS and elemental analysis involving isotopes and ratios through ICPMS has also been used to identify the rhino species (white or black), determine the country of origin and even pinpoint the specific park from which rhino horn hair originated.[44]

3.2.6 Recent Applications and Future Areas of Application – Bioplastics, Microplastics and PFAS

There are three areas of environmental analysis that have recently gained a lot of general interest and where IRMS and/or ICPMS analysis can aid by providing additional intelligence. Previous research had been used to identify the age and compare the potential source of oil spills from wrecks,[45] from a range of ships in busy shipping channels and from natural oil seepages. Metal isotopes of lanthanide, neodymium, and samarium by GC-ICPMS were also used to identify contamination sources of marine sediments.[45]

3.2.6.1 Bioplastics

Carbon and hydrogen isotope ratios have been employed in attempts to discriminate between different bioplastics (originating from C3 or C4 feedstock) and petroleum-based plastics. A collaborative effort between New Zealand and China focused on investigating C and H isotopes using a rapid and low-cost technique to 'compliment' radiocarbon (^{14}C) and FTIR techniques;[46] radiocarbon measurement can show if the carbon source was old – low or below detection limits due to radioactive decay over time – or contained recently grown material. The aim of the publication was to investigate if plastics were being correctly labelled, and, in particular, to ascertain whether products being labelled and sold as bioplastics genuinely conformed to their claimed composition. The authors produced a decision tree that set boundaries for classification as petroleum, C4 bioplastic, C3 bioplastic and mixed petroleum and bioplastics.

The core problem was illustrated by research by Suzuki[47] who showed that, although the carbon isotopic signal varied depending on the source, there were significant overlaps that made discrimination of plastics more difficult. A 2018 study[48] reported that C4 plants (corn ∼ −10‰ and sugarcane ∼ −13‰) had carbon isotopic ratios that overlapped with plastics made from corn, or sugarcane, but C3 plants (rice ∼ −27‰, potato ∼ −27.5‰ and tomato ∼ −30‰) had isotopic values that overlapped with plastics made from polyethylene derived from petroleum ∼ −29 to −27‰. As is the case with plastic analysis, the signal may be a mix, especially if an item is made from recycled plastics which had a signal that ranged from ∼ −24 to −30‰, thus overlapping with petroleum-based plastics and genuine bioplastics from C3 plants.

3.2.6.2 Microplastics

Microplastics in the environment are defined as having a diameter of less than 5 mm whereas plastics with a diameter greater than 5 mm are classed as macroplastics. There is growing concern that these microplastics are impacting the environment. Isotopic analysis is well-suited to contribute to the recent increase in investigations on bioplastics and detecting and cataloguing the prevalence of microplastics in the environment by identifying potential sources and discriminating between oil-based and renewable plastics.

Microplastics have been described as a 'Trojan horse'[49] because of their susceptibility to ingestion by living beings and their ability to act as vectors for persistent organic pollutants (POPs) and heavy metals. Being hydrophobic, plastics are attractive to pollutants in water (they are also hydrophobic); hence plastics can absorb persistent organic pollutants, POPs, and toxic metals. Because microplastics are easily ingested, they can then act as a vector for POP, heavy metals, 'pathogenic and/or antibiotic resistant microorganisms' to pass into organisms.

Berto[48] used elemental analyser IRMS to characterise plastic polymers in the marine environment. They estimated a production of 280 million tonnes in 2016 with 29% coming from China and 40% used in packaging which will end up either in recycling, land dumps or in the ecosystem. With an increase in concern for environmental plastics and the ability of plastics to adsorb POPs and heavy metals, this is an area for future development of combined IRMS and ICPMS techniques to provide forensic intelligence into environmental pollution studies and litigation cases.

3.2.6.3 PFAS

Perfluoroalkyl and polyfluoroalkyl substances (PFASs) are considered to be persistent organic pollutants as they 'never break-down'. There are a large number of PFASs, all of which have a chain of carbons with fluorine atoms – some have the carboxy acid group – others sulphates, phosphates, acrylate, amine or alcohols groups, some with ether or sulphide linkages and some with chloride or hydrogens constituents. These are therefore ideal candidates for isotopic comparison using IRMS.

One study using IRMS determined the prevalence of different PFASs in ecosystems and assessed trophic levels through the analysis of stable isotopes of nitrogen, showing that not all PFASs bioaccumulate in accordance with predictions.[49] Sixty PFASs were studied in 21 species in the St. Lawrence River food web. They concluded that emerging PFASs – including PFECSH, PBSA and 6 : 2 FTAB were detectable in water and biota. Different PFASs behaved differently – C10-13 PFCAs, PFOS, and PFDS were found to be very bio-accumulative, PFOA and EtFOSAA were bio-diluted, FBSA and PFECHS were biomagnified. Invertebrates accumulated short-chain PFCSs and FTSAs differently compared to teleost fishes. For example, PFOS made up 82% of the PFASs found in fish. The authors proposed two mechanisms to account for these discrepancies – some PFASs are two big to be bioaccumulated or some – like C13-14 PFCAs strongly adsorb onto sediment organic matter making them less bioavailable. In addition to the obvious potential for the use of isotope ratios of PFASs to discriminate between different sources of PFASs, it appears that, without using ^{15}N IRMS for determining trophic levels, analysis of PFASs in ecosystems will not take into account the different behaviours – very bioaccumulating, bioaccumulated or bio-diluted – of PFASs. IRMS should be used for these investigations to provide appropriate levels of contamination for different PFAS in different species in an ecosystem.

3.3 Conclusions

Analysis of isotope ratios of the light stable elements, C, H, N, O, and S by isotope ratio mass spectrometry and analysis of elements by inductively coupled plasma mass spectrometry have a long and illustrious background and have been applied to many areas of forensic environmental investigations. These include (i) profiling of food and drink/water to verify their stated origin and (ii) investigations of gold, diamond, silver and other natural commodities that can attract forgeries or questions about their origin. Applications have also been extended to synthetic compounds – especially drugs, either to identify where they were grown – for example, heroin – or from what and how they were synthesised – for example, methamphetamine. These techniques have been successfully developed over time and will continue to develop into established and accepted techniques. Areas where they may play a significant role in the future include the analysis of bio-plastics to prove they are made of bio-material, microplastics – in determining origin by isotope ratio and contamination by POP and heavy metals by IRMS and ICPMS, and analysis of PFAS by IRMS to determine sources and also to determine the extent of contamination with trophic levels through δ^{15}N analysis. The challenge for all analytical techniques is to gain approval for presentation of results as evidence in a court of law. In order to achieve this, techniques need to prove that they meet Fry and/or Daubert criteria for evidence: (i) they are accepted in the scientific area, (ii) errors and uncertainties can be adequately handled and (iii) they produce reproducible results. ICPMS has reached some level of acceptance. IRMS faces some more challenges. This may be due to it being a specialised technique and susceptible to more variations, with delta notation representing a difference in ratios related to a standard rather than a straightforward concentration. For example, the difference between isotope ratios of −29 and −27‰ is as significant as the difference between +0.5 and −1.5‰, although one is positive and one is negative.

Both techniques have provided forensic intelligence in several different areas where other analytical methods fall short and, hopefully, will continue to be embraced more widely in the future of elemental and isotope environmental forensic investigations.

Acknowledgements

The author acknowledges the editors for helpful suggestions for revision. He also acknowledges the crucial role played by his high school chemistry teacher in shaping his interest in chemistry – Dr Shelagh Walker (14/02/1931–07/12/2022).

References

1. M. E. Wieser and M. Berglund, *Pure Appl. Chem.*, 2009, **81**(11), 2131–2156.
2. J. Meija, T. B. Coplen, M. Berglund, W. A. Brand, P. De Bièvre, M. Gröninig, N. E. Holden, J. Irrgeher, R. D. Loss, T. Walczyk and T. Prohaska, *Pure and Appl. Chem.*, 2016, **88**, 293–306.
3. B. Fry, *Rapid Commun. Mass Spectrom.*, 2007, **21**, 750–756.
4. S. Mambelli, P. D. Brooks, R. Sutka, S. Hughes, K. M. Finstad, J. P. Nelson and T. E. Dawson, *Rapid Commun. Mass Spectrom.*, 2016, **30**, 1743–1753.
5. H. Fu, L. Wei, H. Chen, X. Yang, L. Kang, Q. Hao, L. Zhou, Z. Zhan, Z. Liu, J. Yang and L. Guo, *J. Food Comp. Anal.*, 2021, **102**, 103972.
6. E. Leyden, J. Farkas, S. Gilbert, J. Hutson and L. M. Mosley, *Talanta*, 2021, **235**, 122708.
7. H. Dong, K. Xiao and D. Luo, *RSC Adv.*, 2017, **7**, 18946–18952.
8. X. Wang, C.-Q. Liu, N. Zhang, S. Xu, Z. Pang, S.-L. Li, H. Ding, J. Chen and Z. Xie, *J. Anal. At. Spectrom.*, 2023, **38**, 186–196.
9. Kimberly Process and Certification Scheme, https://www.kimberleyprocess.com/en/system/files/documents/KPCS%20Core%20Document.pdf. (accessed May 2023).
10. T. Stachel, P. Cartigny, T. Chacko and D. G. Pearson, *Rev. Mineral. Geochem.*, 2022, **88**(1), 809–875.
11. S. Brill, N. Lilimu and D. Chetty, *Heliyon*, 2020, **6**, e05592.

12. E. M. Smith, K. V. Smit and S. B. Shirey, *Gems Gemol.*, 2022, **58**(3), 270–288.
13. X. Wang, T. Yang, B. Huang, Y. Zhu, H. Li, J. Shen, L. Li and M. Li, *IOP Conf. Ser.*, 2020, **555**, 012052.
14. K. Sowada, R. Newman, F. Albarède, G. Davis, M. R. Derrick, T. D. Murphy and D. B. Gore, *J. Archaeol. Sci. Rep.*, 2023, **49**, 103978.
15. S. M. Gerber and R. Saferstein, *Chemistry and Crime*, American Chemical Society, Washington D.C., 1997, p. 140.
16. B. Beard and C. Johnson, *J. Forensic Sci.*, 2000, **45**(5), 1049–1061.
17. F. Feyrer, G. Whitman, M. Young and R. C. Johnson, *Mar. Freshwater Res.*, 2019, **70**, 1689–1697.
18. A. Zitek, J. Irrgeher, M. Cervicek, M. Horsky, M. Kletzl, T. Weismann and T. Prohaska, *Mar. Freshwater Res.*, 2014, **65**, 878–986.
19. Z. A. Doubleday, C. Izzo, S. H. Woodcock and B. M. Gillanders, *Aquat. Biol.*, 2013, **18**, 271–280.
20. F. Monti, A. Robert, J.-M. Dominici, A. Sforzi, R. T. Bagur, A. M. Navarro, G. Guillou, O. Duriez and I. Bentalab, *Oecologia*, 2021, **195**, 655–666.
21. I. Onoka, A. T. Banyika, P. N. Naberjee, J. J. Makangara and L. Dujourdy, *Forensic Sci. Int.: Synergy*, 2020, **2**, 194–205.
22. G. E. David, D. B. Hibbert, R. D. Frew and A. R. Hayman, *Aust. J. Chem.*, 2010, **63**, 22.
23. N. Nic Daeid, S. K. Jayaram and W. Kerr, *Sci. Justice*, 2012, **53**(3), 278.
24. K.-W. Chan, G.-H. Tan and R. C. S. Wong, *Sci. Justice*, 2013, **53**, 73–80.
25. C. Liu, Z. Hua, Y. Bai and Y. Liu, *Forensic Sci. Int.*, 2014, **239**, 37–43.
26. R. J. Watling, G. Lee, C. J. Scadding, T. S. Pilgrim, R. L. Green, A. E. Martin, C. D. Mat and J. L. Valentin, *Open Chem. Biomed. Methods J.*, 2010, **3**, 179–196.
27. A. Schimmelmann, H. Qi, P. J. H. Dunn, F. Camin, L. Bontempo, D. Potocnik, N. Ogrinc, S. D. Kelly, J. F. Carter, A. Abrahim, L. T. Reid and T. B. Coplen, *J. Agric. Food Chem.*, 2020, **68**(39), 10852–10864.
28. K. Katerinopoulou, A. Kontogeorgos, C. E. Salmas, A. Patalas and A. Ladavos, *et al.*, *Foods*, 2020, **9**(4), 489–498.
29. United Nations Declaration of Human Rights, General Assembly resolution 217 A, 10 December 1948, https://www.un.org/en/about-us/universal-declaration-of-human-rights, (accessed 29/06/2023).
30. United Nations 17 Sustainable Development Goals, https://sdgs.un.org/goals, (accessed 29/06/2023).
31. K. Jaouen, M. Beasley, M. Schoeninger, J.-J. Hublin and M. P. Rochards, *Sci. Rep.*, 2016, **6**, 26281.
32. M. A. Campbell, V. Udyawer, T. D. Jardine, Y. Fukuda, R. K. Kopf, S. E. Bunn and H. A. Campbell, *Biol. Lett.*, 2016, **18**, 20210676.
33. IAEA, Assessment of Zinc Metabolism in Humans Using Stable Zinc Isotope Techniques, *Human Health Series No. 35*, IAEA, Vienna, Austria, 2018.
34. M. Bird, S. A. Crabtree, J. Haig and C. M. Wurster, *PNAS*, 2021, **118**, e2024642118.

35. K. Pye and D. J. Croft, *Forensic Geoscience: Principles, Techniques and Applications*, Geological Society, Special Publication, London, 2007, vol. 232.
36. L. A. Chesson, J. E. Barnette, G. J. Bowen, J. R. Brooks, J. F. Casale, T. E. Cerling, C. S. Cook, C. B. Douthitt, J. D. Howa, J. M. Hurley, H. W. Kreuzer, M. J. Lott, L. A. Martinelli, S. P. O'Grady, D. W. Podlesak, B. J. Tipple, L. O. Valenzuela and J. B. West, *Oecologia*, 2018, **187**(4), 1077–1094.
37. M. Arif, G. Chilvers, S. Day, S. A. Naveed, M. Woolfe, O. Y. Rodionova, A. L. Pomerantsev, O. Kracht, C. Brodie, A. Mihailova, A. Abrahim, A. Cannavan and S. D. Kelly, *Food Control*, 2021, **123**, 107827.
38. L. Strojnik, D. Potocnik, M. J. Hudobivnik, D. Mazej, B. Japelj, N. Skrk, S. Marolt, D. Heath and N. Ogrinc, *Food Chem.*, 2022, **381**, 132204.
39. M. P. V. Matos, M. E. Engel, J. B. Mangrum and G. P. Jackson, *Anal. Methods*, 2021, **13**, 3493–3503.
40. I. Gonzales-Martin, C. Gonzales-Perez, J. H. Mendez, E. Marques-Macias and F. S. Poveda, *Meat Sci.*, 1999, **52**(4), 437–41.
41. P. Tan, Honours Thesis, Flinders University, 2017.
42. M. O. Varrà, L. Husáková, J. Patočka, S. Ghidini and E. Zanardi, *Food Chem.*, 2021, **360**, 129790.
43. C. P. Bataille, B. E. Crowley, M. J. Wooller and G. J. Bowen, *Palaeogeogr., Palaeoclimatol., Palaeoecol.*, 2020, **555**, 109849.
44. V. Sahaipal, S. Mishra and D. Bhandari, Forensic Analysis in Wildlife Crime Cases: Microscopy, DNA Profiling and Isotopic Analysis, in *Forensic Analysis – Scientific and Medical Techniques and Evidence under the Microscope*, ed. I. Freckelton, Intech Open, 2021.
45. L. D. Mee and J. W. Readman, Nuclear and isotopic techniques for investigating marine pollution, *IAEA Bull.*, 1993, **2**, 1–8.
46. K. M. Rogers, J. C. Turnbull, J. Dahl, A. Phillips, J. H. Bridson, L. G. Raymond, Z. Liu, Y. Yuan and J. S. Hill, *Rapid Commun. Mass Spectrom.*, 2021, **35**, e9051.
47. Y. Suzuki, F. Akamatsu, R. Nakaashita and T. Korenaga, *Chem. Lett.*, 2010, **39**, 998–999.
48. D. Berto, F. Rampazzo, C. Gion, S. Noventa, M. Formalewicz, F. Ronchi, U. Traldi and G. Giorgi, in *Plastics in the Environment*, ed. A. Gomiero, Intechopen, London, 2019.
49. G. Munoz, L. Mercier, S. V. Duy, J. Liu, S. Sauve and M. Houde, *Environ. Pollut.*, 2022, **309**, 119739.

CHAPTER 4

Mass Spectrometry Methods for Securing Road Traffic Drugs and Casework Toxicology Convictions

M. C. PARKIN* AND S. C. GROSSE

Eurofins Forensic Services, Feltham, UK
*Email: Mark.Parkin@forensicsuk.eurofins.com

Abstract

This chapter describes the development and application of mass spectrometry methodologies used for obtaining evidence in forensic toxicology casework. It covers the evolution of methods from simple colour spot tests to the mass spectrometry approaches used today. The application of mass spectrometry to both road traffic drugs and casework toxicology is discussed in order to provide both context and insight into how evidence is attained in this area of forensic science and how the technique may evolve in the future to meet the demands of the criminal justice system.

4.1 Introduction and Brief History of Forensic Toxicology

Forensic toxicology is a branch of forensic science employing chemical analysis to help determine whether impairment or death has occurred due to drug use, poisoning or exposure to toxic substances. Typically, these poisons

New Developments in Mass Spectrometry No. 14
Applications of Mass Spectrometry for the Provision of Forensic Intelligence: State-of-the-art and Perspectives
Edited by Simona Francese and Stephen Bleay
© The Royal Society of Chemistry 2024
Published by the Royal Society of Chemistry, www.rsc.org

or drugs together with their metabolites are detected and identified in human biological specimens using some form of instrumental analysis. Evidence acquired from this process is used to aid in the legal investigation of a death or serious crime.

Forensic toxicology has a rich history dating back hundreds of years. Many believe that toxicology began with the Swiss lay theologian, physician and alchemist Paracelsus in the 16th century. In 1538, Paracelsus expounded the idea of dose response by stating in his Third Defence that "all things are poison and nothing is without poison; the dosage alone makes it so a thing is not a poison"; this notion is often summarized as "The dose makes the poison". Mathieu Joseph Bonaventure Orfila is considered the modern founder of forensic toxicology.

The 19th-century Spanish chemist and physician conducted pioneering research on poisons and their antidotes using live animals and effectively established several basic principles of modern pharmacology. He was able to develop and improve the accuracy of tests for small quantities of poisons prevalent at that time, including arsenic, cyanide, phosphorus and morphine. Importantly, this involved refining the Marsh test for the extraction of arsenic from organs, tissue and blood. This was the first occurrence of extracting arsenic from human biological samples, whereas previous analyses were conducted directly on the digestive tract. By systematically applying chemical analysis to autopsy samples, Orfila's results and opinions were sufficient to aid the courts of law in solving medicolegal cases and he effectively became forensic toxicology's first expert witness. Also, around this time in the mid 19th-century, further key advances in toxicological chemistry were made which focused on the extraction and isolation of poisons and drugs from the human body. This included a process for extracting alkaloids developed by the Belgian analytical chemist Jean Stas (who also co-discovered the atomic weight of carbon) which was further refined to include a broader range of alkaloids by the German chemist Julius Otto in 1856.[1] Stas was asked to investigate the murder of Gustave Fougnies by Count Hippolyte Visart de Bocarmé. Stas used potassium hydroxide and diethyl ether to extract nicotine from tissue samples of Fougnies and following distillation, he obtained brown residues which were accompanied by a cold tobacco smell.[2] During that time, it was not uncommon to use both taste and smell as a means for detection and preliminary identification. Having successfully extracted nicotine from the biological samples, he was then able to subject the residues to laboratory tests available for pure nicotine. Upon extraction and purification, colour tests, together with precipitation and crystallisation for conclusive identification, were often shown to juries, thus further establishing toxicologists as expert witnesses. Indeed, this process of extraction from the sample followed by purification and then identification remains the fundamental mechanism for obtaining forensic toxicology evidence that is still used today.

Towards the middle of the 20th century, it was the English forensic toxicologist, Alan S. Curry, working at the UK's Home Office, who developed systematic extraction methods for recovering drugs from human biological samples. Curry combined these processes with paper and thin layer chromatography for purification and then he usually used some form of spectrophotometry for identification. Much of his work focused on refining the Stas-Otto solvent extraction, essentially enhancing its efficiency and selectivity so as to encompass an even broader range of detected compounds, especially those which were labile.[3] Curry's work was extensively documented in his popular and highly cited book "Poison Detection in Human Organs" which had three editions, with the first being published in 1963.[4]

4.2 Mass Spectrometry as Used in Forensic Toxicology Casework

4.2.1 Introduction

Currently, the most powerful analytical tool available to forensic toxicologists is mass spectrometry. Following sample preparation, extraction and some form of hyphenated chromatography, drugs that might be present in the sample undergo molecular weight determination along with a form of structural determination. The technique was first established by the physicist Sir Joseph John Thomson at the Cavendish laboratory in Cambridge, where he was investigating the composition of cathode rays. During the Second World War, the technique gained more practical application and had a major role in the production of fissionable uranium and weapon-grade plutonium. High octane aircraft fuels were also needed and mass spectrometry became an important tool used by the petroleum industry for quality control in refining oil for this purpose. After the war, the power of the technique was realised for its ability to identify and characterise organic compounds. In the 1950s, with the development of so-called 'double-focusing high resolution' mass spectrometers, the concept of using accurate mass measurements to help identify compounds began. This was probably the turning point when mass spectrometry could be applied to forensic cases. Prior to the late 1960s, even though mass spectrometry instrumentation could help detect organic compounds in mixtures, it was not used to provide evidence for courts. The reason behind this is thought to stem from the costs and difficulty in operating instruments as well as the complexity involved in interpreting the data. It was John Herbert Beynon who realised that with sufficient mass accuracy it would be possible to obtain enough information on the elemental composition of a molecule such that its identity could be elucidated from a limited number of chemical formula choices.[5] When coupled to a GC *via* electron ionisation, the approach was known as mass fragmentography and by 1973 it was being routinely used by the UK's Home Office forensic science laboratories.[6] Table 4.1 details the distribution of case types following a full year of operation in 1973. It is interesting to see that seven cases involving driving under the influence of

Table 4.1 Distribution and number of case types in 1973. Data from ref. 6.

Type	Number
Illegal possession of drugs	59
Suspicious death	47
Explosives	18
Arson	17
Miscellaneous	10
Administration of noxious substances	8
Driving under the influence of drugs	7
Malicious damage	7
Documents	4
Biology	2

drugs cases were analysed and now, 50 years on, we typically handle tens of thousands of these case types per year in our own laboratory.

4.2.2 Basic Principles of Mass Spectrometry

Detection and molecular characterisation is achieved in mass spectrometry by the generation, separation and detection of gas-phase ions. The main components of the instrument are the ion source, the mass analyser and the detector. Once formed in the ion source, ions are transmitted in the gas-phase through a mass spectrometer by using electric or magnetic fields or, in some instrument types, a combination of both. In order to prevent the ions from undergoing deflection, and becoming neutralised or fragmented due to their collision with atmospheric molecules, the instrument is operated at low pressure under high vacuum (typically around 10^{-4} to 10^{-7} Pa). Upon exiting the mass analyser, component ions are sent to the detector where amplification of the ion's signal occurs.

Upon separating the ions, a complete mass spectrum is generated showing the *m*/*z* values on the *x*-axis of a graph against their signal intensity on the *y*-axis. Compound elucidation and identification involves careful interpretion of the spectra and often involves comparison to certified reference materials that have also been run through the instrumentation or comparison to a reference library of spectra in a database. To determine the amount of drug or poison present in a sample, a comparison of the signal intensity of the sample ions is made against a series of standard solutions containing known amounts of the reference material; this is known as a calibration curve.

4.2.3 Mass Analysers Commonly Used in the Forensic Toxicology Laboratory Today

The different types of mass analysers that are used in commercial mass spectrometry systems have been adequately covered in many textbooks elsewhere. Today, the three most popular mass analysers used in the forensic toxicology laboratory are the quadrupole, particularly in the triple quadrupole (QQQ) format, along with the high mass accuracy instruments, the Orbitrap and the time-of-flight (TOF), and these will be considered here.

4.2.3.1 *The Quadrupole Mass Analyser*

The quadrupole consists of four mutually parallel circular metal or graphitic rods such that the electric field produced between them is hyperbolic (quadrupolar). A dynamic electrical field can be created by the application of a simultaneous radio frequency (rf) which is 180° out of phase to the opposite pairs of rods. Each pair of opposing rods is also supplied with direct current (dc) of equal magnitude but with opposite polarities. The electrical field created then acts as a mass filter as it only allows ions of certain masses to pass down the gap between the four rods. By controlling the ratio of the rf to dc voltages, a narrow band filter can be produced allowing a packet of ions of one *m*/*z* value to take a stable trajectory between the rods to the detector. Ions with different mass to charge ratios will collide with the rods and become neutralised. By increasing the dc and rf voltages, stable trajectories for ions of different *m*/*z* values can be made to pass through to the detector.

Developed by Richard Yost and Chris Enke in the late 1970s,[7] the triple quadrupole instrument remains very much the workhorse mass spectrometer commonly found in forensic toxicology laboratories today. Here, the mass analyser region consists of two quadrupoles, the first (Q1) and the last (Q3), functioning as normal mass filters. The middle (Q2) quadrupole functions as a collision cell with only the rf potential turned on and filled with an inert gas (usually nitrogen or argon). This allows for all ions to pass through to it from Q1 and then get fragmented into smaller constituents by colliding with the gas molecules. All fragments are then passed through to the last quadrupole (Q3) solely through the rf field. As we shall see later, this combination of quadrupole fields enables the instrumentation to be operated in a particular way that benefits the acquisition of forensic evidence that cannot be achieved using a single quadrupole alone.

4.2.3.2 *The Orbitrap Mass Analyser*

The Orbitrap is one of the most recent types of mass analyser to be developed. It was invented by Alexander Makarov[8] and later commercialised by ThermoFisher Scientific in 2005. The Orbitrap is made up of a pair of cup-shaped outer electrodes which, when placed together, have a barrel shape and a central inner electrode in the shape of a spindle. Ions are injected into the space between the outer and inner electrodes, initiating their orbit around the spindle through a balance between the electrostatic and centrifugal forces. Ions then follow a circular orbit around the spindle electrode between the outer electrodes. The shape of the barrel's outer electrodes and the field they induce prompt the ions to oscillate axially along the spindle's x-axis. The frequency of this oscillation is mass dependent and is detected as a time domain signal and transformed into a mass spectra using a Fourier transform algorithm. This conversion from the time (frequency) domain to the mass domain allows the determination of the mass of the ions extending to four or five decimal places. This makes the Orbitrap a high mass accuracy instrument (often known as a high-resolution accurate mass (HRAM) instrument) with a mass accuracy up to four decimal places compared to a quadrupole mass analyser (either single or triple) that can only measure mass down to one decimal place at most.

4.2.3.3 The Time-of-flight Mass Analyser

Time-of-flight (TOF) mass analysers separate ions based on their m/z using time rather than their interaction with an electric field. The basic principle was first described by William Stephens in 1946[9] and then later in 1955, the first commercial instrument was developed by Wiley and McLaren.[10] The principle behind the TOF mass analyser is one of the simplest to follow. After their initial acceleration in an electric field, ions are separated in the TOF analyser according to their velocities when they drift in a field-free region known as the flight tube. Ions of the same kinetic energy are accelerated towards the detector with a constant energy. Due to the relationship between kinetic energy and mass, ions with a greater mass will have a lower velocity. Therefore, the time an ion takes to reach the detector across the known distance of the flight tube can be used to determine its m/z value.

A hybrid instrument that consists of a quadrupole mass analyser coupled to a TOF analyser (QTOF) was first developed by Howard Morris and colleagues and commercialised by Waters Corp in the late 1990s.[11] This instrument consists of a normal mass resolving quadrupole and an rf-only quadrupole serving as a collision cell like Q2 of the QQQ described earlier. The ion beam can be focused out of the second quadrupole and into a TOF that is placed orthogonally to the direction of the ion beam. Ions can then be selected through the filtering mode of the first quadrupole and then fragmented in the second quadrupole collision cell. The fragment ions formed can then be separated and measured in the TOF mass analyser. These instruments have very high ion transmission efficiencies, thereby resulting in very high sensitivity and good mass accuracy, like the Orbitrap.

4.2.4 Modes of Operation of Mass Analysers in the Forensic Toxicology Laboratory

All the mass analysers described earlier can generate the so-called full scan mass spectra. In this mode, all ions that are produced in the instrument are captured across a set range of m/z values as determined by the operator. On quadrupole instruments, full scan acquisition can take longer than other modes as the instrument's electronics needs to ramp the rf/dc potentials across the voltage range required to obtain the full scan spectra. The wider the *m*/*z* range required, the longer this process will take. Chromatographic peaks obtained on the recent generation of ultra-high performance liquid chromatography systems typically now coupled to QQQ instruments are between 3–5 seconds in width. Full-scan spectra may take too long to acquire enough data points across the chromatographic peak to adequately resolve it. For this reason, QQQ instruments are almost always operated in what is known as the selected reaction monitoring (SRM) or multiple reaction monitoring (MRM) mode. In this mode, Q1 is set at a fixed rf/dc potential to selectively filter only the *m*/*z* of the target analyte. This is known as precursor ion selection. The precursor ions then undergo fragmentation in Q2. Q3 is then set to only allow one, specific, fragment ion through to the detector. This is known as product ion selection. This combination of precursor and product ion selection is known as an SRM transition. Given that quadrupole mass analysers are low mass accuracy instruments, it is reasonable to consider that at that degree of mass accuracy, other ions that share both the same precursor and product ions could interfere with the SRM transition. In order to prevent this from occurring, other product ions characteristic of the drug molecule are selected for further SRM transitions. This combination of multiple SRM transitions is then known as MRM. This way of operating the QQQ allows for much faster data acquisition. As an illustration, the extracted ion chromatogram for our laboratory's road traffic drugs method is shown in Figure 4.1. As can be seen, the entire process to

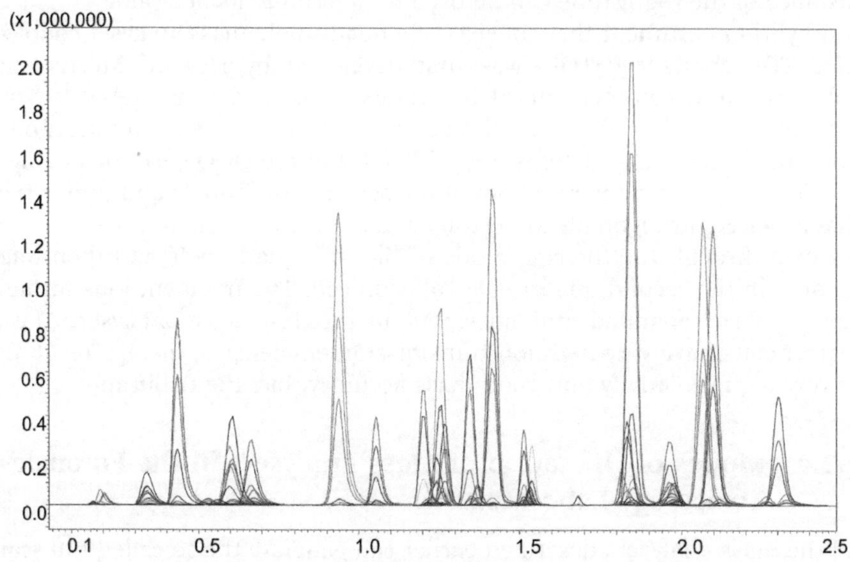

Figure 4.1 Extracted ion chromatogram from the authors laboratory showing the 2.5 minute run capturing 34 compounds for their road traffic drug method.

capture 34 compounds is complete in 2.5 minutes. Apart from facilitating rapid chromatographic run times, the SRM mode also decreases the noise obtained in the spectra from interfering precursor ions and from the smaller precursor fragments that would be present in the full scan acquisition. Since the overall sensitivity of a method is determined from the amount of signal *versus* the noise, the MRM mode can be seen as one of the most sensitive modes of operation for a QQQ instrument. When we know what analytes we need to target, as we will discuss when covering road traffic casework later in this chapter, operating the instrument in this way is known as targeted analysis. In the forensic toxicology laboratory, low mass accuracy instruments are always operated in this way. For untargeted approaches, full-scan modes of operation are used in high mass accuracy instrumentation. Since all ions are collected in this way, whether through precursor or product ion scanning, no data are lost. This allows the operator to revisit the data to interrogate it further if required. This approach is used in the forensic toxicology laboratory when little is known about the drugs or poisons involved, for example, in cases such as drug facilitated sexual assaults or the use of novel psychoactive substances.

4.3 Biological Matrices Used in Forensic Toxicology

During Orfila's time, one of the most common sample types from humans, that was used to obtain evidence, was tissue from the digestive tract. During that period, this approach was logical as sampling this region followed the path of a poison from entry to exit. Forensic toxicology, practiced in the past, would focus on detecting the presence of poisons in these samples. Modern forensic toxicology not only focuses on the presence but also asks the question of how much is present? Blood, either as anti- or post-mortem is now the preferred sample type closely followed by urine. Oral fluid, vitreous humour, tissue, hair and gastric contents are also, although less commonly, used. Table 4.2 details the advantages and disadvantages of each biological matrix presented here.

4.3.1 Whole Blood

In a living person, it is typically whole blood (as opposed to plasma or serum) that is the preferred matrix. The main advantage of using blood is that any drug or poison detected and measured in the sample reflects substances currently affecting the individual at the time when the sample is taken. Concentrations of drugs as measured in whole blood, plasma or serum give a direct correlation with the pharmacological effect of the drug and hence are useful matrices for interpretive purposes as discussed latter in this chapter. Whole blood is primarily used as opposed to serum or plasma as, historically, much of the forensic toxicology literature documents concentrations of drugs and poisons detected in whole blood. Following administration of a drug, detection windows range from a few minutes and up to

Table 4.2 Typical detection windows and characteristics of biological matrices used in forensic toxicology.

Matrix type	Detection window	Advantage	Disadvantage
Whole blood	Up to 24–36 hours	Useful for recent administration. Accurate correlation between dose taken and concertation detected	Invasive collection. Rapid elimination of most drugs. Requires significant sample preparation prior to analysis.
Urine	Up to 72 hours	Contains metabolites of drugs. Can indicate administration has taken place	Lack of parent compound. Concentrations measured of drugs present cannot be used for interpretation.
Oral fluid	Similar to blood – 24 to 36 hours	Useful for recent administration. Non-invasive. Drug concentration can correspond to those in blood	Rapid elimination. Some drugs found at low levels. Limited information in the literature to assist with interpretation.
Hair	Weeks to months depending on length	Very long window of detection. Segmental analysis can provide further information	Cannot detect recent exposure. Environmental contamination a problem, may not indicate administration.
Vitreous humour	Hours – days	Longer sample stability. Ease of sample preparation	Limited sample volume. Not all drugs pass through the blood/retinal barrier.
Tissue – liver	Hours – days	Useful when no other sample types are available	Limited interpretive value. Laborious sample preparation involving several steps.

24–36 hours. Concentrations of drugs vary between whole blood, serum and plasma depending on the partitioning of the drug between the fatty components of the blood cells and the more aqueous component of the plasma between the cells. This needs to be considered when any interpretation of dosage is made based on concentrations measured in these blood components. Furthermore, the centrifugation of whole blood to obtain either serum or plasma introduces an additional step in the sample preparation stage. Post-mortem blood provides limited plasma and serum due to the extent of putrefaction that may have occurred. Additional factors also need to be taken into consideration when using post-mortem blood for analysis, as changes in drug distribution occur following death. This is known as post-mortem redistribution and is covered in greater detail elsewhere.[12] The extent of post-mortem redistribution can significantly impact the interpretation, and therefore the conclusions drawn, of a case.

The practitioner would likely consider using QQQs in the targeted MRM mode on whole blood samples. The reason behind this is that the blood sample is valuable for quantitative analysis and so it is likely the suspected compound or compounds are known to the analyst. Using a QQQ in the MRM mode can provide superior levels of sensitivity when compared to operating other instruments in the full scan mode. This additionally allows the analyst to work with very small amounts (typically around 100 μL) of the blood sample and thus conserving the sample for any further analysis if required.

4.3.2 Urine

When whole blood is not available or if there has been a delay in obtaining samples following suspected drug administration, urine is the preferred matrix. Urine is liquid waste comprising predominantly of water (>95%) and is produced by the kidneys and stored in the bladder. The kidneys serve to regulate salt and water balance and filter waste and metabolic excretion products for elimination from the body *via* the bladder. In a healthy individual, the bladder can hold up to 500 mL from 2 to 5 hours or longer. Sampling of urine unlike blood is non-invasive although privacy is a concern as well as adulteration. Drugs undergo metabolism, primarily in the liver, to enhance their water solubility for passage through the kidneys and out of the body *via* the bladder and urination. This means that it is the metabolites of drugs that are primarily detected in urine. These will accumulate over time alongside a decreasing amount of the parent drug. As water is reabsorbed back into blood as urine passes through the kidney, drug concentrations can be 10 to 100 times more concentrated than in circulating blood. Variations in urine production both through age and diet yields differences in bladder volume. This makes the correlation between dosage and concentration detected somewhat meaningless. However, drugs and their metabolites may be detected in urine for longer periods than in blood. This provides a longer detection window and can be particularly useful for determining administration following drug facilitated crimes. Urine may not always be available as a post-mortem sample as in around 50% of deaths, the bladder is voided.[13]

For urine analysis, this matrix is particularly useful if there are delays in obtaining samples following suspected administration. Examples of such scenarios could arise after drug facilitated sexual assault where the victim may have been incapacitated at length and recollection or reporting to the police occurs some time after the assault. The use of HRAM instrumentation may be the preferred route for these types of samples as urine is likely to contain multiple metabolites of the compounds administered. It is unlikely that the MRM transitions will be known, and so full scan untargeted analysis or data driven (sometimes called data dependent) MS/MS analysis could be useful here. Furthermore, the use of GC–MS to acquire fragmentation rich electron ionisation spectra could also provide supportive information to any HRAM data acquired. Urine is also being used more frequently in the so-called dilute-and-shoot sample preparation approaches as discussed later. Here, any MS instrumentation used should have ion-optics/ion transfer regions that are robust to fouling and are readily accessible for cleaning.

This reduces the risk of loss of sensitivity over time due excessive build-up of matrix materials inside the instrumentation.

4.3.3 Oral Fluid

Oral fluid or saliva is produced by secretion from three salivary glands (parotid, submandibular and sublingual) in the mouth.[14] It predominantly comprises water and contains a variety of salts and electrolytes together with some proteins and peptides. It has a lower protein content compared to blood, making it a cleaner matrix in comparison. The concentration of drugs found in oral fluid has been reported to be similar to that of whole blood.[14] However, some drugs such as benzodiazepines are only found at very low concentrations. Detection of drugs in oral fluid can be significantly affected by the route of administration, the pK_a of the drug and the degree of protein binding.[15] Oral fluid is rarely used in forensic toxicology except, however, for road traffic drugs cases. In this instance, it is used as a matrix for drug screening at the roadside using immunoassay-based testing devices if the police suspect a driver is under the influence of drugs and is impaired. This is discussed in more detail later in the chapter.

4.3.4 Hair

Hair is often used in forensic toxicology cases particularly for its very long window of detection (weeks to months) as drugs are incorporated into the hair shaft and remain stable. Hair is a useful matrix when the windows for detection in blood and urine have passed and for practical purposes, its use can often be complimentary to blood and urine. Hair is primarily composed of protein (keratin), water and lipids. Since hair grows at a rate of 1 cm per month, segmenting the hair into millimetre sections can help determine a timeline of administration. Due to the low amounts of drugs that are incorporated into the hair shaft and its availability, analytical sensitivity is key and presents an ongoing challenge. There is a greater variability in drugs and metabolites that can be found in hair in comparison to blood. Contamination of the hair shaft by the environment and the external breakdown of some drugs (*e.g.*, cocaine) to their main metabolite can hinder and complicate the interpretation of the results obtained from using hair as a matrix.[16] Additional complications arise due to the fact that the deposition of drugs into hair is influenced by its melanin content. Dark pigmented hair seems to bind greater amounts of drug and metabolites then lighter pigmented hair. This is believed to be as a result of drugs binding more effectively to melanin. Since melanin is acidic and hydrophobic in nature, drugs that are basic and lipophilic are more readily incorporated into the hair shaft. Products used for bleaching, dyeing and perming hair often contain strong bases and when used on an individual, they can affect the stability of the drug incorporated into the hair.

Due to the small amounts of drugs that are incorporated into hair, highly sensitive mass spectrometry methods are preferred. Once again, the use of

QQQs and targeted analysis would be the preferred approach. Further sensitivity gains are also achieved by using micro- or nano-flow liquid chromatography coupled to miniaturised electrospray sources. A practitioner would likely consider having a dedicated instrument for hair analysis that can provide high levels of sensitivity without being hampered by the contamination from other biological matrices.

4.3.5 Vitreous Humour

Vitreous humour is the clear, colourless and gelatinous fluid that is found between the lens and retina of the eye. It is made up of approximately 98% water together with a mixture of collagen, proteins salts and sugars. It makes an excellent post-mortem forensic sample as it is one of the last sites in the body to undergo putrefaction. Its relative isolation in a closed compartment within the body makes it less susceptible to post-mortem redistribution and contamination. This preserves drug and metabolite concentrations for much longer periods in comparison to post-mortem blood. This feature makes it a useful matrix for when a body has not been discovered for days or weeks following death. It also simplifies the sample preparation process for analysis. However, like other matrices with high protein content, compounds that exhibit strong protein binding are less likely to be detected due to the small amount of protein found in this matrix. Sample volumes obtained can be limited, with a maximum of 2–3 mL available for collection from each eye. Due to this, a QQQ would be the instrument of choice when analysing this sample type.

4.3.6 Stomach Contents

Stomach contents have been proven to be useful in casework, particularly where visual examination for tablets, capsules or herbal remedies is possible. It can be an important matrix where oral intoxication has occurred and where drug material may still be present. If significant degradation has not taken place, then identification can help guide further analysis in other specimens that may have been taken. The gastric content can be homogenised and drugs extracted using methods described later on in this chapter. Subsequent analysis would focus on the parent drug as metabolites would not usually be expected to be present.

4.3.7 Tissues

Liver and muscle are still used frequently in post-mortem toxicology. Since the liver is the primary organ for the metabolism of drugs and poisons, it can provide some information when blood and urine are not available. However, it is limited in its usefulness for interpretive purposes. Muscle, particularly if taken from a suspected injection site, can be a useful sample type. There is a paucity of information on drug levels in this type of tissue, which again limits its interpretive value.

4.4 Sample Preparation and Extraction Techniques Used in Forensic Toxicology

Before a sample can be analysed and potential evidence obtained, it usually requires some form of preparation before being subjected to the instrumentation. This is primarily to separate potential compounds of interest from other interfering components that make up much of the matrix. This helps to avoid damage to the instrumentation and helps make the overall analytical method more robust. Sample preparation can also serve as a preconcentration step, allowing the analytes to come within the detection range of the MS instrumentation used. This approach also offers further benefits of using less sample. One of the features of samples acquired for any forensic analysis is that it is often not possible to obtain another sample. Each sample and the evidence contained within it is unique and it is desirable to obtain as much information as possible from it. The less sample that is required per test means that a greater number of tests can be carried out. Furthermore, in forensic science, it is not desirable to use the sample in its entirety. Only a portion of the sample is used to obtain a valid analytical result. This portion of the sample must be representative of the entire sample.

Owing to the complexity of the different biological matrices described in the preceding section, a variety of sample preparation techniques are available that vary in their intricacy as the complexity of the matrix increases from simple fluids (urine and vitreous humour) to blood and tissue (Figure 4.2). Most sample preparation techniques involve two key stages; (i) sample clean up and (ii) analyte extraction and concentration. Since the focus of this book is on mass spectrometry, only those methods that are used in forensic toxicology casework and subsequently analysed on mass spectrometers will be considered here.

4.4.1 Protein Precipitation

As we have seen earlier, whole blood remains the most useful and one of the common sample types used in forensic toxicology. Due to its very high protein and lipid content, these components need to be depleted in order to (i) unbind any drug bound to them, (ii) prevent them from fouling the instrumentation and (iii) suppressing the signal from the target analytes when undergoing introduction to the mass spectrometer. Traditionally, whole blood samples can be treated with agents to break up the blood cells (a process

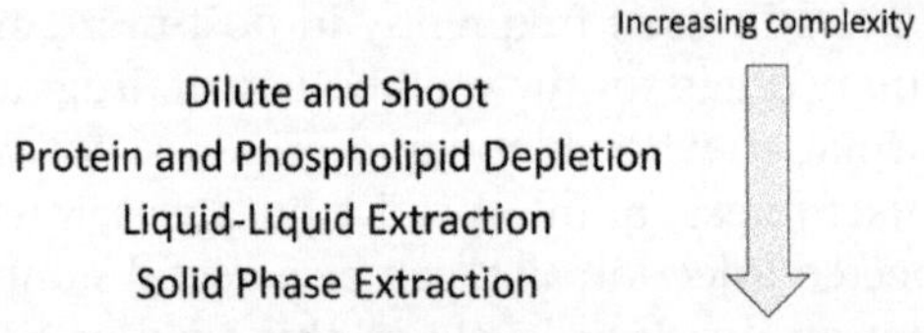

Figure 4.2 Complexity of the different sample preparation procedures.

known as haemolysis) followed by a precipitation step using organic solvent to separate the protein and lipid content.

4.4.2 Extraction Steps

Following protein precipitation, one of the two commonly used extraction methods is then used on blood: liquid–liquid extraction (LLE) or solid-phase extraction (SPE). If a biological sample does not contain large amounts of protein, then the samples are directly subjected to this extraction stage. LLE is simply the process of moving the analytes from one liquid phase to another, where the two liquids are immiscible but in contact with each other. The sample phase is the aqueous phase and is mixed with an organic solvent such as hexane or ethyl ether. The two solutions are vigorously mixed and then separated by centrifugation. The top, organic solvent, layer is then recovered and dried and reconstituted (concentrated) in solvents compatible with the chromatography-mass spectrometry system being used for analysis. There are significant drawbacks of LLE; in particular, the approach is time consuming with limited capacity to automate. Furthermore, large amounts of environmentally unfriendly solvents are used that are expensive to procure and dispose of. SPE is an alternative to LLE and is currently the preferred extraction technique in forensic toxicology laboratories. SPE works on the concept that the analyte can selectively absorb onto the active surface of a solid phase or be distributed into the liquid phase passing through the cartridge to which the solid phase is bonded. Modern SPE cartridges typically consist of purified silica particles (20–50 μm i.d.) that have hydrocarbon chains (C_8–C_{18}) and/or anion or cation exchange sites covalently attached. Recently, silica has been replaced with more pH resistant polymers that allow the utilisation of a wider range of buffers for targeting analytes.

4.4.3 Dilute-and-shoot

In recent years, when forensic toxicology laboratories handle urine samples, these are diluted with water and injected onto the LC–MS systems directly without any further sample preparation. This so called 'dilute-and-shoot' approach allows for any matrix interferences to be diluted to levels that are acceptable, thereby eliminating their interference with the analyte signal. This approach is far quicker than subjecting the sample to an extraction process and allows for increased sample throughput on the instrumentation. However, there is a greater risk of the instrumentation fouling due to a build-up of matrix materials. So, a balance has to be struck between throughput and downtime for additional cleaning and replacement of consumable components. The latest generation of mass spectrometers often have modifications in their ion-transfer optics, sample introduction and ionisation components to accommodate dilute-and-shoot methods. This provides a little more robustness and sensitivity, rendering such approaches achievable and this aspect would be particularly beneficial if urine is the preferred matrix for the laboratory.

4.4.4 Automation

Both SPE and dilute-and-shoot methods are amenable to automation. This involves the use of liquid handling robotics to dispense and process the samples alongside preparation of buffers and extraction solvents. Multiple samples are prepared simultaneously with minimal involvement from the laboratory analyst. In recent years, SPE cartridges have been miniaturised and produced in the 96-well microtiter plate format. This format can be used in liquid handling robotic systems for very high throughput of samples in parallel. This gives the forensic toxicology laboratory much greater capacity and helps reduce sample turnaround times. Other benefits with this type of automation are that human error is reduced and sample continuity is better managed through the use of robotic sample tracking and barcoding.

4.5 Quality Management of Forensic Toxicology Casework

The analytical results and interpretation provided by a forensic toxicology laboratory plays a fundamental role in solving criminal cases. Therefore, the mass spectrometry-based evidence provided to the criminal justice system must be reliable and robust. Every forensic toxicology laboratory should be strictly controlled and regulated by clear and defined guidelines. All activities, whether administrative or laboratory-based, should be outlined in standard operating procedures (SOPs). Methods must be validated and competent analysts must perform the work.

For forensic science providers (FSPs) in England and Wales, the forensic science regulator (FSR), the United Kingdom Accreditation Service (UKAS) and independent groups such as the United Kingdom and Ireland Association of Forensic Toxicologists (UKIAFT) are all sources that laboratories can use for guidance.

4.6 Forensic Casework – Driving Under the Influence of Drugs (DUID)

4.6.1 The History and Law (England and Wales)

Despite the commencement of British Motoring history in the late 19th century, it was not until 1930, with the introduction of the 'Road Traffic Act 1930', that it became on offence in England and Wales to drive, attempt to drive or be in charge of a motor vehicle on a road or any public place while being under the influence of drink or a drug to the extent of being incapable of properly controlling the vehicle. Over thirty years later, in 1967, a legal limit for driving under the influence of alcohol was set and it became an offence to drive, attempt to drive or be in charge of a motor vehicle with a blood alcohol concentration that exceeded the maximum prescribed limit (80 milligrams of alcohol per 100 millilitres of blood and 107 milligrams of alcohol per 100 millilitres of urine). At that time, the act referred to alcohol only, and for drugs, it remained necessary to prove impairment before a charge could be brought. In 1988, legislation was updated in the form of 'The Road Traffic Act, 1988', where according to Section 4, it is an offence to drive or attempt to drive a mechanically propelled vehicle while being unfit due to the influence of drink or drugs. Unlike the 'Road Traffic Act, 1967' for alcohol where Section 5 outlines the specified limit for alcohol, there was no such parallel act set for drugs, and Section 4 of the 'Road Traffic Act 1988' continued to stipulate that police officers must prove that the 'driver' was unfit. This is usually based on the arresting officer's observations at the scene (typically through field impairment test) and the medical practitioner observations while obtaining the individual's biological sample. The forensic toxicologist's role is to provide an expert opinion on any drugs found in the biological specimen, as to whether the drug and the amount of drug present could have a potential effect on psychomotor skills. This is highly challenged in the court and the prosecution success rate is relatively low. This was noted in a review conducted by Sir Peter North in 2010, who suggested that a new procedure should be introduced for drug driving that would not solely depend on the offence of impaired driving under Section 4 of the Road Traffic Act 1988.[17]

An expert panel of scientists was formed to assess the suitable legal limits for a set panel of 17 drugs.[18] The report which was produced by the expert panel, together with the 'North Report', led to the introduction of a new offence in England and Wales. The Crime and Courts Act 2013 inserted a new Section 5A in the Road Traffic Act 1988, which makes it an offence to drive, attempt to drive or be in charge of a motor vehicle with a concentration of a specified controlled drug in the body above a specified limit. A legal limit was set for the panel of 17 drugs, as outlined in Table 4.3, which meant it was no longer required to demonstrate impairment. The offence came into force on 2nd March 2015.

4.6.2 Analysis of Whole Blood Samples

Once an individual has been arrested for a DUID traffic offence, a blood sample can be taken and seized by the police. A person can be arrested for a Section 4 offence or Section 5A in England and Wales. For Section 5A, whole blood samples are collected in glass tubes containing potassium fluoride and EDTA as the preservative and anti-coagulant, respectively. This sample is then stored, packaged and transported appropriately to a forensic toxicology laboratory for analysis. This is the same for a Section 4 offence, but a defendant can refuse to give a blood sample, and a urine sample must be supplied instead.

Table 4.3 List of all substances covered under Section 5a of the Road Traffic Act 1988 and the specified limits in blood.

Substance	Specified limits ($\mu g\,L^{-1}$ of blood)
Amphetamine	250
Benzoylecgonine	50
Clonazepam	50
Cocaine	10
Delta-9-tetrahydrocannabinol (THC)	2
Diazepam	550
Flunitrazepam	300
Ketamine	20
Lorazepam	100
Lysergic acid diethylamide (LSD)	1
Methadone	500
Methylamphetamine	10
Methylenedioxymethamphetamine (MDMA, ecstasy)	10
6-Monoacetylmorphine (6-MAM)	5
Morphine	80
Oxazepam	300
Temazepam	1000

Historically, preliminary screening tests of DUID Section 4 samples in forensic toxicology laboratories were performed using immunoassay screening methods such as enzyme-linked immunosorbent assay (ELISA). A 2020 survey, completed by 65 forensic toxicology laboratories in the United States reported that 51% of the laboratories used ELISA for preliminary screening.[19] Papers from European forensic toxicology laboratories analysing DUID samples also report the use of ELISA for preliminary screening.[20] This was probably a much higher percentage of laboratories in earlier years, and in England and Wales, until recently, it is likely that most of the forensic laboratories screening DUID samples used such a technique.[21,22] However, with the advancement of mass spectrometry, there has been a shift in the approach to toxicological analysis for DUID cases. Despite the advantages of using ELISA for screening, such as low sample volume, minimal sample preparation and its ability to be automated for high throughput, limitations do exist with the technique. The main disadvantages include the technique's lack of specificity and the numerous ELISA kits required to test the full range of drugs and metabolites required for DUID analysis.

As described earlier, modern mass spectrometers, and QQQs in particular, have faster MRM acquisition and polarity switching speeds, allowing for the acquisition of many transitions in one method in both positive and negative ionisation modes without data loss and with acceptable quantitation since sufficient points across a chromatographic peak are achieved. Furthermore, mass analysers with larger dynamic response ranges can increase the sensitivity of methods by enabling detection of both very low and very high concentrations of target analytes within the same analysis. This has led to the analysis of a wide range of drugs, covering central nervous system (CNS) depressants, CNS stimulants, hallucinogens, narcotics and in certain cases cannabinoids in a single method.[23]

The Section 5A offence in England and Wales requires the analysis of whole blood samples for 17 specific drugs as listed in Table 4.3. THC and cocaine (with its metabolite benzoylecgonine) are the most encountered drugs, partly because of the initial roadside testing used to screen an individual for these compounds, and partly due to the popularity of the drugs over the past decade. It was noted in a report which evaluated the new drug driving legislation one year after its introduction that over 80% of tested blood samples contained THC.[24] This is a trend that forensic laboratories in England and Wales have continued to see.

The introduction of the new law gave forensic toxicology laboratories the opportunity to review the workflow of DUID samples through the laboratory. It used to be a common practice to use two independent analytical techniques before a substance could be identified and quantified in a sample. However, with the expansion of capabilities of instrumental techniques and by using a combination of hyphenated mass spectrometry techniques, the DUID samples can be screened for the required panel of drugs, and at the same time, any drug present can be confirmed and quantified. In line with guidelines such as 'Standard for Identification Criteria in Forensic Toxicology' ANSI/ASB Standard 113, 2023,[25] a low mass accuracy LC–MS system is sufficient to achieve identification of an analyte. To ensure reliability of any quantitative results, two aliquots from the same sample are independently taken through the extraction process and analysed, and sufficient number of SRM transitions should be monitored. This means that any laboratory wishing to undertake DUID analysis can use QQQ instrumentation at the levels of sensitivity typically offered as the "entry level" by MS vendors. There is no necessity to use highly sensitive instrumentation or HRAM technology.

4.6.3 Analysis of Urine Samples

Although whole blood is the preferred specimen for the analysis in DUID cases, urine is an option covered by Section 4 of the Road Traffic Act 1988. Section 7 (4) of The Road Traffic Act 1988 (provision of specimens for analysis) gives the arresting police officer the authority to choose whether the specimen required is blood or urine. Blood samples must be taken by a medical practitioner or health care professional and only if the medical practitioner (or health care professional) is of the opinion that for medical reasons a blood samples cannot or should not be taken, a urine sample will be collected for analysis.[26] A specimen of urine shall be provided within one hour of the requirement for its provision, as outlined in s.7 (5) of the act. However, two specimens must be provided in this time slot, with the first sample being discarded and the second specimen being used for analysis. This ensures that the urine sample tested is in equilibrium with the blood.

Of note, with regard to the *per se* law in England and Wales, although Section 5A (2) RTA 1988 states that the specimen can be blood or urine, there are no specific limits set for drugs in urine and therefore the sample must be blood and is reflected in police procedures. The reasonings for this are set out in the 2013 report 'The Drug Driving Expert Panel Report' which concluded that urine samples cannot be used due to the interpretative limitations of urine to be an indicator of the effects of drugs on an individual's ability to drive a motor vehicle safely.[18]

The analytical technology used for screening urine samples, like blood samples, was historically based on immunoassay techniques. However, as discussed above, the desire to move to multi-class and multi-analyte methods has prompted more laboratories to use mass spectrometry methods, like those set up for blood samples. However, simpler and quicker extraction methods can often be implemented, such as dilute-and-shoot, as urine, unlike blood, often does not require protein and lipid removal.

4.7 Casework Toxicology

Forensic casework toxicology covers a wide range of case types and sample types. It covers the analysis of biological samples from the living and the deceased (post-mortem toxicology). Although blood and urine remain the main focus of many forensic toxicology investigations, as discussed in Section 5.3, vitreous humour, hair, bile, liver, muscle and brain tissues can be useful specimens to aid investigations.

Casework toxicology can involve the detection of drugs used in sexual or other assaults, known as drug facilitated crimes, the determination of drugs in biological samples from those being accused of a crime, ranging from murder to grievous bodily harm (GBH) as well as alleged poisoning. Post-mortem toxicology may include looking for large concentration of drugs in overdose cases, or low therapeutic doses to establish if an individual was taking their prescribed medication. Ultimately, the forensic toxicologist is supporting coronial and criminal investigations and court procedures in determining 'how' and 'what' may have influenced an individual or contributed to the cause of death.

Historically, casework analysis would start with the screening of one or more biological samples using an immunoassay technique to screen the most common drugs of abuse such as amphetamines, opiates, cocaine, methadone, 3,4-methylenedioxymethamphetamine (MDMA,) a limited number of benzodiazepines, barbiturates and cannabinoids. Although as previously discussed, this screening technique is quick and requires minimal sample volume, these initial screening results could only suggest the class of drug that was potentially present in the sample. Post-mortem samples would often produce false positive amphetamine results, due to the cross-reactivity of putrefactive substances present in the samples.[27] Other false positive results were also seen on other kits, such as the antidepressant drug trazodone causing false positives on MDMA kits, and the antidepressant sertraline producing false positives on benzodiazepine kits.[28] This could lead to unnecessary further testing, using both case sample volume and extending the analysis time of the case. In addition to this, the scope of the testing was limited and although further individual immunoassay kits were available, to include drugs such as ketamine, buprenorphine and tramadol, a GC–MS screen was often used for the screening of a wider range of compounds and compliment the original immunoassay results. The GC–MS screening method could cover a larger scope of drugs, however, in certain case types, such as drug facilitated crimes, several other methods were needed. The extraction step for the general screen by GC–MS did not include a derivatisation step and therefore low-dose benzodiazepines and GHB were extracted and analysed using separate GC–MS methods. The advancement of LC–MS/MS has led to the phasing out of the immunoassay techniques. A multi-class drug method through LC–MS/MS allows a comprehensive screen to be conducted, yielding more specific results than the previously used immunoassay techniques, and it can provide a quicker turnaround time than the combination of an immunoassay screen and a GC–MS screen. Although further quantitation work is required following the initial LC–MS/MS screen, the initial analysis can give an indication of the direction a case may take at an early stage in the investigation.

The introduction of screening through HRAM into routine casework analysis also allowed a further comprehensive screening to be completed and consolidated a number of separate analytical methods, improving both the scope of analysis and reducing the length of time to achieve this more comprehensive screening of samples.

Traditional analytical workflows in forensic toxicology were often performed based on drug classes, for example, a laboratory would have a protocol for the analysis opiates and opioids, and a different method for analysing benzodiazepines. GC–MS was historically regarded as the 'gold standard' equipment for a forensic toxicological laboratory. However, the advances in analytical instrumentation using a mass spectrometer in conjunction with liquid chromatography has led to the quantitation of multi-drug classes in one method.

The biological sample size obtained in many forensic cases can be low and therefore requires a strategy stage to prioritise which analysis should take precedence. Decision making at this stage must be based on the information that is available on the circumstances of the case, at the time of receiving the sample, which is often limited. Improvements in the sensitivities of analytical techniques have enabled laboratories to minimise the sample volume required for poly-drug group analysis, resulting in analysis being conducted on as little as 0.1 millilitres of matrix, whereas in the past, this would have required ten times as much sample. This is a significant advantage for cases which arrive at the laboratory with minimal volume, when previously workflow could only offer very limited forensic toxicology evidence.

In 2021, there were 4859 deaths related to drug poisoning in England and Wales, which is the highest number of deaths since records began in 1993, and 6.5% higher than the previous year.[24] Approximately half of the drug deaths (45.7%, 2219 deaths) involved an opiate or opioid. Heroin and morphine continue to be the most frequently mentioned drugs, with 1213 drug poisonings mentioning either one of these substances. Cocaine and methadone are also increasingly being involved in drug deaths. These drugs are easily quantified on the 'workhorse' QQQ LC–MS instruments that are found in most modern forensic toxicology laboratories.

There has also been a significant increase in deaths involving new psychoactive substances (NPS) and modern analytical instrumentation have transformed and expanded the capability of a forensic toxicologist to support the everchanging field of new illicit designer drugs and powerful psychoactive medication that presents itself in modern casework.

The use of high mass accuracy mass spectrometry to perform screening on a wide range of forensic toxicology samples is well documented. High mass accuracy mass spectrometry allows both targeted and untargeted analysis, giving laboratories enhanced scope of testing. The rapidly changing trend in NPS provides a complex, albeit fascinating, challenge for forensic toxicology laboratories. Traditional analytical techniques cannot provide the specificity needed to identify these chemically similar compounds. The use of high mass accuracy mass spectrometry and libraries has helped in identifying compounds in casework scenarios where previously they might have gone unnoticed or been difficult to identify. The use of high mass accuracy mass spectrometry has enabled the development of comprehensive screening methods which detect a significant number of analytes, in some cases greater than 800 analytes per method, as well as allowing the potential for retrospective data analysis.[29–31]

A significant time interval between an incident and the collection of samples is not uncommon in forensic casework and is especially encountered in drug facilitated crime. For example, there may be a delay in reporting incidents to the police, which could be due the victim being unclear of events that led up to the incident and potentially reluctant to report the crime. In other situations, individuals may require emergency treatment due to life threating situations and it is not uncommon that very little or indeed none of the original biological samples taken on first admission to hospital would be remaining following critical clinical testing on the samples. Modern sensitive mass spectrometers can aid in forensic interpretation in such cases. Improvements in ion transmission efficiency, selective enrichment of targeted ions, improvements in ion utilization rate and the improvement in signal-to-noise have continued to improve sensitivity.[32] This has resulted in the detection of sub therapeutic concentrations of drugs in low sample volumes, where previously this may not have been achieved, potentially limiting the toxicological evidence available in cases where there is a time interval between the incident and the collection of the sample. Urine samples are generally the specimen of choice to conduct initial screening of drugs in cases where a time interval is involved. The longer detection window of drugs and their metabolites in urine enables the possibility of uncovering potential toxicological evidence several days after the alleged incident. However, the interpretation of drugs and metabolites in urine is limited and therefore the confirmation of the drug in urine is more pertinent, as opposed to the quantitation of the amount of drug present. Although dilute-and-shoot extraction techniques can often be employed for urine samples, as discussed in Section 5.4.3, more involved clean up steps are also used to ensure the attainment of detectable levels of of analytes for these types of crime.

4.8 Emerging Trends and Future Developments

It is likely that in future HRAM instrumentation will get faster in terms of data acquisition and scan speeds potentially reaching levels comparable to those of QQQ instrumentation. When this occurs, the role of QQQ instrumentation becomes questionable. A laboratory may be better off using HRAM and untargeted analysis as a standard approach rather than using targeted MRM methods. In order to reduce turnaround times, some laboratories are now combining both screening and quantification into single methods on HRAM instrumentation.[33] QQQs may still have an important role to play with limited sample volumes, in particular with post mortem blood. With smaller amounts of sample, smaller amounts of drugs are recovered and sensitivity becomes key.

For DUID work, miniaturisation of QQQ instrumentation may also play a role. A number of vendors are offering much smaller QQQ instruments which can now be stacked on top of the LC pumps. This offers the opportunities of moving the technology out of the traditional forensic toxicology laboratory and out into the field, either as mobile roadside units or in police stations.

Automation, particularly in the use of liquid handling robotics capable of managing the entire sample handling and extraction process, is increasingly being used by forensic toxicology laboratories. Here, fast and sensitive MS instrumentation is crucial as in order to process samples using 96-well extraction plates (that are amenable to automation), sub-100 μL sample and solvent volumes are used.

4.9 Conclusions

The acquisition of evidence in forensic toxicology cases has steadily advanced over the past 150 years, from simple visual methods but with high degrees of ambiguity to instrumental methods that have much greater certainty. The rapid commercialisation of mass spectrometry instrumentation over the past 50 years has greatly facilitated this evolution. This has not only given the courts more confidence in the scientific circumstantial evidence presented but also supports a system that now has to deal with tens of thousands of exhibits on an annual basis. Recent developments in high mass accuracy instrumentation have broadened the range of drugs and poisons that can be detected through the use of untargeted approaches. This together with sensitivity improvements across most mass spectrometry instrumentation now means that the forensic toxicology laboratory can do much more with much less.

References

1. K. D. Watson, *Acad. Forensic Pathol.*, 2020, **10**, 35–46.
2. R. Wennig, *Drug Test. Anal.*, 2009, **1**, 153–155.
3. R. Singh, *Forensic Sci. Int. Rep.*, 2022, **5**, 100275.
4. A. S. Curry, *Poison detection in human organs*, Charles C. Thomas, Springfield, Illiniois, USA, 1st edn, 1963.
5. J. H. Beynon, *Advances in Mass Spectrometry*, 1959.
6. J. A. Zoro and K. Hadley, *J. Forensic Sci. Soc.*, 1976, **16**, 103–114.
7. R. A. Yost and C. G. Enke, *J. Am. Chem. Soc.*, 1978, **100**, 2274–2275.
8. A. Makarov, *Anal. Chem.*, 2000, **72**, 1156–1162.
9. M. M. Wolff and W. E. Stephens, *Rev. Sci. Instrum.*, 1946, **24**, 616–617.
10. W. C. Wiley and I. H. McLaren, *Rev. Sci. Instrum.*, 1955, **26**, 1150–1157.
11. H. R. Morris, T. Paxton, A. Dell, J. Langhorne, M. Berg, R. S. Bordoli, J. Hoyes and R. H. Bateman, *Rapid Commun. Mass Spectrom.*, 1996, **10**, 889–896.
12. D. J. Pounder and G. R. Jones, *Forensic Sci. Int.*, 1990, **45**, 253–263.
13. G. R. Jones, in *Clarke's Analytical Forensic Toxicology*, ed. S. Jickells and A. Negrusz, Pharmaceutical Press, London, 2008, pp. 191–217.
14. O. H. Drummer, *Clin. Biochem. Rev.*, 2006, **27**, 147–159.
15. A. G. Verstraete, *Ther. Drug Monit.*, 2004, **26**, 200–205.
16. C. Holden, *Science*, 1990, **249**, 1099–1100.
17. Department of Transport, https://www.ias.org.uk/uploads/pdf/Drink%20driving%20docs/NorthReview-Report.pdf, April 2023.
18. Department of Transport, https://assets.publishing.service.gov.uk/government/uploads/system/uploads/attachment_data/file/167971/drug-driving-expert-panel-report.pdf, April 2023.
19. A. L. D'Orazio, A. L. A. Mohr, A. Chan-Hosokawa, C. Harper, M. A. Huestis, J. F. Limoges, A. K. Miles, C. E. Scarneo, S. Kerrigan, L. J. Liddicoat, K. S. Scott and B. K. Logan, *J. Anal. Toxicol.*, 2021, **45**, 529–536.
20. J. Maurer, E. Vergalito, A.-F. Prior, N. Donzé, A. Thomas and M. Augsburger, *Forensic Sci. Int.*, 2021, **329**, 111081.
21. H. J. Burch, E. J. Clarke, A. M. Hubbard and M. Scott-Ham, *J. Forensic Leg. Med.*, 2013, **20**, 278–289.
22. B. Rooney, G. J. Gouveia, N. Isles, L. Lawrence, T. Brodie, Z. Grahovac, M. Chamberlain and G. Trotter, *J. Anal. Toxicol.*, 2016, **41**, 140–145.
23. M. Farley, H. Tran, S. Towler, J. Gevorkyan, S. Pearring and L. N. Rodda, *J. Anal. Toxicol.*, 2021, **46**, 658–669.
24. Office for National Statistics, https://www.ons.gov.uk/peoplepopulationandcommunity/birthsdeathsandmarriages/deaths/bulletins/deathsrelatedtodrugpoisoninginenglandandwales/2021registrations#:~:text=4%2C859%20deaths%20related%20to%20drug,(79.5%20deaths%20per%20million), April 2023.

25. American Academy of Forensic Sciences, https://www.aafs.org/asb-standard/standard-identification-criteria-forensic-toxicology, April 2023.
26. U.K. Government, https://www.legislation.gov.uk/ukpga/1988/52/contents, April 2023.
27. T. Kupiec, L. DeCicco, V. Spiehler, G. Sneed and P. Kemp, *J. Anal. Toxicol.*, 2002, **26**, 513–518.
28. A. Saitman, H.-D. Park and R. L. Fitzgerald, *J. Anal. Toxicol.*, 2014, **38**, 387–396.
29. A. N. Kimble and A. P. DeCaprio, *Forensic Chem.*, 2019, **16**, 100189.
30. J. B. Stephenson, M. L. Flater, J. Austin, L. T. Bain, L. A. Holt and J. M. Mehan, *J. Anal. Toxicol.*, 2020, **45**, 243–251.
31. E. Partridge, S. Trobbiani, P. Stockham, C. Charlwood and C. Kostakis, *J. Anal. Toxicol.*, 2018, **42**, 655–660.
32. C. Li, S. Chu, S. Tan, X. Yin, Y. Jiang, X. Dai, X. Gong, X. Fang and D. Tian, *Front. Chem.*, 2021, **9**, 813359.
33. E. Rab, S. Martin, A. Freemont, S. Maslin, R. Marr and M.-S. Kanu, *J. Anal. Toxicol.*, 2023, **47**, 317–323.

CHAPTER 5

Gas Chromatography–Mass Spectrometry Based Approaches for the Analysis of Seized Drugs

ENOCH ADEWALE,[a] MIKE DAY[b] AND KAREN CHAN*[b]

[a] Eurofins Forensic Service, 1 Dukes Green Ave., Feltham TW14 0LR, UK;
[b] United Kingdom Accreditation Service, 2 Pine Trees, Chertsey Lane, Staines-upon-Thames TW18 3HR, UK
*Email: Karen.Chan@ukas.com

Abstract

Gas chromatography–mass spectrometry (GC–MS) is a powerful analytical tool with an extensive range of applications in the analysis of seized substances. In this chapter, several such applications will be outlined and discussed, including the use of GC–MS to carry out drug identification within a forensic environment and its impact on the criminal justice system. Furthermore, a discussion on the challenges arising from the utilisation of this technique, and how forensic laboratories tackle these challenges, will be presented. Other operational difficulties encountered by forensic laboratories will also be critically appraised along with their impact on the criminal justice system.

New Developments in Mass Spectrometry No. 14
Applications of Mass Spectrometry for the Provision of Forensic Intelligence: State-of-the-art and Perspectives
Edited by Simona Francese and Stephen Bleay
© The Royal Society of Chemistry 2024
Published by the Royal Society of Chemistry, www.rsc.org

5.1 Introduction – Drug Identification

When suspected drugs are seized, one of the primary aims is to ascertain the presence and identify any controlled compounds present, ensuring that (i) they are present at a justifiable level and (ii) are identified robustly enough to withstand intense scrutiny if challenged within a court of law.

There are multiple routes to analyse and identify drugs of abuse, and, over a period of years, the Scientific Working Group for the analysis of seized Drugs (SWGDRUG) has produced and updated guidelines to assist in developing effective procedures to identify unknown substances. These guidelines include categorising commonly used techniques depending on their ability to discriminate between compounds (Table 5.1[1]).

Based on the above categories, the scientific working group recommends that a minimum of two different techniques be used to determine the identity of an unknown substance. Due to the discrimination power of the category A techniques (Table 5.1), when they are used as part of the identification procedure, the secondary technique can be chosen from any of the remaining categories, and when the results of each technique are corroborative, this should be considered sufficient to confirm the identification of the suspected drug in the sample.

While some illicit drug samples are sold in a relatively pure state, others can be heavily mixed with adulterants and diluents, meaning that it is useful to be able to separate the different component parts of a mixture.

Table 5.1 Analytical techniques based on selectivity.

Category A (selectivity through structural information)	Infrared spectroscopy Mass spectrometry Nuclear magnetic resonance spectrometry Raman spectroscopy X-ray diffractometry
Category B (selectivity through chemical and physical characteristics)	Capillary electrophoresis Gas chromatography Ion mobility spectrometry Liquid chromatography Microcrystalline tests Supercritical fluid chromatography Thin layer chromatography Ultraviolet/visible spectroscopy Macroscopic identification (cannabis only) Microscopic identification (cannabis only)
Category C (selectivity through general or class information)	Colour tests Fluorescence spectroscopy Immunoassay Melting point Pharmaceutical identifiers

5.1.1 Drug Identification Using Presumptive Test and GC–MS

Gas chromatography coupled with mass spectrometry provides a rapid and effective method for both the separation of compounds and the subsequent identification of the individual components based on the *m*/*z* values of their parent ion and/or fragments generated, providing structural information. This makes it an extremely useful analytical tool for the identification of controlled drugs as each technique independently fits into one of the two categories; therefore the combination of these analytical techniques covers both category A and B techniques (Table 5.1) within a single analytical process.

However, it should also be noted that when the sample size allows, it is recommended to conduct a simple presumptive test (*e.g.*, a colour test or immunoassay) prior to GC–MS analysis of a sample.[2]

These tests are useful as they are very quick and can give an indication of which drugs or families of drugs may be present within a sample; for example, the Marquis reagent yields vivid colour changes to indicate the presence of various groups of drugs including phenethylamines, amphetamine/methylamphetamine, other amphetamine-type stimulants such as methylenedioxymethylamphetamine or MDMA (commonly known as ecstasy), and opiates such as diamorphine (heroin[3]).

Depending on the instrumental setup and method used, the results of these 'spot tests' provide the analyst with useful information to make appropriate next steps in the analysis of drugs of abuse using GC–MS. Some drug molecules are highly volatile or produce a mass spectrum with very little characteristic information. These compounds may benefit from a derivatisation step prior to GC–MS analysis. This step ensures that they transition to the gas phase and elute after the solvent front thereby generating more detailed structural data which ultimately can distinguish between similar, non-characteristic mass spectra.

The identification of suspected seized drugs by GC–MS is primarily carried out by comparison of the acquired electron ionisation (EI)-generated mass spectrum with known spectra generated either through the analysis of a known standard, or by comparison to a mass spectral library.

Mass spectral libraries usually fall into one of three categories, each with its own advantages and disadvantages. Many companies choose to develop their own in-house library of mass spectra. These work well as they can be tightly controlled to adhere to industry standards and can be highly specific and targeted to specific drug types; for example a company may choose to have a specific in-house library for synthetic cannabinoids, or cocaine alkaloid impurities. These libraries, however, demand significant time from the analyst to develop and are reliant on the availability of reference material; as such, in the fast-moving world of drug analysis, they can quickly become outdated and key new and emerging drugs could be missed. By the same token, provided they can be suitably characterised, new drugs that are encountered can be quickly and easily added to in-house libraries so that some gaps can be closed quickly. Libraries created by industry research groups/bodies are often ideal to fill some of these gaps. Those provided by the European Network of Forensic Science Institutes (ENFSI) or the Centre for Forensic Science Research and Education (CFSRE) often contain EI mass spectra of many of the newer compounds in the drug market and are generated with the specific reference to the sector considered. The main disadvantage of these libraries is that they tend to be updated and distributed on a fixed schedule, and therefore if a new drug emerges shortly after an update, it might take several months before the corresponding mass spectrum becomes available. The third main source of libraries is commercially available libraries, such as the NIST library (by the National Institute of Standards and Technology, US). These libraries often contain many thousands of compounds and are regularly updated. However, they can entail a significant cost for purchase and implemention, and due to the size of the libraries, there could be a number of irrelevant compounds/matches identified within an impure sample.

GC–MS acquisition software usually either has a built-in spectral search function or is bundled with a secondary software designed to fulfil this function. This software allows the comparison of an acquired spectrum with that of a known compound or specified libraries of compounds.

Each mass spectral comparison software will work in a slightly different way and return slightly different results for the same matches; however, they all work in accordance with the same set of principles. Each software package compares the fragmentation ions present, relative abundances and ratios between major peaks. Depending on the software, a higher or lower significance is assigned to factors such as the presence or absence of a certain mass ion at the expected *m*/*z* ratio. This comparison results in the generation of a correlation or match scores which are typically expressed as a number, with either 99 or 999 as the maximum match score.

Table 5.2 illustrates an example of an interpretation of match score quality provided by mass spectrum search software. In the example system, a match score of 999 is the highest possible match.[4]

However, the match score alone should not be used as a definitive identification of the unknown compound but rather as an indicator. In addition to the match quality, the unknown mass spectrum and the reference mass spectrum should be visually compared to ensure that the peaks present and

Table 5.2 MS library comparison match score and potential interpretation.

Match factor/match score	Interpretation
>900	Excellent
800–900	Good match
700–800	Fair match
<600	Poor match

ion ratios are similar. While this approach may enable the differentiation of similar compounds in the library, or allow the recognition of a true unknown (*i.e.*, a compound not present in the library), this might not always lead to a definitive identification, as will be discussed later in this chapter.

5.2 GC–MS Analysis and Interpretation

While mass spectrometry often serves as the primary component of GC–MS used to identify a compound, it is important to carefully consider the chromatographic aspect in order to ensure a robust analysis of the submitted sample.

First, it is important to ensure that the method profile is capable of achieving the desired results. The optimisation of the chromatographic method is paramount. First, the inlet temperature must be optimised; depending on the solvent, the inlet temperature must be high enough at the time of injection to ensure the vaporisation of all compounds at the point of injection.

Some drug compounds are highly volatile; therefore the starting temperature of the GC oven should be low enough to facilitate the retention of the compound on the column until after the detection of the solvent front to avoid masking. As a result, it is often recommended to start methods at temperatures of around 70–80 °C. This is particularly important when analysing low mass compounds such as amphetamine. Once the solvent has eluted out of the column, it is important to assess the temperature needed to allow the separation of any high mass compounds present on the column, and the length of time needed to ensure that all compounds of interest have eluted out of the column. While it is often beneficial to have one method set up which is able to separate and identify the majority of seized drugs encountered, it may also be necessary to have a number of different methods available, with (i) either adjusted temperature gradients or varied carrier gas flow to separate selected drug peaks, or (ii) a lengthened run time when high mass compounds are suspected to be present (or shortened run times for samples where it is known that all possible drugs present will elute early within the run).

5.2.1 Retention Time

Retention time (RT) data can provide additional information to support an identification, or to dismiss a possible false identification. When a certified reference material (CRM) is run on the same system as the suspected drug to be analysed, the RT of the peaks of interest would be expected to match within a margin of uncertainty (which is often determined at the point of validation), if the sample peak is in fact the same compound as the CRM. If the RT of the CRM and sample peaks exhibit significant differences, this could indicate either the presence of a similar compound not present in the mass spectral library, or the existence of a positional isomer of the compound. Although it is possible for multiple compounds to share the same or similar RT within a given system, as mentioned previously, by varying the parameters, changing the column, the separation technique or the method, it may be possible to increase selectivity and chromatographic resolution, thus resulting in separation of similar compounds. When multiple compounds generate a very similar mass spectrum, a CRM should be run for each possible match in order to confirm or eliminate each one individually. Any CRMs used as standards for RT comparison should be run using the same method parameters, and re-run if any significant maintenance is carried out on the instrument, such as column trimming, as this will affect the results obtained.

Comparison of a sample RT against that of a CRM can be a very useful additional tool; however, when the range of compounds encountered becomes very large, running a CRM standard for each compound on a regular basis can be very time consuming and costly. To counteract the inefficiencies of constantly running a certified standard, a database of RT of compounds relative to a set of standards, known as linear retention indices (such as Kovats) can be developed. To do this, a standard mixture of alkanes (typically C8–C40 chains) is injected into the instrument along with the compound standards. The RT of the compound relative to the alkanes that elute before and afterwards can then be calculated to derive an LRI reference value. This can be completed/calculated for each compound of interest within an organisation. Daily quality controls with compounds of known retention such as one of the alkanes used in the LRI calculation could be run daily with an expected RT based on the previously run alkanes compared against the daily RT. A variability limit must be set based on validation. When a change is made to the system that warrants re-running the standards, the alkane mixture can be run alone, and the RT of each compound is calculated using the RT of the relevant alkanes and the LRI values previously calculated.

5.2.2 Unknown Samples and Sample Preparation

One of the most important considerations that can affect the success of the GC–MS analysis is sample preparation. For the majority of 'simple' powder-based samples (powders, tablets and solids), it is sufficient to select the correct solvent based on likely solubility for the task; methanol, ethanol and chloroform are commonly used as solvents because the majority of drugs are highly soluble in these solvents, or at least they exhibit a solubility level which will produce a good enough peak to illicit further investigation. This step may need to be revisited depending on the results of the initial analysis.

For example, diamorphine (heroin) is poorly soluble in ethanol (approximately 1 : 37 w/v) but highly soluble in chloroform (approximately 1 : 1.5 w/v); therefore, a powder containing a low percentage of diamorphine may produce a poor peak if run in ethanol, but will produce a higher peak response if run in chloroform. Interestingly, paracetamol, one of the most common adulterants in illicit heroin powders is highly soluble in ethanol (approximately 1 : 7 w/v), but poorly soluble in chloroform (approximately 1 : 50 w/v).[5]

Alternative sample matrices may require even more preparation stages before being analysed by GC–MS. Certain medical preparations containing commonly abused drugs present themselves as a linctus. These commonly contain a large proportion of sugars, which can interfere with the compound extraction process and the resulting peaks obtained, and cause mechanical issues with the instrument, such as sticking and bending of the injection syringes. However, using a suitable solvent (or series of extractions), the drugs within these liquids can be extracted without the sugars being carried across.

5.2.3 Use of Derivatisation

Alongside solvent selection, it may be necessary to undertake a derivatisation step as part of the sample preparation process. Derivatisation in the analysis of seized drugs fulfils a number of functions. First, it can be used to increase the mass of a compound, which, as a result, can enhance the distinctness of the resulting mass spectrum. This is achieved through transformation of unstable hydrogens into amides, esters.[5] This is particularly useful for amphetamine-type stimulants, which often have a scarcely fragment-rich spectrum when un-derivatised. For example, the spectrum of amphetamine becomes significantly more fragment-populated and characteristic when derivatised with *N*-methyl-bis(trifluoroacetamide) (MBTFA), making its identification more reliable[6] (see Figure 5.1).

Another benefit of derivatisation is that it can increase the retention of analytes within the column. This increased RT may result in better separation of analytes but can be particularly useful for smaller and more volatile compounds, which may otherwise elute very early within an analytical run. Once again, using amphetamine as an example, underivatised,[7] the RT of amphetamine is observed at approximately 1.21 minutes. This RT can be close to the solvent front from the injection and, as such, the peak may be masked, especially in a low abundance (typical illicit amphetamine samples contain less than 10% amphetamine). In contrast, the trifluoroacetamide (TFA) derivative of amphetamine elutes at a much delayed RT of 1.65 minutes, using exactly the same instrumental parameters and temperature profile.

Another instance in which derivatisation may be useful is for enabling differentiation between some isomers and even diastereomers. For example, ephedrine and pseudoephedrine yield different chromatograms when derivatised, giving rise to a different numbers of peaks, due to the limitations on how TFA can bind to the two diastereomers; as TFA can have a preferential interaction with one diastereomer more than another due to the spatial structure of functional groups, depending on the isomer arrangement, derivatisation can results in two different chromatograms.

Chiral derivatisation enables the differentiation of compounds that are otherwise indistinguishable, such as enantiomers, for example, levomethorphan and dextromethorphan. Although the use of chiral derivatising agents can be time consuming and expensive, it may at times be required,

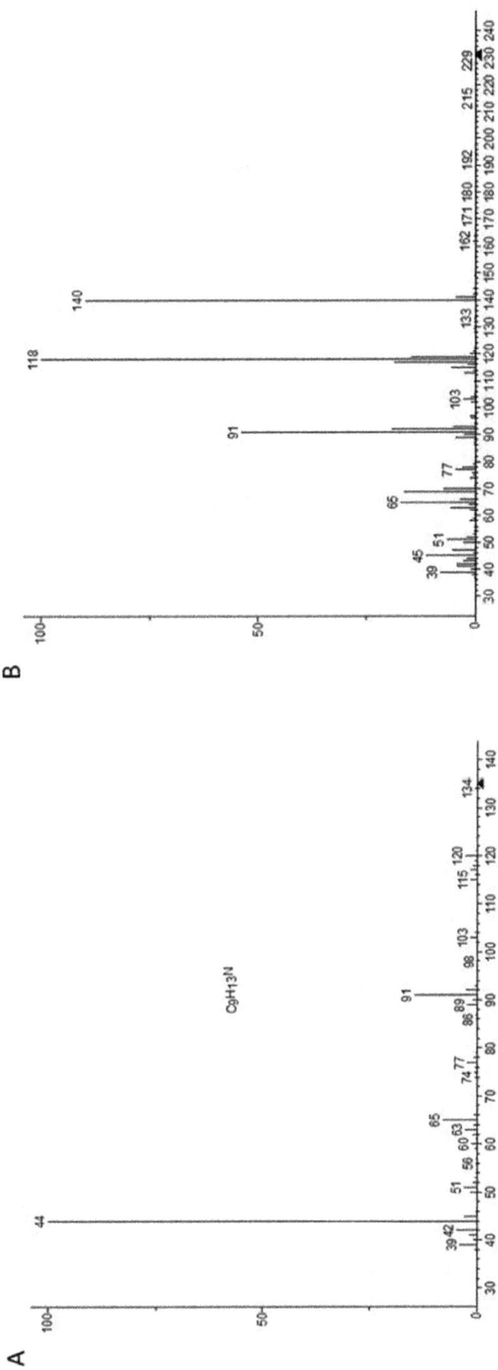

Figure 5.1 Mass spectrum of amphetamine (panel A) and amphetamine TFA (panel B).

due to the different effects and therefore different legal controls associated. With the example of methorphan, levomethorphan (a strong opioid analgesic) is a class A controlled drug under current UK legislation, while the dextro-enantiomer (dextromethorphan, a common antitussive) is not controlled, so distinction between the two can be important for law enforcement purposes.

Finally, derivatisation can be used to impart higher thermal stability to compounds, thus making them amenable to GC–MS. One particularly good example of this is the compound gamma-hydroxybutyric acid (GHB). When subjected to sufficient heat, such as within the injection port, or column of a GC–MS instrument, GHB thermally decomposes to form gamma-butyrolactone (GBL). Therefore, if GHB is not derivatised using a suitable agent such as *N*,*O*-bis(trimethylsilyl)trifluoroacetamide (BSTFA), it can lead to an incorrect identification.

5.3 Technical Challenges Related to Identification of Suspected Illicit Drugs

While analysing seized drugs by GC–MS, any number of difficulties might arise. In the next sections, a number of scenarios will be discussed along with ways in which they could be resolved, with reference to specific drugs as examples.

5.3.1 Differentiation of Isomers

Positional isomers – 2, 3 and 4-methylmethcathinone (mephedrone). These molecules share a similar structure with just the position of the methyl group attached on the benzene ring being the differentiator.

As expected, given the similarity between the structures, the mass spectra of each of the compounds are also similar as shown in Figure 5.2. However, these structural isomers can be differentiated relatively easily using an appropriate method that produces different enough retention times on the column. Within the system used to collect the data for this chapter, 2-MMC (2-methylmethcathinone) has a RT of approximately 2.35 min, 3-MMC has a RT of approximately 2.41 minutes, and 4-MMC has a RT of approximately 2.46 minutes. Although the variation is still quite small, it is significant enough that it can be measured, especially when standards are run alongside the sample.

Two structural isomers-α-methyltryptamine (AMT) and 5-(2-aminopropyl)-indole (5-IT). As with the methylmethcathinone compounds, the structural similarities lead to almost indistinguishable mass spectra (Figure 5.3).

However, in this case, an added complication arises: under the standard operating conditions used to identify drugs, the two compounds also elute with the same approximate RT. In this case, the identification of the isomer present is important as under the provisions of the Misuse of Drugs Act, AMT is a class B controlled drug, while 5-IT is not controlled. In order to

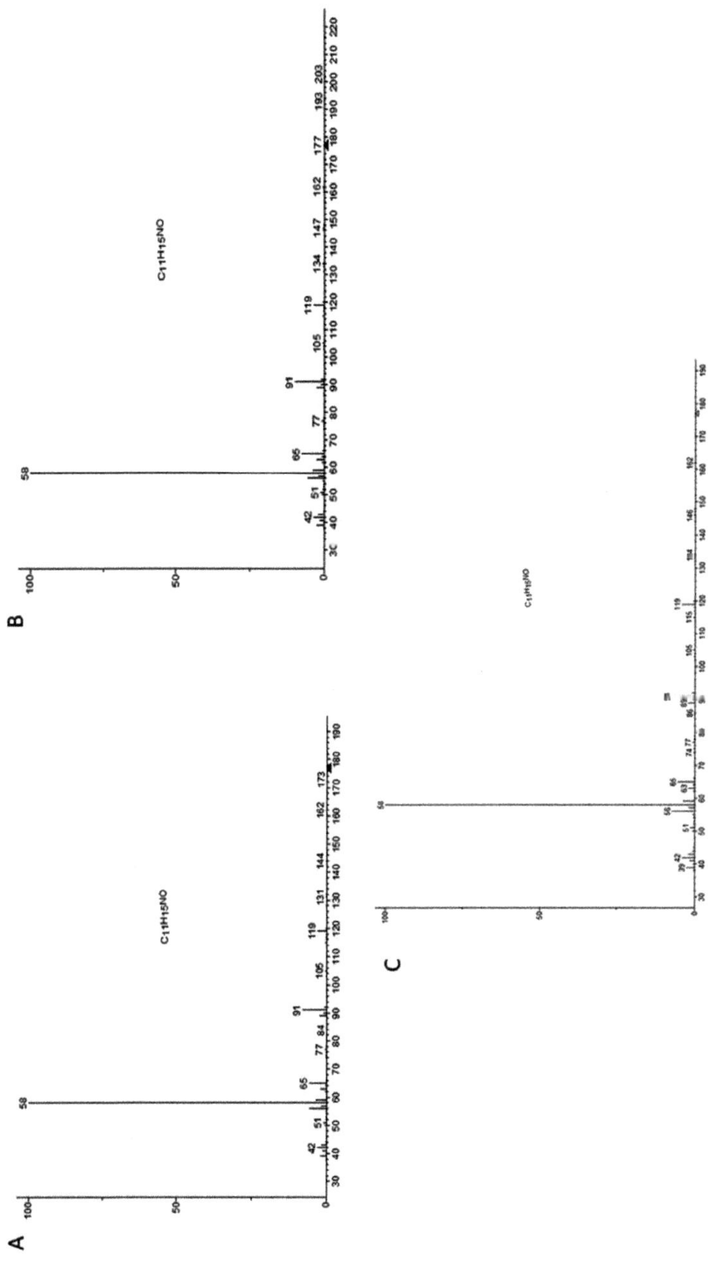

Figure 5.2 Mass spectrum of 2-MMC (panel A), 3-MMC (panel B) and mephedrone (panel C).

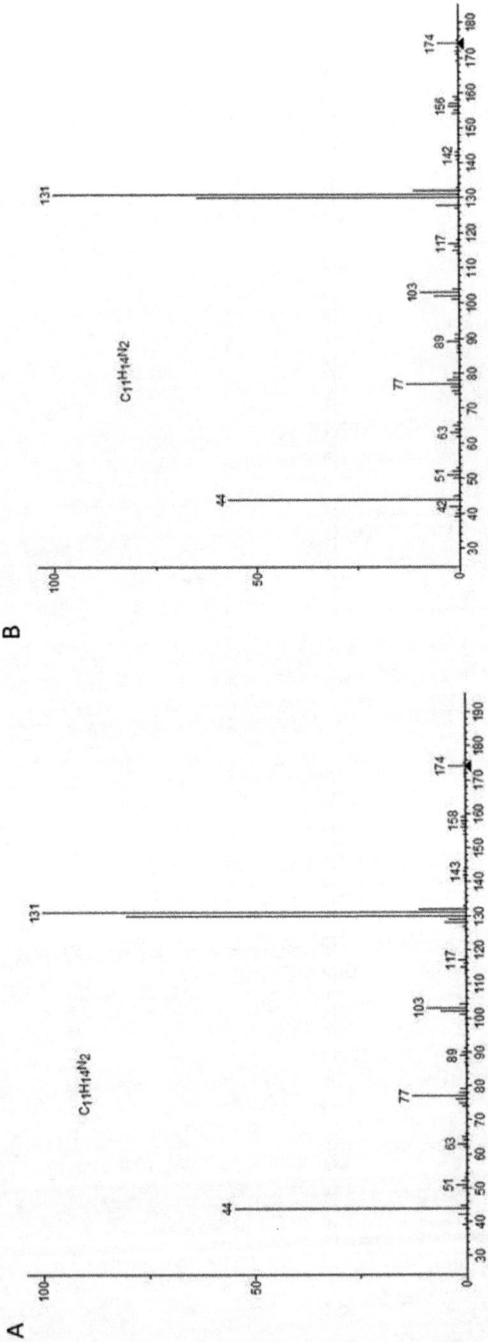

Figure 5.3 Mass spectrum of alphamethyltryptamine (panel A) and 5-(2-aminopropyl)indole (panel B).

achieve separation and be able to definitively identify the sample, a lower starting temperature gradient of the method would extend the retention of the compounds within the column and result in two different RT for AMT and 5-IT.

5.3.2 Co-eluting Compounds

Some compounds may still share high levels of structural similarity, while not actually being isomers. The two synthetic cannabinoids, 4F-MDMB-BINACA and MDMB-4-en-PINACA, are not isomers and it is possible to exploit their structural differences to differentiate them.

The difficulty with this particular pair is that (i) they have been observed both as individual compounds, and in combination within the same drug sample, and (ii) under the standard system (defined as a system that uses a consistent method for analysing the two drugs), they share the same RT under particular conditions of chromatographic methods. Consequently reproducibility of results can be difficult when a sample is repeatedly analysed due to the two compounds sharing the same RT. This means that it is not possible to use RT to confirm which compound is present. However, knowing the ions present in the individual mass spectra of these compounds (Figure 5.4), characteristic ions can be targeted, such as those at m/z 219, 275 and 307, for 4F-MDMB-BINACA, and at 213, 301 and 357 for MDMB-4-en-PINACA.

Using ion extraction tools available in most GC–MS software suites, a selective search can be conducted for these characteristic ions within a spectrum, enabling differentiation and identification of the two compounds, even if one is at a much lower concentration and is masked by the presence of the other.

Mis-identification of compounds with minor differences in the structure – methylphenidate and ethylphenidate. These two compounds have identical structures, except for the presence of an extra carbon chain. This means that the differences between the mass spectra obtained are minimal (Figure 5.5) and these can frequently be mis-matched using GC–MS library matching software. These compounds highlight the importance of checking that the mass spectrum obtained from the sample visually matches that provided as a match by the software; indeed, the software may produce matches within the acceptance range required, due to the presence of similar ions in both compounds and hence a visual assessment by the analyst allows for the identification of distinguishing ions.

5.3.3 Evolving and Emerging Drugs

As legislation expands to cover the full range of drugs seen in the illicit marketplace, so does the range of drugs produced to circumvent legislation.

The newly synthesised drugs often have broadly similar structures to existing controlled drugs and are designed to produce similar effects, but are outside of the range of substances included within generic controls.

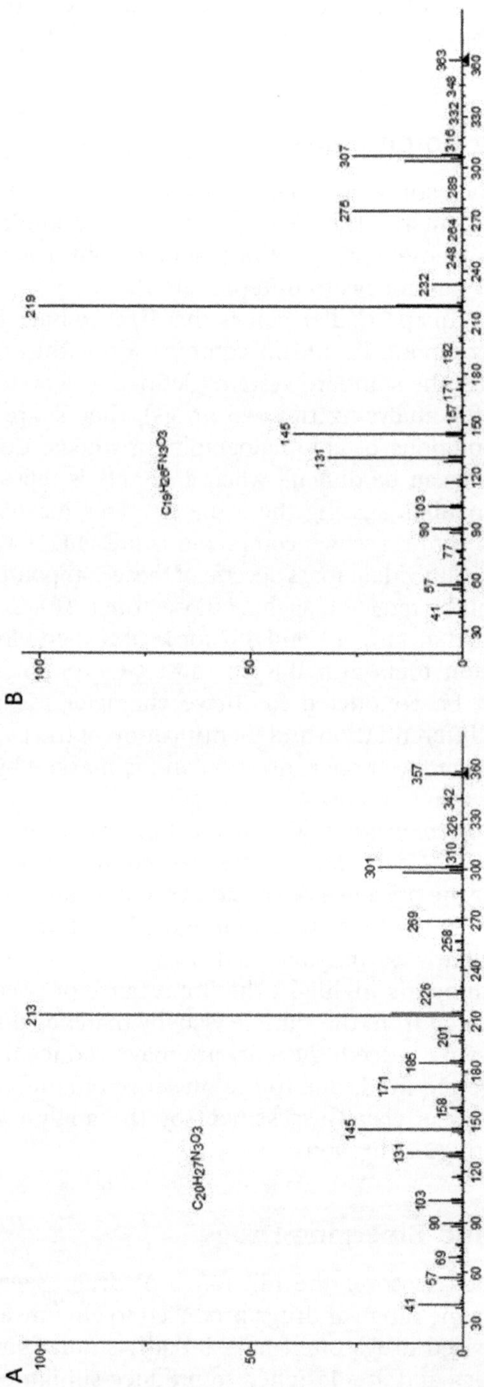

Figure 5.4 Mass spectrum of MDMB-4EN-PINACA (panel A) and 4F-MDMB-BINACA (panel B).

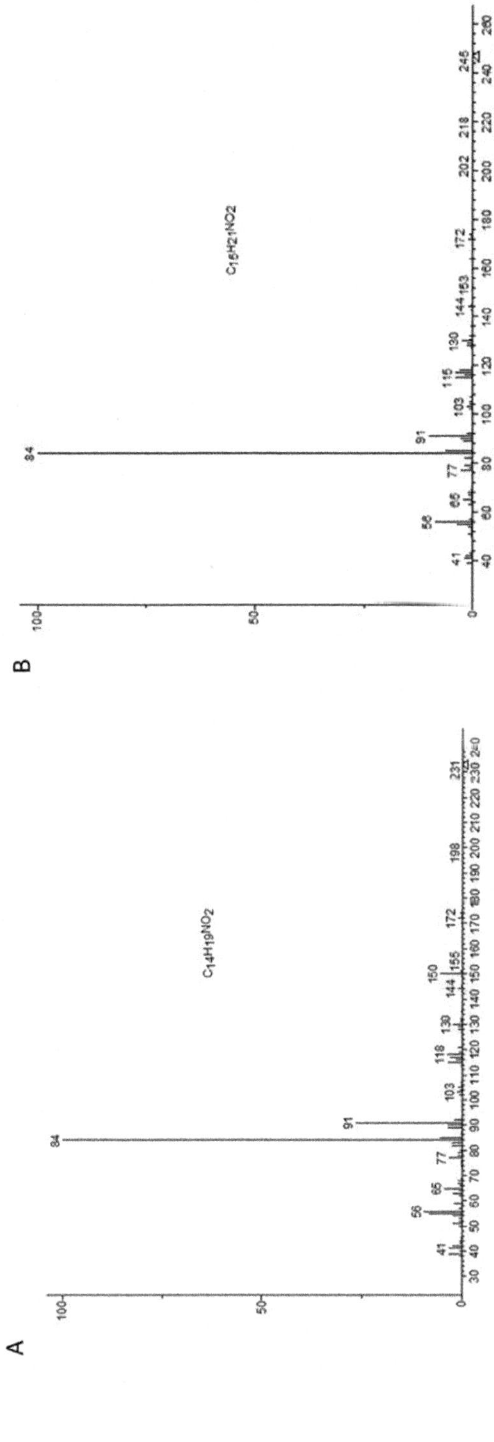

Figure 5.5 Mass spectrum of methylphenidate (panel A) and ethylphenidate (panel B).

Legislation is therefore increasingly focussed on controlling the generic groups of chemical structures and is designed in such a way that minor tweaks (for example, replacing a chlorine atom at the end of a chain with a fluorine atom) would still result in a drug controlled under the same legislation. Currently, this is particularly common with synthetic cannabinoid receptor agonists (SCRAs), which are drugs designed to mimic the effects of cannabis. Whenever a country or a group of countries regulates a set of SCRA drugs, a new raft of similar drugs, which fall outside of the legislation, emerge. Often these new drugs are first encountered within routine casework. In these scenarios, it is often not known or suspected that a new drug is present; therefore the first time a new drug is analysed, the compound may not be present in any mass spectral library. Consequently, during analysis, it would lead to inconsistent results such as a match to the library but with a RT that is not consistent with known values.

When a new compound is encountered, there are a number of possible routes for identifying and confirming its identity. The simplest is to check against alternative databases and online sources, such as CFSRE and others mentioned earlier in this chapter. CRM manufacturers will also often have reference spectra available for their products. However, while these sources may have a reference spectrum for the compound, it is often observed that it has not been encountered widely enough, thereby lacking an available reference. Under this circumstance, the use of alternative techniques, such as nuclear magnetic resonance (NMR), may be beneficial to characterise the sample and provide an identification. Even after its identification *via* these means, it is recommended to obtain an alternative reference spectrum, or a CRM to ensure robustness and reliability of the identification.

Earlier in the chapter, the importance of solvent selection has been mentioned, in particular when extracting drugs from sugary linctus preparations, in order to avoid interference of sugars with the GC–MS analysis, particularly with the injection and damage to the syringe. In recent years, there has been an increase in the number of samples seized in the form of foodstuffs claiming to contain tetrahydrocannabinol (THC), or cannabidiol (CBD) which are active components of cannabis. The majority of these come in the form of gummy sweets, in small branded pouches. These sweets pose a similar problem to the linctus, wherein it is important to extract the active drug, while avoiding the presence of sugary residue within the sample aliquot for analysis. However, unlike medicinal linctuses (which will usually be labelled), with these illicitly made edible preparations, it is often difficult to know which compounds are present and at what concentration. Empty printed pouches are available online, which can be filled with any kind of product; in the case of 'cannabis' edibles, these may contain THC as listed (although this could be at a much lower level than advertised), but they have also been found to contain SCRAs and even with no drug content at all. Therefore, it is important to know that the extraction method is efficient enough to detect any compounds present when analysing the sample by GC–MS.

5.4 Non-identification Applications of Seized Drug Analysis Using GC–MS

In addition to being a powerful tool for use in the identification of seized drugs, GC–MS is also used, to a lesser extent, in a number of ways that are less focussed on the identification of a primary drug. Once a drug's identity has been confirmed, there are other forms of analysis that can be conducted using GC–MS in order to provide more context to a case or for intelligence purposes. These are discussed in the following sub-sections.

5.4.1 Quantitative Analysis

GC–MS can be used for qualitative and quantitative purposes simultaneously and this can be advantageous in a commercial forensic environment, or in situations where a speedy result is required.

Quantification of seized drugs by GC–MS is based on measuring the peak area of a compound, identified through the presence of one or more qualifier ions, which is then compared to that of a standard (or set of standards) of known concentration. As with any quantitative analysis, the accuracy of the method will be increased by increasing the number of calibration points used for the comparison. However, while having more calibration points will enable a more accurate quantification, it will also take more time to set up and run; therefore, when conducting quantitative analysis of seized drugs, it is important to establish the accuracy of the method required by the end user. Here too, ensuring the use of the correct solvent is critical. If not all of the compound is extracted, its quantity will be an under-estimation. Some compounds are also more stable in particular solvents depending on the chemical form of the compound. For example, cocaine hydrochloride is readily soluble in water whereas the base form of cocaine (most commonly encountered as 'crack' cocaine in forensic drugs laboratories) is not very soluble in water.[3] As it may not always be immediately obvious in which chemical state a drug is present, it is important to select a solvent which can be used to analyse all of the different chemical states.

Something which has not yet been mentioned, as it is not critical for identification, is injection volumes. Although it is possible to set an injection volume for each analysis method, there will always be an amount of variability between the exact amount injected, and therefore the amount of solvent and analyte present. In order to mitigate this source of error within quantification, an internal standard of known concentration should be used within quantitative GC–MS (in both the sample being extracted and the standards) to compensate for injection variability, volatility, *etc.* Quantification is undertaken by using and plotting the ratio between the internal standard response of the standard injection and that of the sample injection *versus* the concentration of the analyte.

5.4.2 Drug Profiling

The majority of casework requests for forensic laboratories are identification and quantification of drugs; however, in addition to routine analysis, forensics laboratories are also sometimes requested to perform comparisons between seizures of controlled drugs to try and determine if they have originated from a common source. While results can be used evidentially, they are often used either to support the existing information regarding supply chains within a case, or, in the case of nationwide or force-wide profiling schemes, as an intelligence gathering tool, where the results may prompt new lines of investigation or highlight previously unknown links between individuals or organised crime groups.[8]

Profiling of drugs using GC–MS analysis is primarily an extension of quantification of drugs. However, instead of a single compound of interest, a larger number of compounds will be identified and quantified. The compounds used include adulterants/diluents and drug related impurities, such as alkaloid impurities in cocaine, or by products of manufacture of amphetamine type stimulants. Once the profile of a sample has been acquired, the ratio of each impurity can be established and compared to that of other samples. This requires analyses to be performed by experienced practitioners for optimum results, and the larger the database of samples, the more meaningful the resulting comparison will be.

5.5 Logistical Challenges for Forensics Laboratories and Impact of Results

Robust forensic drug analysis is paramount and plays a crucial role in the criminal justice system; however private forensic service providers and in-house police force laboratories often face a range of challenges from the lack of funding, lack of expertise in the field to provide drugs analysis results, interpretations and opinions, and the ever-increasing complexity in drugs substances and criminal activities.

5.5.1 Funding and Costs

Funding has always been a significant challenge for forensic science in the UK, as it relies on police budgets, which are constantly under scrutiny. Due to this, there is an intimate link between forensic science and the economic landscape.

The increase in costs often forces providers and in house police force laboratories to find new innovative ways to change the landscape of analysis in search of new ways to retain a high level of quality while increasing efficiency of throughput.

However, this challenge is exacerbated by the lack of funding allocated to forensic science in general. For continuous improvement and development, a research culture that is not restricted by funding is often required as it enables free flowing research and development to occur. This is often not the case as shown by the decrease in funding into forensic science with continuous decrease in funding year after year.[9]

The knock-on effect of insufficient investment in research and development at government levels is that it leaves the research and development to the forensics providers. This means that laboratories independently must push the research and development in various fields within their budgets. As a result of the funding for these research and development comes from within, providers may be forced to delay immediate instrumentation upgrades until there is a significant improvement in the analytical instrument market or the introduction of features which improves efficiency significantly, whether that is a new column which enables fast elution of all compounds and better separation or a new model of GC–MS that minimises maintenance.

This is not a big problem, provided everything is working as it should be. A huge drawback of this circumstance is the challenge in sourcing parts and lack of manufacturer's support when the instrument has been discontinued.

Another aspect of drug's analysis affected by the cost and funding is the ever-increasing cost of CRMs, often used for confirming the presence of new or uncommon compounds; the sourcing and availability of reference materials can vary significantly from compound to compound with some compounds costing over £700 for five milligrams. These ever-increasing costs reduce the profitability of the industry for private providers while also delaying results as they await delivery of these compounds, with some compounds having a lead time of two to three months.

All these expenses are in addition to usual running costs and the costs required to operate a drug department, such as Home Office licensing and accreditation costs.

These factors mean that analysis of seized drugs is mainly conducted by a small number of larger commercial providers. This leads to additional complexity as issues within a single provider can lead to significant build up of work and backlogs across the criminal justice system.

5.5.2 Drug Complexity

The other challenge that forensics laboratories face is the precision and accuracy that are required for drug analysis. Laboratories invest a lot of time and resources to ensure that the analytical techniques they use are accurate and reliable, and while well-established analytical methods and increased sensitivity of the instruments used provide a high level of accuracy, on occasions where very low levels or unusual mixtures of drugs are detected, forensic practitioners can rely on their knowledge and experience to determine whether a particular drug is present, and when looking at traces of drugs recovered from an item, the relevance of the results becomes important. Although the quality of the GC–MS spectrum obtained and the number of compounds within the library significantly influence the interpretation, it is important for forensic practitioners to spend sufficient time when completing analysis and interpreting data so that they are confident that no false-positive or false negative results will be reported. As referenced earlier, this is particularly important when unidentified peaks are present, which may be as a result of the presence of novel psychoactive substances (NPSs).

Novel psychoactive substances (NPSs) are compounds designed to mimic the effects of 'traditional drugs' which are controlled under the Misuse of Drugs Act 1971. As discussed earlier in the chapter, SCRAs are designed to operate beyond current legislation and as such are often referred to as 'legal highs'. As mentioned previously when these new drugs emerge, it is often forensic providers who first 'discover' that they are present as there are unidentifiable peaks within the GC–MS analysis results. The identification of these compounds often presents a challenge for forensics laboratories; this is due to the requirement of a relatively large amount of experience to competently and confidently interpret the mass spectra and identify a compound (not to mention to then establish whether or not it falls within the current legislation). The number of experienced drugs experts is limited within the UK, which can lead to delays in this type of interpretation.

As forensic practitioners may be challenged on their results by defence scientists and in courts of law, the reliability of the results produced must be provably robust. Most established forensic laboratories demonstrate this in a number of ways, but most commonly by using accredited analytical methods which are compliant with the requirements set out in international standards, such as ISO/IEC 17025 as well as showing that they are compliant with the Forensic Science Regulator's Code of Practice. Having accreditations provide additional reassurance to the criminal justice system; however, due to the increasing number of drugs and the complexity around them, it is not often possible to gain accreditation for all method activities within the laboratory (due to cost and timescales involved with gaining accreditation) and again, the responsibility rests on the practitioners and the laboratories to provide the confidence to the court that the analytical methods are fully validated and that the practitioners have sufficient knowledge and experience to analyse and interpret results and data.

Forensic laboratories can also provide confidence in their results through regular participation in proficiency testing and inter-laboratory comparison exercises. These exercises, which are either independent of the laboratories taking part, or are set by a group of laboratories, set benchmarks and act as blind tests of a laboratory's ability to identify unknown compounds. As with accreditation, these exercises should cover the full range of compounds tested for within a given laboratory; however, in practice, proficiency schemes are not available to cover all possible compounds and to do so *via* interlaboratory comparison would be too time consuming for all involved. As such, most laboratories participate in a number of these exercises throughout the year, which will target all of the major groups of drugs encountered, even if not each specific drug.

5.5.3 Handling and Storage of Drugs

Seizures of drugs frequently weigh multiple kilograms, and some may even weigh tonnes. This can pose a problem due to the sheer weight of these samples and the requirement of adequate storage space for physical examination, all of which need to be suitably secure.

One of the most fundamental requirements in forensic science is maintaining the integrity and the security of exhibits and ensuring that the chain of custody is unbroken. This requires accurate records of the (i) individuals who have or have had any handling or interaction with the evidence, (ii) the movements of the evidence and (iii) its storage.

Secure storage under correct conditions is an important part of this, particularly with samples containing herbal materials (such as cannabis), which can go mouldy, or samples which are susceptible to degradation if stored incorrectly. For example, when diamorphine is exposed to water it begins to break down to monoacetylmorphine (6-MAM). If a sample is submitted damp, or becomes damp and is not dried, there is a risk that the diamorphine present will break down to 6-MAM and when analysed it may be possible that the expected diamorphine peak does not provide a suitable match to the library. This is another example of a scenario requiring knowledge and experience to correctly interpret the GC–MS results.

During method validation, factors such as improper storage, which may affect the results should be factored in when assessing the uncertainty associated with the method. This is best carried out by testing samples which have been stored under the full range of conditions expected to be encountered in the laboratory.

The other facet to this is the handling of suspected drugs. While the small amounts required for analysis by GC–MS are usually (but not always) safe to handle, working with larger quantities of powdered drugs can present significant risks. As such, ventilation, careful handling and appropriate protective equipment, such as fitted face masks, are essential in order to protect staff from the risks of seized drugs. The safe handling and analysis of drug exhibits has been particularly significant in recent years with the increased prevalence of extremely potent opioid analgesics, first in the form of fentanyl analogues, and more recently with nitazene analogues. These drugs are highly toxic and can cause adverse reactions and even death at a sub-milligram level. An additional challenge with these particular compounds is that they have often been seen in powdered form resembling heroin powders. Within these powders, these compounds are present at relatively low concentrations, which, while being enough to imbue significant potential harms to drugs users, they can be difficult to detect, especially when the drug has recently emerged and has not been analysed/identified freely; this is the case for the nitazene compounds and many of the newer fentanyl analogues. In addition to analytical challenges, these samples are frequently submitted due to public health incidents, which can include sudden death cases or, severe adverse reactions to the drugs taken. In these scenarios, because of the significant impact to public health, there is an urgency to correctly identify all components of the drug sample in a very short period of time, so that notifications to users, first responders and medical staff can be issued.

5.5.4 Turnaround Time Expectation and Limitation

The majority of sample submissions received by forensic laboratories have a standard turnaround time, of between one week and one month depending on their perceived complexity. In addition to standard submissions, forensic laboratories are also required to provide urgent analysis services, which, while they may have a turnaround time of 24 hours, in reality a result is often provided to officers within four hours. This is determined by the operational needs of the submitting officers, based, for example, on the amount of time they have available to charge a suspect if they wish to keep them in custody. They may also need intelligence from the seizure to generally progress their investigation or to ensure the safety of officers or the public.

This short turnaround time means that laboratories are required to offer an out-of-hours on-call service in order to ensure that samples can be analysed when required. These services will typically be staffed by a small number of scientists, adding further challenges, if there are unexpected results that need interpretation. In order to help meet the expectations, practitioners work with the submitting officers in order to determine the best strategy for analysis, targeting certain exhibits or samples only in order to provide the evidence needed as efficiently as possible.

5.6 Conclusions

GC–MS analysis of seized drugs focuses on the identification of drugs using a combination of gas chromatography and mass spectrometry. The combination of these two techniques allows for an efficient and rapid confirmation of compounds, with multiple-layered result assessment that ensures a high level of confidence in the generated outcomes.

With the application of correct sample preparation processes, whether that is choosing the correct solvents or ensuring derivatisation is carried out when required, GC–MS is of great value to assist investigations and support judicial debates in the context of drugs of abuse identification.

Other considerations also need to be taken into account when handling seized drugs, including employing the correct health and safety measures and storage conditions, to ensure the safety of the practitioners and maintain the integrity of the sample without compromising it.

Acknowledgements

The Eurofins Forensics Services Drugs department is gratefully acknowledged for providing all spectra used within this chapter.

References

1. Supplemental Document SD-7 For Part III B – Methods of Analysis/Analytical Scheme for Identification of Drugs or Chemicals, https://swgdrug.org/Documents/SD-7_FINAL_ForPostingRev_101719.pdf, (accessed 3rd Jul 2023).
2. D. R. Christian and S. Bell, Seized Drug Analysis, in *Forensic science: An introduction to scientific and Investigative Techniques*, ed. S. H. James, J. J. Nordby and S. Bell, CRC Press, Boca Raton FL, 4th edn, 2014, pp. 282–298.
3. A. R. W. Jackson, J. M. Jackson, H. Mountain and D. Brearley, *Forensic science*, Pearson, Harlow, 4th edn, 2017.
4. NIST/EPA/NIH Mass Spectral Library Compound Scoring: Match Factor, Reverse Match Factor, and Probability, https://jordilabs.com/wp-content/uploads/2017/07/Whitepaper-NIST-EPA-NIH-Mass-Spectral-Library-Compound-Scoring.pdf, (accessed Jul 2023).
5. A. C. Moffat, M. D. Osselton and B. Widdop, *Clarke's analysis of drugs and poisons*, Pharmaceutical Press, London, 2004, vol. 2.
6. C. Schummer, O. Delhomme, B. M. R. Appenzeller, R. Wennig and M. Millet, *Talanta*, 2009, **77**(4), 1473.
7. T. Kumazawa, K. Hara, C. Hasegawa, S. Uchigasaki, X.-P. Lee and H. Seno, *et al.*, *Int. J. Spectrosc.*, 2011, 318148.
8. R. Ahmed, M. J. Altamimi and M. Hachem, *Molecules*, 2022, **27**(19), 6602.
9. R. M. Morgan and E. A. Levin, *Int. J. Spectrosc.: Synergy*, 2019, **1**, 243.

CHAPTER 6

AccuTOF-DART for Rapid Presumptive Screening of Drugs of Abuse

MARK BENNETT[a] AND BRIAN MUSSELMAN*[b]

[a] Virginia Department of Forensic Science, 6600 Northside High School Road, Roanoke, Virginia, USA; [b] Bruker Corporation, 40 Manning Road, Billerica, Massachusetts, USA
*Email: Brian.Musselman@Bruker.com

Abstract

Methods described in this chapter enable rapid and confident drug identification in the forensic laboratory. The use of direct analysis in real time ionization with high-resolution time-of-flight mass spectrometry to generate accurate mass and fragment ions is described with examples covering the detection of both major and minor compounds present in samples. The development of publicly available drug libraries and spectrum search technology is also reported to complete this chapter.

6.1 Introduction

The development of direct analysis in real time (DART) as an open-air ionization method configured with a time-of-flight based mass detector, the JEOL AccuTOF, has opened a new avenue for forensic scientists. The device itself, introduced at the 2005 Pittcon where it was awarded the highest prize for innovation,[1] drew immediate attention from a wide audience of potential

New Developments in Mass Spectrometry No. 14
Applications of Mass Spectrometry for the Provision of Forensic Intelligence: State-of-the-art and Perspectives
Edited by Simona Francese and Stephen Bleay
© The Royal Society of Chemistry 2024
Published by the Royal Society of Chemistry, www.rsc.org

users. Positioned in the JEOL booth at the conference and operating without use of any sample preparation devices, the instrument was an instant draw, with attendees presenting all manner of hastily gathered samples. The most demonstrated of those objects being the $20 US bill, from which protonated cocaine was routinely desorbed and ionized in seconds per exposure. Samples of orange peels, coffee with and without cream, tea, and various pharmaceutical tablets supplied from the pockets of visitors to the booth invariably yielded high-quality, high-resolution mass spectra, in which compounds including various fungicides on the peels, caffeine in tea and coffee (with or without lipids depending on the presence of cream), and in the case of tablets, the expected active pharmaceutical ingredients along with excipients were detected in seconds per sample by visual inspection of the data. Perhaps the most notorious moment occurred when the co-inventor, Dr. Robert "Chip" Cody, was presenting a sample of an explosive into the ionization region, just as the bomb-sniffing canine was passing the booth. While the signal generated by the desorption ionization was relatively abundant, the canine simply wandered off; this was a true demonstration of the sensitive nature of DART-TOF-MS.

In fact, DART had been in use for several years in the laboratory at JEOL USA and the US Army laboratory at Aberdeen Proving Grounds in Maryland, as recounted in the seminal publication describing a "versatile new ion source," which appeared in early 2005.[2] The ion source ushered in the era of ambient ionization with the promise of reducing requirements for sample preparation and elimination of time-consuming chromatography separations which had been standard practice in forensic laboratories for the previous 35 years. Since its inception, DART-TOF-MS has evolved into a high throughput sampling tool enabling confident seized-drug identification along with the capability for problem solving in a wide variety of analyses including psychoactive substances in plants, active pharmacological components of traditional medicines, and a demonstrated capability to rapidly characterize a plethora of synthetic cannabinoids, bath salts, and increasingly complex opioids as well as those modified fentanyl analogs currently of international concern because of their risk to life. A recent review in forensic chemistry[3] provides a clear description of the technology, including over 130 references specific to DART-MS; these authors point to it as an excellent reference for those intending to utilize this powerful instrument.

Our interest here is to describe the validated methods initially developed in the laboratories of the Virginia Department of Forensic Science (VDFS or the department) as first described in the literature in 2009[4] and their continuing efforts to expand and improve on those methods as published in subsequent years. Inevitably, as the technology has evolved, the need to make the method more robust and expand its use beyond those laboratories with advanced instrumentation has been addressed with the efforts of the VDFS to produce initial general high resolution mass spectrometry libraries of common drugs and pharmaceuticals. Those libraries, starting with the publication of a collection of reference spectra posted to the NIST website, and now fully curated as the updated NIST DART-MS forensic database, are described later in this chapter. Elements of the database construction and its use for seized drug analysis will follow the description of the analytical method.

6.2 Establishing the Capability for DART Source Utilization

VDFS received its first AccuTOF-DART instrument in late 2006 and immediately began the process of validating it for use in casework within the Controlled Substances Section. The department was drawn to the instrument's ability to screen suspected controlled substances samples quickly and reliably with minimal sample preparation to reduce overall analysis and turnaround time. By late 2007, the instrument was validated and online for use in controlled substances casework for the general screening of unknown samples.[4] It was not until a couple of years later when the VDFS further expanded the instrument's utility in casework by validating a method to confirm the active component(s) of pharmaceutical preparations.[5] These two uses remain the primary function of the instrument within the department today.[6] The number of AccuTOF-DART instruments across the VDFS has grown to five, and the department is continuing its efforts in exploring other scenarios for acceptable use in casework. To date, this technique has been used on the order of hundreds to thousands of cases that have been adjudicated within the Commonwealth of Virginia; cementing its acceptance for use within the field. The subsequent sections will provide further context and considerations for the use of AccuTOF-DART in suspected controlled substances samples in forensic casework.

6.3 General Screening of Unknown and Suspected Controlled Substances Samples

The analytical scheme for general unknowns at the department, including weighable materials and residues within the Controlled Substances Section, generally involves a progression of analytical techniques of increasing specificity.[6] For the majority of casework falling into this category, the final confirmation of the component(s) present within an unknown sample requires the analysis of standard(s), blank(s), and sample(s) on a traditional gas chromatograph/mass spectrometer (GC/MS) system. Due to emerging trends in certain classes of compounds, such as cannabimimetic agents, synthetic cathinones, and fentalogues, it became increasingly more common to use this analytical scheme without established standards since certified standards were often not yet available for synthetic analogs of the substances. The analysis of samples containing novel synthetic compounds thus required longer screening methods than those frequently being used in the confirmation of general unknowns. With efficiency and a growing backlog in mind, a method for the general screening of unknowns was developed by Steiner and Larson using the AccuTOF-DART.[4]

The software used for instrument control, method setup, and data collection/processing is either Mass Center or MSAxel, both by JEOL, Ltd.[6,7] Data analysis is performed using Mass Mountaineer, a software program developed by Dr. Robert "Chip" Cody of JEOL, Ltd. Samples are prepared for analysis by dissolving them in a suitable solvent (*e.g.*, methanol or chloroform). Every day, quality assurance checks are performed by running a dilute solution of polyethylene glycol (PEG) 600 for internal mass calibration and a quality control (QC) mix solution containing methamphetamine, cocaine, and nefazodone. These QC mix components have calculated protonated mass-to-charge (*m*/*z*) ratios of ($[M+H]^+$) of 150.128, 304.155, and 470.232, respectively. The acceptance criteria chosen for these controls were within ±5 millimass units of these calculated values. These three components were selected because they span the low-, mid-, and high- mass regions of the method used for routine controlled substances casework. Many specific instrumental parameters vary by instrument based on its individual tuning, but some common shared parameters used at the VDFS in casework are summarized in Table 6.1.

The instrument is almost exclusively operated in the positive-ion mode. A low mass cutoff of *m*/*z* 66 was chosen in order to avoid collection of protonated methanol dimers (*m*/*z* 65), which can be generated when that solvent is used for sample extraction or dilution. Additionally, both PEG 600 and the QC mix spectra are acquired within each data file collected for casework. Due to the near exact mass output of this instrument, any slight fluctuation that may occur due to environmental conditions can result in mass assignment variations within sample runs throughout the day. Running PEG 600 for internal mass calibration and QC mix as a set of positive controls within each data file addresses this issue and assures the correct functioning of the instrument at the time of data collection for casework. While the method used for data collection includes function switching regardless of the intent to screen or confirm, it is either the 20 V or 30 V data that are typically used for screening purposes. Function switching refers to the rapid cycling of voltages within the ion optics portion of the instrument and can provide mass spectra from both soft and hard ionization processes. If the results show the presence of empirical formula isomers, it is recommended that the

Table 6.1 Summary of general AccuTOF-DART parameters for controlled substances casework at the VDFS.[6]

Parameter	Setting
DART source temperature	350 °C
Orifice 1 temperature	100–120 °C
Scan range (*m*/*z*)	66–600
Sampling interval (ns)	0.25 ns
Function switching cycle	20, 30, 60, and 90 V

60 V or 90 V data be examined for fragmentation patterns of the sample. Data generated from higher voltages are discussed in more detail in the Pharmaceutical Confirmation section (6.4).

Single use glass melting point tubes are used for sample introduction into the instrument.[6] The melting point tubes are cleaned prior to use to remove dioctyladipate, a common plasticizer from the melting point tube packaging that is routinely detected at *m/z* 371.316.[6,7] The closed end of the tube is dipped into the sample and placed into the sample gap of the instrument. If the open end is used, the heat provided by the source causes the sample drawn up within to release and fall out. The sample-coated tube should stay within the sampling gap for a period of several seconds to allow ionization to occur across all components present in the sample. Replicate "wandings" for each specimen within a sample run are recommended. "Wanding" refers to the motion of passing the sample-coated tube through the heated ionizing gas exiting the DART source for brief intervals of time in order to generate a series of peaks with sufficient relative abundance so that they can be discriminated from the background, making peak integration more practical. It is acceptable, on occasion, to run samples in their neat form. The PEG 600 calibration spectrum is collected with each group of wandings and applied to the collected data, after which background subtraction is completed to generate mass spectra for the samples with that spectrum being saved using Mass Mountaineer/MSAxel.

Saved spectra are searched in Mass Mountaineer against lists made in-house at the VDFS. These lists are regularly updated to include novel and/or emerging compounds, the identities of which can be obtained from information shared within the controlled substances field and other analytical literature. These lists simply include the name of drug compounds and their corresponding chemical formula. Mass Mountaineer software calculates the exact mass of each compound using the formula provided while adding the mass of a proton and uses that as the basis of comparison to the mass values determined from the sample spectra. Sample components should fall within ±5 millimass units of these calculated values to be considered positive for this singular method. This concept works exceptionally well for screening because it eliminates the need to have a standard on hand to run for comparison purposes. While a standard needs to be available for actual confirmation of sample components, the identity of compounds present in a sample is often uncovered with this screening method with few exceptions. However, list results provided by Mass Mountaineer software should not be overemphasized in the event of multiple matches at the same exact mass. Even with the near-exact mass output of the instrument, there are several commonly encountered examples in casework of compounds present in a list with the same exact mass $[M+H]^+$. Examples of this include naloxone and 6-monoacetylmorphine ($[M+H]^+$ of *m/z* 328.155), cocaine and scopolamine ($[M+H]^+$ of *m/z* 304.155), and fluoxetine and a prominent heroin fragment ion of $[\text{heroin-}C_2H_4O_2]^+$, both with an exact mass of *m/z* 310.144 (discussed below). In any of these instances, differentiation of these compound pairs is easily discerned within the analytical method(s) used for confirmation. The basis of this screening method places significant importance and continued emphasis on the lists used for reference. The exchange of information within the field of forensic drug analysis is crucial for the sustained success of this method.

There are a few things to consider when looking at the data generated by AccuTOF-DART as it applies to suspected controlled substances. Unless coupled with additional instrumentation, there is no chromatography involved with this technique. The resulting mixed mass spectra can include all the individual components in a sample, atmospheric compounds, and even dimers/trimers of sample components. This can complicate data interpretation and lead to masking of some components present in the sample. Successful drug detection is largely predicated on the proton affinities of sample components relative to each other. The DART gas stream can preferentially ionize certain sample components while excluding others to varying degrees. As previously described by Song *et al.*, this can affect the ionization of sample components in a few different ways.[8] First, a component not initially ionized by the gas stream may finally receive its proton transfer when other components in the sample with higher proton affinities are exhaustively ionized. Another scenario exists in which components in a sample are preferentially ionized based on their proton affinities, subsequently serving as a source for the ionization of other sample components. In either situation, there is a likelihood for portions of a multi-component sample to be ionized to a lesser extent relative to their respective concentration within the sample. This sort of ionization competition makes it difficult to determine relative sample concentration of individual components present within a mixed sample. This belabors the recommendation to keep samples within the sample gap of the instrument for several seconds during data collection.

A practical instance in which this occurs in casework is in samples that contain both fentanyl and heroin. Casework samples have frequently shown the presence of fentanyl to significantly mask or eliminate the detection of heroin altogether at lower voltages. In these scenarios, the department has noted that further observation of higher voltage spectra (60 V and 90 V) can reveal the presence of heroin *via* breakdown fragments resulting from the exposure to higher voltages within the ion optics portion of the instrument (Figure 6.1). Even then, these breakdown fragments at higher voltages are still relatively minor in abundance. These fragments can include *m/z* 328.154 $[\text{6-monoacetylmorphine} + H]^+$, *m/z* 310.144 ($[\text{heroin-}C_2H_4O_2]^+$), and *m/z* 268.133 $[\text{apomorphine} + H]^+$.

There are additional uses of lower voltage data that have been found useful for casework in a supporting and supplemental role. The softer ionization process taking place at 20 V and 30 V allows for data collection that includes molecular ion information for compounds that may otherwise break down on a traditional GC/MS instrument. It is important to remember that the molecular ion appears at lower voltages on the AccuTOF-DART

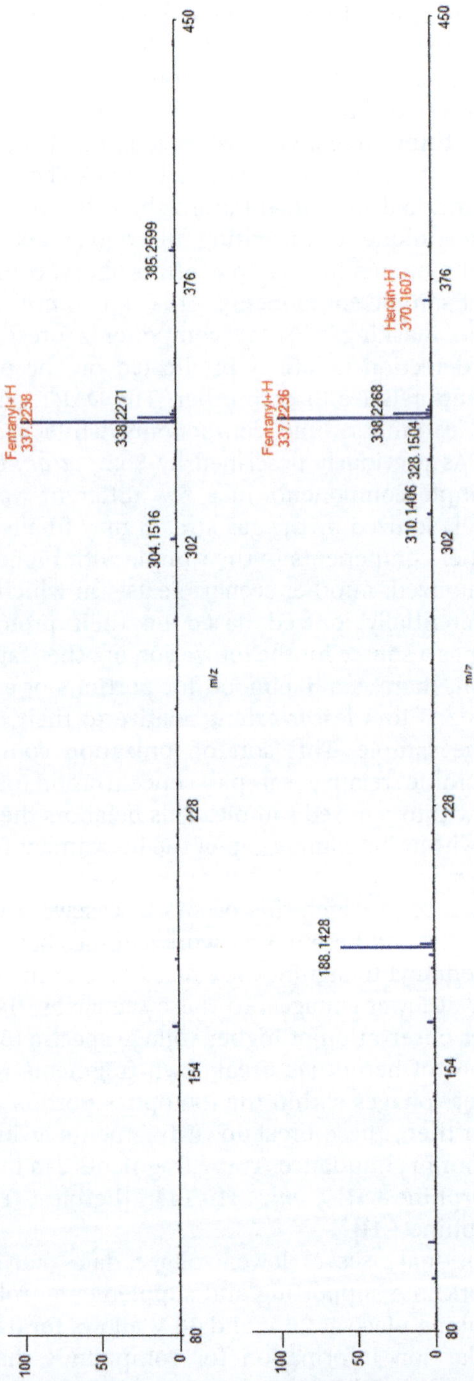

Figure 6.1 Heroin and fentanyl mixed sample at 30 V (top) and 60 V (bottom).

system in its protonated $[M+H]^+$ form. Compounds that the department sees in casework with some regularity that fit this scenario are lorazepam and gabapentin. Lorazepam has a nominal molecular weight of 321 Da. In an underivatized sample containing lorazepam, the molecular ion is observed at *m*/*z* 302 on a traditional quadrupole instrument due to rearrangement and water loss after ionization. Gabapentin has a nominal molecular weight of 171 Da. However, it also loses water upon ionization on a GC/MS system, thus also generating mass spectra without molecular ion information. Both of these compounds, when analyzed using the AccuTOF-DART system, are detected as their $[M+H]^+$ ions at *m*/*z* 321.020 for lorazepam and *m*/*z* 172.134 for gabapentin in either 20 V or 30 V data, corresponding to their intact parent compound. To this effect, DART data at lower voltages have been used at the VDFS to help strengthen the support of a conclusion for a given piece of evidence. This has become increasingly helpful for aiding in the identification of (i) compounds that do not readily present molecular ions in their mass spectra, (ii) compounds that require higher sample concentrations to exhibit ^{13}C peaks associated with their molecular ions, and (iii) compounds that are thermally labile and would otherwise need to be derivatized on traditional GC/MS systems.[6]

The VDFS also successfully utilizes the AccuTOF-DART in the analysis of samples from clandestine labs, in particular for the illicit synthesis of methamphetamine using a "one pot" method.[9] This name is appropriately given as commonly sourced household goods may be readily obtained and mixed within a single vessel ("one pot") to produce methamphetamine.[9] A specific section of "The Code of Virginia"[10] designates a list of chemicals, where possession of two or more from the list can serve as grounds for prosecuting the manufacture of methamphetamine. While the majority of these chemicals are most appropriately analyzed within the department's Trace Evidence Section, pseudoephedrine is listed within this code section and best analyzed within the Controlled Substances Section. Pseudoephedrine, a decongestant, is not specifically controlled within the Commonwealth of Virginia. Its purchase, however, is regulated and tracked due to its role as a precursor in the synthesis of methamphetamine. The identification of methamphetamine is also certainly of value in these sample types as it is the desired end product of the reaction/synthesis.

The challenge in analyzing samples of this nature lies in their complexity where incomplete synthesis means that the composition of the sample is not uniform and may or may not contain end products. These samples are often submitted to the department in various forms, from multi-layer liquids to sludges. General preliminary testing is necessary to gain a better understanding of the sample that can be applied to an appropriate extraction and subsequent analytical scheme. Light petroleum distillates, such as Coleman® fuel, are often used as an organic solvent for extraction within this synthetic route. Both methamphetamine and pseudoephedrine, if present in the organic layer of a sample, will elute in a GC/MS method within the alkane-heavy region of the light petroleum distillate. This makes the analysis of weaker samples more difficult due to the matrix contribution. Acetyl derivatization of both methamphetamine and pseudoephedrine pushes their derivatized counterparts to later retention times and away from the alkane heavy region of the matrix. This provides a better opportunity for identification. Analogous to using DART in a supplemental fashion as previously described, both derivatized and non-derivatized samples from suspected clandestine labs have been subjected to analysis on the AccuTOF-DART to help strengthen the overall analytical data packet to support a conclusion. Acetyl-derivatives can be made using traditional methods as provided by the manufacturer or even in-source on the AccuTOF-DART given the high temperatures provided within the sample gap. Methamphetamine has a calculated mass $[M+H]^+$ of m/z 150.128 while acetyl-methamphetamine has a calculated mass $[M+H]^+$ of m/z 192.139. Pseudoephedrine has a calculated mass $[M+H]^+$ of m/z 166.123 while acetyl-pseudoephedrine and diacetyl-pseudoephedrine have a calculated mass $[M+H]^+$ of m/z 208.134 and 249.137, respectively. The m/z 208.134 $[M+H]^+$ ion for acetyl-pseudoephedrine is often minimized and manifested on a spectrum with the loss of water at m/z 190.123.

6.4 Pharmaceutical Confirmation Using AccuTOF-DART

Increased specificity with this instrument is due to the near-exact mass nature of its data output. Data output from the AccuTOF-DART is accurate enough to permit mass determination up to three decimal places, whereas a single quadrupole-based mass spectrometer instrument is limited to determination of the nominal mass of an ion. However, due to the existence of molecules which, when ionized, produce isobaric mass ions (those with the same nominal mass but different elemental composition), this does not necessarily mean that a molecule that is listed as a match from an in-house list is in fact that specific molecule. This redundancy can apply to intact protonated molecules, any corresponding fragment ions arising from the degraded condition of the sample, or fragment ions that are generated from function switching within the optics portion of the instrument. With function switching, AccuTOF-DART allows for near-simultaneous collection of mixed spectra at varying but consistently repeating low and high voltage(s). Intact protonated species are detected using lower entrance orifice lens voltages (20 V and 30 V) while mixed spectra showing more characteristic fragmentation are detected using higher entrance orifice lens voltages (60 V and 90 V). This versatility allows for more specificity of data generated by the instrument, and it is a feature utilized for a method of confirmation of suspected pharmaceutical preparations.[5]

Pharmaceutical preparations have long represented a significant portion of controlled substances casework for the VDFS. With the increasing prevalence of counterfeit tablets within items of evidence, the ability to screen suspected pharmaceutical preparations efficiently and accurately for legitimacy or confirmation purposes was a worthy endeavor for the department. Easter and Steiner validated a method for pharmaceutical preparation confirmation using AccuTOF-DART that is still in use today.[5,6] In this method validation, sensitivity (lower limit of detection), selectivity, and reproducibility were all considered. The selectivity aspect of that validation is the focus of discussion here, although it is worth mentioning that each newer generation instrument currently at the department showed a sensitivity exceeding 0.01 $mg\,mL^{-1}$ during instrument verification. In this scheme, pharmaceutical identifiers are incorporated and combined with corroborating low and high voltage data from the AccuTOF-DART to form the basis by which the legitimacy of any active component(s) present in a suspected pharmaceutical exhibit may be confirmed and reported. Pharmaceutical identifiers refer to the process by which the physical characteristics of a suspected pharmaceutical preparation are noted and used as the basis for making a preliminary identification of the active component(s) present in that dosage unit. These characteristics include, but are not limited to, size, color, shape, and marking/imprint of the preparation. With these characteristics noted, a variety of references exist that can be used to make a preliminary identification of the drug component(s) present. References accepted by the VDFS for a positive pharmaceutical identifier include the Drug ID Bible, Epocrates.com, Drugs.com, DailyMed, and manufacturer's websites.[6]

The Scientific Working Group for the Analysis of Seized Drugs (SWGDRUG) provides a minimal recommendation guideline for the analyses of suspected drug samples.[11] Categories A, B, and C are established and used to identify analytical techniques based on their selectivity. Mass spectrometry is included within category A, the category representing the highest level of selectivity for a technique. Within this scheme, a category A technique may be used in tandem with any one technique from the other categories for identification. SWGDRUG further qualifies pharmaceutical identifiers as a category C technique, which means that with a positive result for a C level technique in hand, data from the AccuTOF-DART at both low and high voltage(s) are then sufficient for confirmation.[5,6,11] Furthermore, it is also acceptable, under the procedures of the department, to use pharmaceutical identifiers in tandem with a traditional GC/MS instrument for confirmation. In having both instruments as options for pharmaceutical identification, the VDFS promotes flexibility in analytical approach as well as efficiency in analysis.

Within the ion optics region of the instrument, rapid switching of the voltage potential between orifice 1 and the ring lens occurs, allowing for simultaneous collection of data at 20, 30, 60, and 90 V. As previously discussed within the screening method, low voltage data (typically 30 V used by the VDFS) provide information on the intact, protonated molecule of any components present in a sample. These data are compared to the in-house generated lists already described in the general screening section. Data generated at the higher voltages of 60 and 90 V provide fragmentation patterns for a sample in a process known as collision-induced dissociation (CID). These data must compare favorably to a standard spectrum run at the same voltage as the saved spectrum. "Compare(s) favorably" means that the overall fragmentation pattern of the sample will match that of the standard spectrum from the library. The department initially created an in-house standard library that encompassed all single component standards as well as any independently confirmed pharmaceutical standards compiled over years' worth of casework or local procurement. Data collected from all four voltages used by the instrument were included in these libraries. Some of these pharmaceutical standards included mixed preparations containing multiple active components. These libraries were shared across the department, and any additions made by individual laboratory locations, upon the receipt of a new standard, have subsequently been distributed accordingly. Additionally, all prominent and individualizing peaks present in higher voltage data must also be within ±5 millimass units of their calculated molecular weight. Of note, most controlled substances in pharmaceutical preparations produce adequate fragmentation data at 90 V for confirmation purposes. Lower molecular weight compounds, such as amphetamine ($[M+H]^+$ of *m*/*z* 136.113), dissociate to produce only one low mass fragment ion at 90 V. This makes confirmation from only using spectra acquired at 30 and 90 V unrealistic. In these instances, 60 V data are used as they generally allow the intact $[M+H]^+$ species to remain detectable along with other higher mass fragments.[5]

Because of the possibility of empirical formula isomers within pharmaceutical preparations, using DART as a confirmation tool required the demonstration of increased selectivity beyond just the near-exact mass, lower voltage data. Easter and Steiner showed that in such instances, both unique fragment ions and the relative abundance of some shared fragment ions between isomer pairs could be used for differentiation (Table 6.2, and Figure 6.2).[5] Examples of compounds that are seen in suspected pharmaceutical preparations with potential isomers at some frequency include codeine/hydrocodone (Figure 6.2A), both having formula $C_{18}H_{21}NO_3$ with their $[M+H]^+$ detected at *m*/*z* 300.160, hydromorphone/morphine (Figure 6.2B)

Table 6.2 Summary of empirical formula isomer differentiation.[5]

Isomer pair	Unique ion(s) at higher voltages (*m*/*z*)	Shared ions at higher voltages with different abundances (*m*/*z*)
Codeine	267.1259, 282.1494	199.0759[a], 225.0916, 300.1600[a]
Hydrocodone[a]	257.1178	
Hydromorphone	243.1021	179.0861[a], 211.0759[a], 286.1443
Morphine[a]	238.0994, 253.1103, 268.1338	
Methamphetamine[b]	119.0861	None
Phentermine[b]	133.1017	

[a] Higher abundance shown in the spectrum.
[b] 60 V data listed.

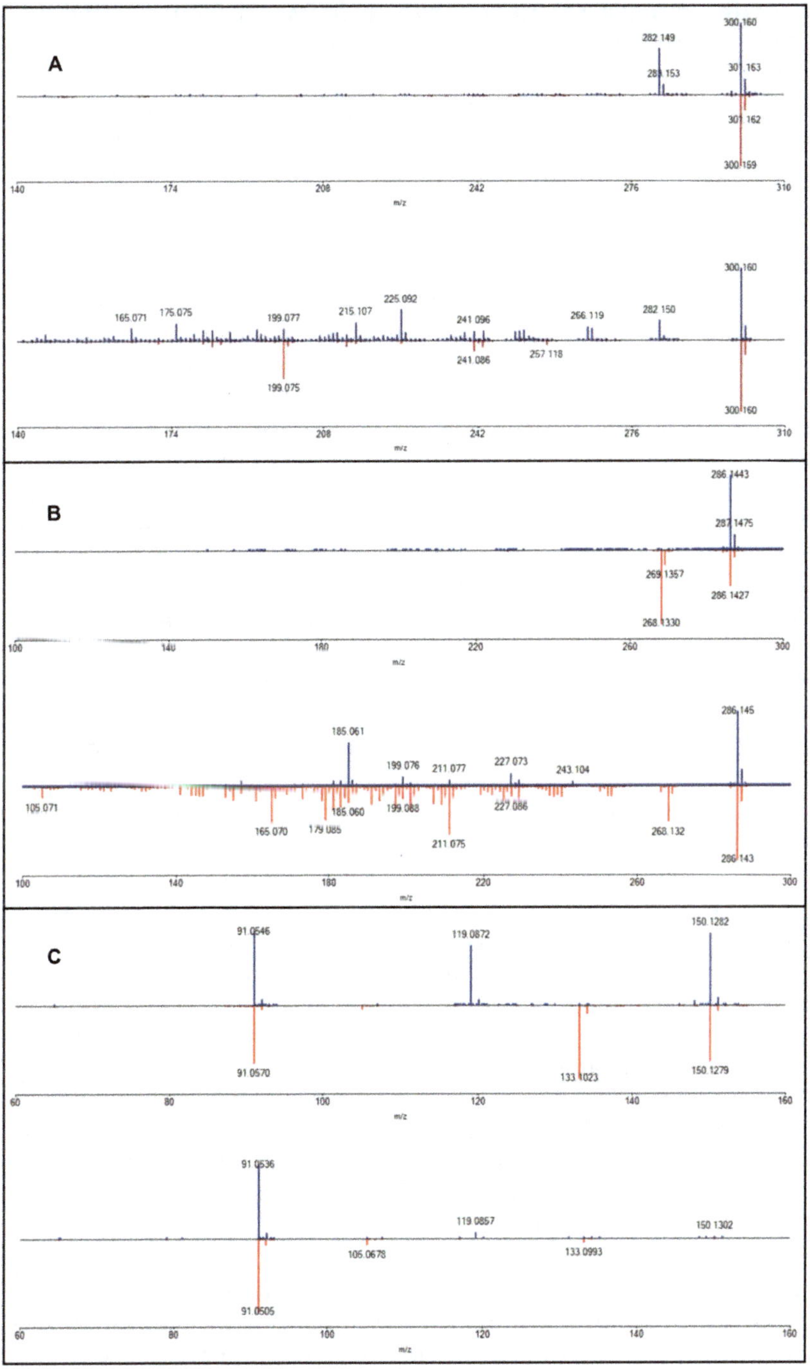

Figure 6.2 Empirical formula isomer pair differentiation at 30 V and 90 V for (A) codeine (top)/hydrocodone (bottom) (B) hydromorphone/morphine and (C) methamphetamine/phentermine.

both having formula $C_{17}H_{19}NO_3$ with their $[M+H]^+$ ion detected at *m/z* 286.144, and methamphetamine/phentermine (Figure 6.2C) both having the formula $C_{10}H_{15}N$ with their $[M+H]^+$ detected at *m/z* 150.128.

A discussion of the potential for using fragmentation to confirm the presence of one isomer *versus* another is warranted, given the lack of chromatographic separation in the DART method. There are multiple significant isomer pairs in the seized drug list. These include the six listed above. Using the mass spectra and fragmentation spectra of the structural isomers methamphetamine and phentermine, it is relatively easy to distinguish between them by acquiring the mass spectra at 30 V and verifying the presence of the confirming ion at *m/z* 91 in the spectra acquired at 60 V and 90 V. These relatively low mass molecules (<160 Da) sometimes produce only a few fragment ions; however presence of these two compounds can be confirmed by combining the data. The detection of only one of these isomers is also viable where only one of them is present in the sample. Using data from the fragment ions generated at 30 V shown in Figure 6.2C permits the confirmation of the presence of methamphetamine using the ion at *m/z* 119 ion or phentermine using the ion at *m/z* 133, where the fragment ion for either one is detected in the absence of the other.

As the complexity of the molecules increases with their molecular weight, additional fragmentation tends to occur, as observed in the data. In particular, spectra for both codeine and morphine show a general fragmentation pattern that is much more extensive at higher voltages than their respective isobaric counterparts. "Compares favorably" was mentioned before as a requirement for a positive result at higher voltages (Figure 6.3).

Any other major peaks present in the sample mass and fragmentation spectra that are not included in the library mass and fragmentation spectra must be further investigated or explained. Often, pharmaceutical preparations contain different binders or fillers in their formulations that can give rise to some differences in ions detected in the mass spectral data. It is important to recall that this instrument, as outfitted at the VDFS, does not include any chromatography prior to mass spectral data collection. Mass spectra obtained at any voltage apply to the entirety of the sample. Additionally, it has been observed that a library match can incorrectly register as the top result simply because of the extensive fragmentation that some compounds offer at higher voltages. In this scenario, many peaks may be shared between the sample and library match spectra, but careful inspection of the two will show a stark difference between their overall fragmentation pattern(s) and relative abundance values of shared fragments.

To reiterate, matrix effects of mixed component pharmaceutical preparations should be considered during solvent selection/extraction to prevent any components from being completely masked or so concentrated as to impact their mass assignment by the software. The importance of proton affinity previously discussed in this chapter should be considered as the role of acetaminophen in pharmaceutical preparations is discussed. Acetaminophen is a weakly acidic compound commonly seen in casework with

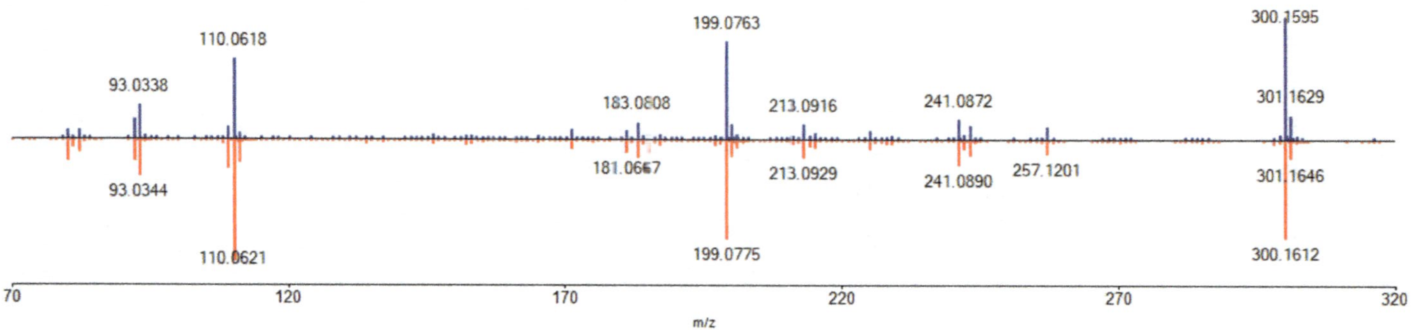

Figure 6.3 Example of the case sample (top) and library spectrum (bottom) that "compare favorably".

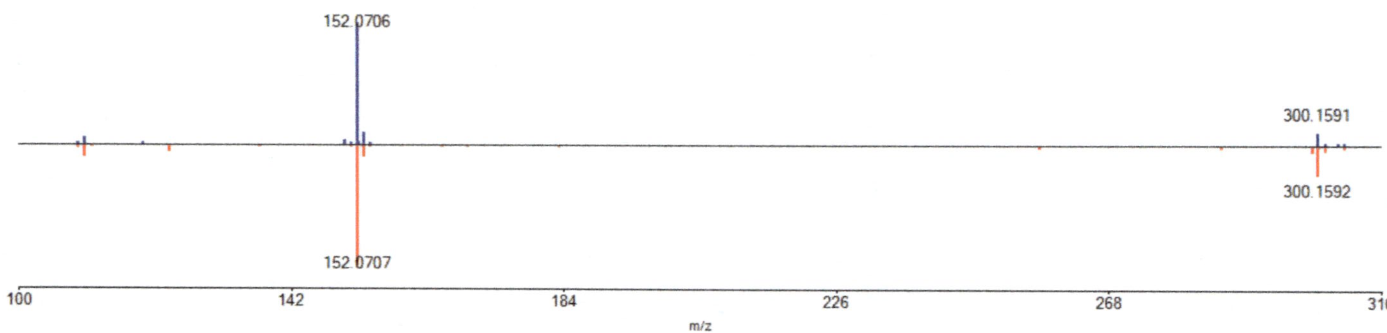

Figure 6.4 Hydrocodone and acetaminophen tablets in methanol (top) and ammoniated chloroform (bottom) at 30 V.

pharmaceutical preparations. Often, it is encountered in pharmaceutical formulations with low doses of semi-synthetic opioids such as oxycodone and/or hydrocodone. Due to its chemical nature and overwhelmingly large dosing in these preparations, a significant amount of acetaminophen must be removed prior to analysis to avoid masking of other sample components. Within the department, these general types of preparations are often dry extracted using ammoniated chloroform (an ANOR (alternate non-aqueous organic ratio) solvent) that will retain some acetaminophen while not completely overwhelming and excluding the slightly basic semi-synthetic opioid (Figure 6.4).[12] Completing this relatively simple extraction improves the potential for detection of components that might be present in low percentages in a given preparation. Reducing the level of this legal pharmaceutical within a sample spectrum also serves to reduce its relative abundance, contaminating the mass spectrometer inlet and potentially degrading the capability of the instrument to preform accurate mass assignment.

6.5 Enabling the Method: A New Library and DATABASE Search Program

Initial work in the Virginia Forensic laboratories involved curating a database of mass spectra using standards and the function switching method to generate a collection of four mass spectra per compound as described here. In 2012, a DART-MS database containing 3217 spectra of 828 compounds was created and released by the National Institute of Standards and Technology (NIST).[13] As a matter of clarifying nomenclature, the function switching mass spectra described in this work became in-source collision induced dissociation spectra (is-CID) when described in the NIST publication. Data analysis previously performed using Mass Mountaineer, the proprietary software program described earlier, could use the spectra in the NIST library when conducting searches, thus providing a means for the method to be used in other laboratories with similar instrumentation.

The growing problem of seized drugs and the occurrence of increasing backlogs in forensic laboratories throughout the country led to a reinvestment of time and thought into the NIST library. In 2021, NIST published the first improvement in the 2012 database[14] as well as the second paper describing the library search algorithm.[15] The new DART-MS database described in the publication contained 1989 spectra of 663 compounds with 3 in-source CID spectra per compound, instead of the previously used four with the change being to eliminate the 20 V mass spectra from the database. The 2021 NIST database utilizes 30 V, 60 V, and 90 V CID spectra. Along with this incarnation of the database, the NIST team developed a computer program to "enable evaluation" of the mass spectra. With this generation of the search program, the NIST team introduced a subset of rules including an indicated preference for the precursor ion in the spectra acquired at +30 V to constitute at least 60% of the base peak with a mass accuracy of 0.0005 Da, and rules for protonated dimer abundance, fragmentation consistency, and noise evaluation. Most significantly, the NIST DART database can be downloaded, thus providing a means for forensic scientists to begin reducing their sample backlog number.

6.6 Conclusions

The work described here provides a method for rapid identification and confirmation of seized drugs, synthesis intermediates from clandestine laboratories, new psychoactive chemicals and characterization of drug related material. The use of DART for ionization is shown to reduce the requirements for sample preparation and use of time-consuming chromatographic analysis. This helps streamline the forensic analysis workflow, reducing both time and expenses. In the 2022 calendar year, the Western Regional Laboratory for the department averaged 385 case samples run on the instrument per month; a number showing both the necessity of its use by examiners and general acceptance of the technique in a court of law. Finally, the work of NIST to implement a true MS/MS database and search algorithms is described as the availability of a validated library supported by the federal government serves to permit wider application of the method, thus enabling a wider application of the method.

Acknowledgements

The author(s) would like to express gratitude to the Virginia Department of Forensic Science, including Director Linda Jackson, Chemistry Program Manager Robyn Weimer, and the Controlled Substances Section in its entirety as well as across the specific examiners of Charles Fishel, Rebecca Hutchison, Cristina Chilcott, Jacob Easter, Thomas Simpson, Erin Corcoran, Neil Vowels, and retired examiner Robert Steiner. Additionally, appreciation is extended to Dr. Robert "Chip" Cody of JEOL, Ltd., for his continuous availability as a resource and for his support in the Department's validation endeavors related to the AccuTOF-DART.

References

1. S. Borman, *Chem. Eng. News*, 2005, **83**(13), 39.
2. R. B. Cody, L. A. Laramee and H. D. Durst, *Anal. Chem.*, 2005, **77**, 2297.
3. E. Sisco and T. P. Forbes, *Forensic Chem.*, 2021, **22**, 100294.
4. R. R. Steiner and R. L. Larson, *J. Forensic Sci.*, 2009, **54**, 617.
5. J. L. Easter and R. R. Steiner, *Forensic Sci. Int.*, 2014, **240**, 9.
6. https://dfs.virginia.gov/wp-content/uploads/221-D100-Controlled-Substances-Procedures-Manual.pdf, (accessed May 2023).
7. AccuTOF-DART Training Guide, 200706, 18th edn, JEOL USA, Inc., 2006.
8. L. Song, S. C. Gibson, D. Bhandari, K. D. Cook and J. E. Bartmess, *Anal. Chem.*, 2009, **81**, 10080.
9. E. C. Person, L. A. Knops and D. M. Northrop, *J. Clandest. Lab. Invest. Chem.*, 2004, **14**(2), 14.

10. https://law.lis.virginia.gov/vacode/title18.2/chapter7/section18.2-248/, (accessed May 2023).
11. Part III B Methods of Analysis/Analytical Scheme for Identification of Drugs or Chemicals, SWGDRUG, 4th August 2018.
12. A. R. Adir, F. T. Noggle Jr., M. S. Odom and M. A. Rhodes, *Microgram*, 1983, **1**, 220.
13. https://chemdata.nist.gov/mass-spc/ms-search/downloads/Readme_NIST_Forensics_DART.txt, (accessed May 2023).
14. E. Sisco, A. S. Moorthy and L. M. Watt, *J. Am. Soc. Mass Spectrom.*, 2021, **32**, 685.
15. A. S. Moorthy and E. Sisco, *J. Am. Soc. Mass Spectrom.*, 2021, **32**, 1725.

CHAPTER 7

Emerging Technologies: Use of Matrix Assisted Laser Desorption Ionisation Mass Spectrometry for the Analysis of Fingermark and Blood Evidence

S. FRANCESE* AND C. HEATON

Centre for Mass Spectrometry Imaging, Sheffield Multimodal Imaging Centre, Sheffield Hallam University, Sheffield, UK
*Email: s.francese@shu.ac.uk

Abstract

For over a decade, a considerable body of knowledge has been built, demonstrating both the versatility and operational feasibility of Matrix Assisted Laser Desorption Ionisation Mass Spectrometry (MALDI MS) for the analysis of fingermarks (molecular fingerprinting). Originally developed for the analysis of biological tissues, this technology, applied to fingermarks, has evolved for the provision of both chemical intelligence (in profiling modality) and biometric information, through the reconstruction of the ridge pattern (in imaging modality). Several applications have been developed linking the biometric and contextual information by visualising the presence of forensically interesting molecules directly onto the identifying ridge

New Developments in Mass Spectrometry No. 14
Applications of Mass Spectrometry for the Provision of Forensic Intelligence: State-of-the-art and Perspectives
Edited by Simona Francese and Stephen Bleay
© The Royal Society of Chemistry 2024
Published by the Royal Society of Chemistry, www.rsc.org

pattern. Furthermore, the compatibility between MALDI MS-based approaches and several fingermark enhancement techniques, as well as sequential processes has been demonstrated. The recognition of this technique as category B in the Home Office/Dstl Fingermark Visualisation Manual, along with its deployment in some police/forensic provider casework, makes MALDI MS an emerging technique moving towards full operational deployment, albeit additional validation and identification of sustainability models are still pending. This chapter describes the applications, the operational capabilities and the limitations of MALDI MS for molecular fingerprinting.

7.1 Introduction

A fingerprint is commonly defined as an impression made by the friction ridges upon contact with a surface. In the UK, the forensic community further distinguishes between the terms fingerprint and fingermark; specifically the former is defined as a reference, a ridge impression deliberately taken under controlled conditions (for example, upon arrest or at a border crossing) and the latter as a ridge impression accidentally left on a surface. Crime scene investigators (CSIs) and fingermark enhancement laboratories are tasked with visualising fingermarks on surfaces of interest by applying fingermark enhancement techniques (FETs) that involve both physical and chemical methods, in order to subsequently compare them to fingerprint records stored in national databases. As such, and given that no two fingerprints/fingermarks have ever been found to be identical, they have been used as a means of biometric identification for over 130 years.

It is now commonly known that fingerprints/fingermarks are not just a pattern of lines, but a pattern of lines made up of molecules. However, while this knowledge dates back to 1978,[1] the analytical community only started exploiting the fingermark chemical composition, for purposes other than developing or refining enhancement techniques, in 2007. A few studies have been published reporting the composition of fingermarks.[2–5] Molecular compositional studies were and still are of great importance as they provide the baseline knowledge to develop applications of fingermark analysis. Building upon these studies, to the best of the authors' knowledge, the first ever report exploiting fingermark chemical composition and demonstrating an alternative and complementary use of this evidence was presented by Asano *et al.* in 2002.[6] In this study, the authors explored the use of laser desorption ionisation mass spectrometry to detect lipids in fingermarks and determine the sex of the individual. Although eventually unsuccessful, this paper could be considered a landmark in introducing the concept that fingermarks could be potentially used to also gather intelligence about the person who left the impression. Subsequently, the analytical community has worked to develop ways to extract chemical intelligence and the reports published fall into two categories of analytical techniques, namely, mass spectrometry and spectroscopy, with the former group outnumbering the latter in terms of publications.[7] Mass spectrometry (MS) techniques proposed include laser desorption ionisation (LDI), surface assisted laser desorption ionisation (SALDI), desorption electrospray ionisation (DESI), matrix assisted laser desorption ionisation (MALDI), secondary ion mass spectrometry (SIMS), silver-assisted LDI (Ag-LDI) and direct analysis in real time (DART).[8] Among these techniques, MALDI MS has generated the most significant body of knowledge[7] in the context of molecular fingerprinting. First reported in 2009[9] for this application, MALDI MS has been used in both 'profiling' (MALDI MSP) and 'imaging' (MALDI MSI) modes.[8,10] The profiling modality provides access to the knowledge of the chemical content of a fingermark in a matter of seconds, and this particular aspect is better clarified in Section 7.2.1. As such, it could provide useful contextual intelligence complementing the biometric information (ridge detail) recovered by the CSI/fingermark enhancement laboratories. In the imaging mode, this technique enables the visualisation of molecules directly within the ridge pattern. This capability enables the reconstruction of fingermark images from the molecules detected within them, thus (i) simultaneously providing chemical intelligence and (ii) linking contextual information to the identity of the individual. While the underpinning principles of working for this technique are briefly discussed in Section 7.2.1, it is worthwhile to highlight at this point that MALDI is a soft ionisation technique. As such, it generates, in the vast majority of instances, intact, singly charged ions, making the spectral interpretation very simple compared to other techniques like SIMS or even DESI (for higher molecular weight species). It is considered a semi-destructive technique for three reasons: (i) it is a surface technique, meaning that only a few microns of molecular layers are removed during analysis, (ii) the sample remains otherwise intact and can be inserted into the mass spectrometer as it is (except for certain requirements reported in Section 7.3.1), (iii) a chemical is needed to spot or coat the sample; however, that chemical can be easily removed without any ridge merging. Wolstenholme *et al.* demonstrated that a fingermark is still present after the chemical's removal, as it could be subsequently re-enhanced.[9] MALDI can be operated under vacuum or at atmospheric pressure (AP). However, all of the publications available in the context of molecular fingerprinting have featured vacuum MALDI. While others have reported potential issues relating to molecular degradation under high vacuum conditions,[11] our group has not experienced them in our research, or at least they did not hinder the development of methods for recovering chemical and biometric information. Furthermore, it remains to be seen whether AP MALDI has comparable sensitivity to vacuum MALDI. Comparable sensitivity would be a very welcome specification as AP MALDI would then lend itself to on-site use, at crime scenes, as opposed to vacuum MALDI which can only be used remotely in specialist laboratories.

Within Sheffield Hallam University (SHU), fingermarks remained the main focus of the fingermark research group (FRG) for the first 6 years of MALDI-based application development, with >15 publications in this area. However, since 2014,[12] in an effort to demonstrate versatility of the technique in the forensic domain, MALDI MSP and MSI were further developed as confirmatory tests for the presence of blood. Indeed, a considerable body of work has been developed since then, demonstrating the operational potential to confirm the presence of blood, determine its species of origin and recover additional intelligence when present in both stains and fingermarks. This has included demonstrating compatibility with the prior chemical enhancement of stains/blood marks.[13–22] Furthermore, additional papers have been published on the use of MALDI for detection and mapping of drugs and metabolites in hair.[23–27]

While the potential of being a multi-evidence forensic analysis platform is now clear, this chapter will solely focus on the use of MALDI based methods for the analysis of fingermark and blood evidence, followed by an assessment of the technology readiness level and how this technique is being used in the operational space. A personal perspective on the future developments and full integration of this technique will be also offered.

7.2 Working Principles and Performance of Modern Instrumentation

7.2.1 Working Principles

Matrix assisted laser desorption ionisation (MALDI) is a surface and soft mass spectrometry ionisation technique producing intact ions in the gas phase and it was first brought to the analytical community by Karas and Hillenkamp in 1985.[28] This technique was originally invented to facilitate the ionisation of large non-volatile and labile molecules. With further hardware development, it has been possible to detect protein species ranging from several hundred thousand Daltons[29] to a million Daltons, as reported by Colvax.[30] However, contrary to the perception among non-experts, (UV-)MALDI MS is also capable of analysing small molecules, typically from 100 Da. Although the use of a chemical (matrix) to assist ionisation may lead to the production of a chemical background in the 100–1000 Da region that could overlap with ion signals from small molecules, this phenomenon is very much matrix- and analyte-dependent. Furthermore, in cases where this signal overlap does happen, instrumental mass resolution and ion mobility often positively counteract it. As such, UV-MALDI has expanded its versatility in the analysis of a wide range of molecule classes including low molecular weight drugs, metabolites, polysaccharides and lipids, in addition to biopolymers and synthetic polymers, peptides, proteins and small nucleic acids.

In (UV-)MALDI MS, a UV-absorbing non-volatile, organic small molecule (generally a weak acid or a weak base) is mixed with the analyte of interest and left to co-crystallise on the MALDI target plate forming matrix–analyte clusters, where the matrix is in large excess and completely surrounds the analyte to protect it from the direct impact of the laser. Upon laser activation (characterised by a wavelength within the UV region of the spectrum), rapid heating on the surface of the sample occurs causing desorption of the matrix-analyte clusters in the gas phase. Two main theoretical models exist to explain what happens next, namely, the gas phase protonation model[31] and the Lucky Survivors model.[32] The former proposes absorption of laser energy by the matrix and subsequent matrix vibrational excitation leading to primary ionisation of the matrix; when operating in the positive mode, the photoexcited matrix molecules are stabilised through proton transfer (or cation attachment) to the analyte, thus generating mainly intact protonated molecules. The second model was initially referred to as peptide and protein ionisation and it assumes that, given their amphoteric nature, they are already incorporated within clusters as charged species, much like they were in the original solution. While a more detailed description is given elsewhere,[32,33] protonated ions are released and called lucky survivors if they escape neutralisation through absorption of photoelectrons or electrons from the stainless-steel target plate.[33]

In the profiling mode (MALDI MSP), the matrix is mixed with an analyte solution, or spotted onto the sample (for example, a fingermark) and mass spectral information is obtained in a handful of seconds by directing the laser to discrete areas within the matrix spots. The imaging modality was first conceived and described in its infancy by Spengler.[34] However, it was Caprioli's group in 1997 who developed it in an operational and comprehensive manner, focussing on the imaging of tissue sections in a clinical context.[35] In the imaging mode (MALDI MSI), the matrix is coated homogenously over the sample, with careful consideration to prevent delocalisation. Different methods of matrix application are available, the most common of which include spray coating and sublimation.[36] In the imaging modality, the laser is automatically fired over a raster of points set *via* software corresponding to specific (x, y) coordinates. An array of mass spectra is acquired, each reflecting the molecular composition of the raster point fired of specific (x, y) coordinates. The spatially closer the points, the higher the image resolution. The software can then be interrogated to visualise any detected ion. Every image pixel in which that ion was detected will be 'painted' in a user instructed colour; thus, an image of the distribution of that particular ion is reconstructed, with the pixel hue and brightness correlating to the ion intensity of each species in the associated mass spectrum. Images can be obtained for both known and unknown ions. A schematic of the application of MALDI MSP and MSI to fingermarks was given by Francese *et al.* in 2013.[10] In the imaging modality, it is therefore possible to reconstruct the image of a fingermark by mapping molecules distributed along the ridges. Indeed, as thousands of molecules can be detected in any one analysis, it is theoretically possible to generate thousands of images of the same mark, with considerable advantages, from a biometric intelligence point of view, as described in Section 7.3.1.

7.2.2 Performance of Modern Instrumentation

As observed across a large number of mass spectrometry techniques, hardware and software advancement has been occurring at a very fast pace in the last two decades. MALDI MSP/MSI has benefitted from coupling with ion mobility spectrometry which increases ion separation and the ability to detect and confirm the presence of both known and unknown targets. The increased mass resolution and accuracy offered by modern instrumentation are crucial when acquiring potentially incriminating or individual-specific chemical intelligence from fingermarks. While the Orbitrap (Thermo) and the FT-ICR (Bruker Daltonics) mass spectrometers have a mass resolution ranging between 500 000 and 2 000 000 FWHM, other configurations such as the MRT SELECT SERIES (Waters Corp.) and the Tims-TOF offer up to 200 000 and 60 000 FWHM, respectively. Mass accuracy on these high-end, modern instruments can reach remarkable levels achieving sub-ppm RMS error with internal calibrants. It is difficult to compare sensitivities as vendors use different parameters. For example, the nominal sensitivity for [Glu1]-fibrinopeptide is 2 fmol on the Waters Corp. SELECT SERIES MALDI MRT system, sub-femtomolar for Angiotensin 2 on the Thermo AP-MALDI Orbitrap Fusion Lumos HRMS system and 250 amols for [Glu1]-fibrinopeptide on the Bruker Daltonics timsTOF system. The laser repetition rate may vary across vendors and instrumentation between 1 to a maximum of 12 kHz.

If, from a chemical intelligence point of view, these specifications are extremely important, high spatial resolution is fundamental to enable the provision of level 1 and level 2 (*minutiae*) fingermark detail and permit fingermark matching against a fingerprint record. Nowadays, within high end instrumentation, it is possible to achieve lateral resolution down to a nominal value of 5 μm as offered by the main vendors, depending on the ion being imaged. While still not reaching the spatial resolution as offered by SIMS Imaging (submicron), MALDI MSI has certainly become very competitive. Certainly, considering the dimension of fingermark ridges, a lateral resolution of 20 μm is more than enough to accurately capture the *minutiae*. Figure 7.1 displays the fingermark images acquired on increasingly better performing instruments at increasing spatial resolution. The acquisition time depends on the spatial resolution set, the size, the laser repetition rate available on the instrument and on the mode of acquisition (stop and go or rastering). The image acquired at 100×100 μm was acquired in 90 minutes (Figure 7.1a), the image at 80×80 μm was acquired in 15 hours (Figure 7.1b), that at 20 μm on the Synapt G2 HDMS (Waters Corp.) was acquired in 20 hours (Figure 7.1c) and finally the image run at 20×20 μm on the MRT SELECT SERIES MALDI qTOF was acquired in 6 hours (Figure 7.1d).

The latter is an 'inverted' image, in that the orange colour displays the distribution of an ion within the valleys of the fingerprint and not the ridges. This ion is also distributed in the sweat pores (the orange full circles visible on the black lines (the ridges)). It is interesting to observe that even if the nominal spatial resolution is the same for fingermarks in Figure 7.1c and d, the actual image resolution is considerably better within the MRT image (Figure 7.1d).

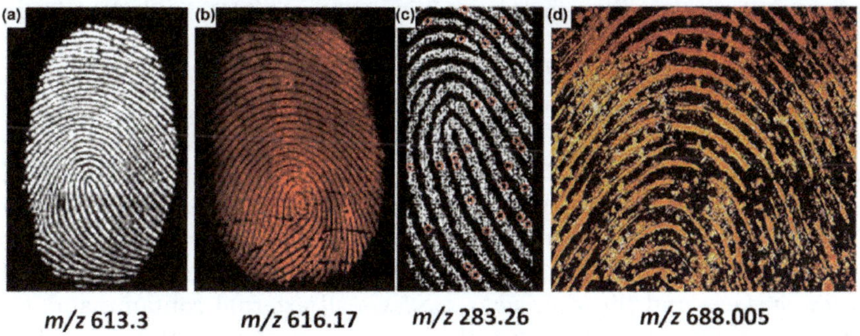

Figure 7.1 MALDI MS images of fingermarks acquired on differentially performing mass spectrometers and at different spatial resolution. (a) MALDI MSI of a PEG-contaminated latent fingermark acquired on a Qstar Elite MALDI qTOF (Applied Biosystems) spectrometer at *m/z* 613.3 and 100 × 100 μm spatial resolution in 2011; (b and c) MALDI MS images of a blood mark (*m/z* 616.17, haem) and of a latent mark at *m/z* 283.26 were acquired on a Synapt G2 HDMS MALDI qTOF (Waters Corp.) at 80 × 80 μm (2021) and 20 × 20 μm (2012) spatial resolution, respectively; (d) the image at *m/z* 688.005 was acquired on the MRT Select SERIES MALDI qTOF system (Waters Corp.), at 20 × 20 μm (2022). Marks in (a) and (c) were prepared with the SunCollect Autosprayer (SunChrom Germany) while marks in (b) and (d) were prepared with the HTX M3+ sprayer (HTX Technologies, US). (a) is courtesy of Dr Robert Bradshaw, Sheffield Hallam University, UK, (b) reproduced from ref. 37 with permission from HTX, Copyright 2020, (c) reproduced from ref. 39 with permission from American Chemical Society, Copyright 2011. (d) is courtesy of Dr Katie Kennedy (formerly from Sheffield Hallam University, now University of Liverpool, UK).

7.3 MALDI MS Applications in Fingerprinting and Technology Readiness Level

In the context of the research of the SHU FRG, the wording "molecular fingerprinting" refers to the process of obtaining intelligence directly from a crime scene mark by exploiting its molecular content. Intelligence spans from biometrics (complementary/alternative/unique visualisation of the ridge pattern and *minutiae*) to the detection of molecules of forensic interest. The latter information can either be used as it is, and complement investigations, or can be used to reconstruct molecular images of fingermarks, thus linking identity to intelligence/circumstance of the crime. In relation to molecules of forensic interest, Francese's group coined the terminology "criminal chemical profiling"[38] to indicate the potential to gather intelligence around an individual's actions, lifestyle, physio/pathological and pharmacological state in order to profile an offender from the molecular composition of his fingermarks. This type of intelligence would be gathered in a complementary way to traditional criminal profiling techniques based on reconstruction of a psychological and personality profile from a modus operandi (MO) and the circumstances of the crime.

MALDI MS-based methods can detect and map the three classes of constituents present in fingermarks, namely, (i) *endogenous species* – produced within the body and potentially excreted in sweat thus ending up in a fingermark upon contact of the fingertip with a surface (ii) *semi-endogenous species* – external substances entering the body (in the form of food, medicines, drugs, drinks, *etc.*) and that are excreted in sweat intact and/or as metabolites, also ending up in a fingermark upon contact of the fingertip with a surface, and (iii) *exogenous species* – all external contaminants that come into contact with the fingertips from the environment and reside on the fingertips for a period of time, before being transferred to a fingermark upon contact of the fingertip with a surface.

Both biometric and chemical molecular profiling capabilities may corroborate other gathered evidence or otherwise inform investigations. Molecular fingerprinting *via* MALDI MS-based methods was pioneered by SHU; as it can be expected from an academic entity, SHU does not hold ISO17025 accreditation (a requirement for forensic laboratories defined by the Forensic Science Regulator). However, the process has undergone various validation studies in addition to intrinsic validation within pseudo-operational and operational trials which have increased the technology readiness level (TRL) of the applications developed and described in Sections 7.3.1–7.3.6.

7.3.1 Visualisation of the Ridge Pattern

Presently, MALDI MSP/I can be applied to fingermarks present on any thin (up to 200 microns) and flat surface; this requirement arises from the necessity to avoid surpassing the height of the target plate, ultimately causing laser de-focussing and loss of data. However, *ad hoc* recessed target plates could be engineered for thicker surfaces. Flat surfaces are also a necessity for the same reason. For any other surfaces, from curved to those large in size, fingermarks can be recovered from the surface for MALDI analysis with a conventional CSI fingerprint lifting tape. Overall, MALDI MSI has been proven to enable fingermark imaging from glass, varnished wood, leather,[39] plastic,[38,39] paper,[39,40] polymer banknotes,[41] metallic surfaces,[39] parcel and clear tape,[43] and marks obtained directly from crime scenes (textured plug sockets and the interior side of a uPVC window frame).[43]

In terms of biometric information, it is possible to improve the ridge pattern continuity by stitching together or superimposing complementary molecular images from many different ion species[10] as well as separating overlapping fingermarks[44,45] as shown in Figure 7.2.[45]

Although stitching/superimposing images increases the chance of a suspect identification against a fingerprint record in the national database (if present), as well as against a fingerprint taken on arrest, separation of fingermarks may be very useful in the process of 'fingermark elimination' and/or to distinguish an assailant's fingermark from the victim's fingerprint.

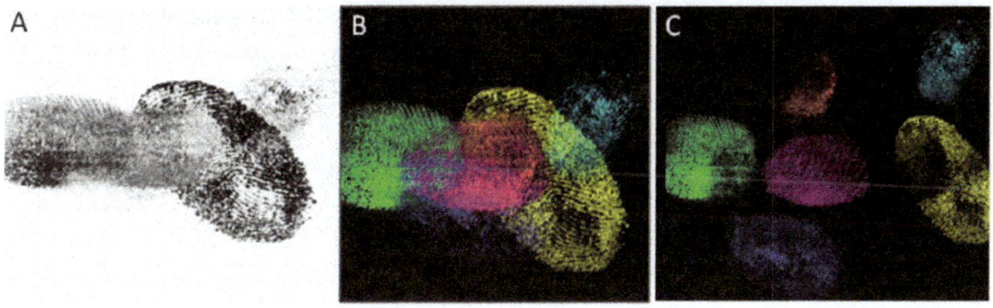

Figure 7.2 Separation of overlapping fingermarks by MALDI MSI. (A) The optical image of overlapping inked marks. (B) The overlapping MALDI MS images of *m/z* signals unique to each mark. (C) The separate recalling of MALDI MS images showing six different fingermarks. Reproduced from ref. 45 with permission from the Royal Society of Chemistry.

7.3.2 Chemical Intelligence – Determination of Sex

One obvious way to conduct criminal chemical profiling from fingermarks would be to determine the sex of the individual. Although Asano *et al.*[6] first suggested this possibility, the utilization of laser desorption ionisation (LDI), rather than MALDI, and lipids as the molecular target, did not permit the verification of this hypothesis. Other groups have attempted to develop a molecular methodology for sex determination from fingermarks but the number of samples was inadequate and the statistical modelling not robust/reliable.[46–48] Francese's group also endeavoured to gather this type of intelligence using MALDI MSP and targeting endogenous peptides/small molecules instead. The first proof of concept study was published in 2012[49] using a cohort of 80 donors and it achieved a prediction accuracy of 75% (85% if less harsh classification criteria were applied). In a subsequent study published by Heaton *et al.* in 2020,[50] the cohort size was increased to 200 donors and more sensitive instrumentation, with higher mass accuracy, was employed, yielding 86.1% accuracy of prediction. While the accuracy of prediction (86%) does not allow suspect elimination, West Yorkshire Police (co-authors of the study) stated the method would be useful in triaging marks at crime scenes. It has also been observed that this method is compatible with the prior application of white granular powder but not aluminium powder; a more comprehensive study on the compatibility with a range of FET needs to be undertaken to assess its feasibility for operational use.

7.3.3 Chemical Intelligence – Detection of Pathologies

In a recent ground-breaking publication, the potential to detect pathologies from fingermarks was shown. Specifically, Russo *et al.* undertook a small proof of concept study with 15 patients aiming to detect breast cancer in a non-invasive screening from fingertip smears which targets, once again, endogenous proteins. This study demonstrated the possibility of discriminating between benign, early cancer and metastatic pathology with

97.8% accuracy of prediction.[51] In particular, every diseased patient was correctly diagnosed. The 2% of fingertip smears derived from healthy donors, which were incorrectly diagnosed (according to the ground truth disclosed), are undergoing further investigation to understand the reasons for the mis-classification. It is recognised that these studies need to be repeated and validated with hundreds of patients; nonetheless the machine learning algorithm employed mitigates, to some extent, over-fitting originating from the low number of the patients investigated. Although in its infancy, this work is at the interface between clinical diagnostics and forensics and it is anticipated that it will boost additional studies from the community to operationally use sweat in both areas. Another example of the potential to diagnose pathologies, useful in both a clinical and forensic environment, has been provided by Heaton *et al.* in 2021 for blood marks.[52] Here, blood marks were generated using silicone rubber fingerprint stamps contaminated with the blood of healthy *versus* patients with haemoglobinopathies originating from the presence of different haemoglobin variants (HBVs). The marks were then analysed by both MALDI MSP and MSI utilising *in situ* bottom-up proteomics to identify HBV proteotypic peptides. These species were then used to reconstruct the ridge pattern. In a scenario where a blood mark is found at the scene, generated by the suspect with their own blood in the context of a violent crime, it may therefore be possible to link the biometric information (suspect identity) with physio-pathological information of the donor, such as disease states. Furthermore, on the basis of the available literature, it would be very interesting to verify the observation that some HBVs are linked to bio-geographical provenance in order to provide additional contextual information.

7.3.4 Chemical Intelligence – Detection and Mapping of Drugs and Metabolites

During the course of an investigation, intelligence on consumption of medications (prescribed or unprescribed) and/or recreational/drugs of abuse may provide important contextual information. This information is routinely recovered from blood or urine (or occasionally from hair) and the gold standard techniques are gas chromatography (GC) and liquid chromatography (LC) MS/MS. However, following the work of Bailey *et al.*,[53] Groeneveld *et al.*[54] and in live casework by Bradshaw *et al.*,[42] it is clear that, at least qualitatively, toxicological information of this kind can also be obtained by MALDI MS from fingermarks, thus enabling the extraction of lifestyle information, physical wellbeing and/or state of mind of the individual. To date, MALDI MS has been successfully tested to detect and/or map seven common drugs of abuse and their metabolites in enhanced and unenhanced ('as deposited') fingermarks.[53–55] Although there are no quantitative studies reporting detection limits in relation to drug and metabolite excretion in fingermarks, in the work by Bailey *et al.*,[53] benzoylecgonine was detected in a fingermark of a drug rehabilitating patient at a level above the 16 ng ml^{-1} cut-off in the corresponding oral fluid, as determined by complementary GC–MS analysis. As for other types of chemical intelligence, adding the imaging dimension (*i.e.*, visualising the drug/metabolite on the fingermark ridge pattern) provides investigators with associative evidence that may further contribute valuable insights into the case. The importance of visualising distribution also contributes more strongly to the hypothesis of non-accidental contact; a few isolated particles could arise from accidental contact but consistent and extensive distribution along the ridges would provide evidence against the first scenario. A simple swab may give a positive result in both instances; therefore chemical mapping is much more useful to discriminate scenarios.

7.3.5 Chemical Intelligence – MALDI MSP/MSI Application for the Detection and Mapping of Pharmaceuticals and Metabolites

Much less explored is the detection of pharmaceuticals and metabolites;[56,57] unpublished MALDI profiling data by Francese's group, acquired through MALDI IMS–MS/MS show the possibility to detect propranolol (a prescribed beta blocker) and its main metabolite (4-hydroxypropranolol) in fingermarks, 2 hours following administration. Despite the paucity of published research on this topic, there is no reason why other medications could not be profiled (and imaged) by MALDI MS to gain additional insights about the individual who has left the fingermark at the scene; this appears to be simply an unexplored area. Bailey's group has already reported the detection of a prescribed anti-psychotic drug and its metabolite from fingerprints using a tandem mass spectrometry approach in the context of monitoring drug adherence.[58]

7.3.6 Chemical Intelligence – MALDI MS Application for the Detection and Mapping of Exogenous Substances (Contaminants)

As highlighted already in this chapter, establishing a link between the biometric information (through the reconstructed MALDI MS image of a fingermark) and the criminal chemical profiling information provides additional associative evidence. This evidence can become corroborative through parallel standard investigations (still being mindful of the unsolved gap in knowledge revolving around ascertaining the most probable origin scenario explaining the presence of a certain substance[59,60] activity level propositions). Research on detection and mapping of exogenous substances such as condom lubricants[59,60] and blood[12–22] have certainly strengthened this proposition.

In the case of sexually motivated crimes, where the offender uses a condom to avoid leaving DNA behind, condom lubricant polymers can be targeted as they are detected and mapped by MALDI MSP/MSI very easily. Bradshaw *et al.*[61,62] were the first to report this application in condom-contaminated fingermarks by reconstructing the fingermark images and generating spectral information showing molecular composition differences for each condom brand/type investigated. This capability has been adapted for MALDI MSP casework commissioned by a European forensic provider (Forensicon BVD). Furthermore, as polymers are persistent substances, their detection and mapping are possible in older marks too, thus potentially counteracting the delayed reporting of this kind of crime.

Blood is a biofluid frequently encountered at the scene of violent crimes and its detection (both in stains and in fingermarks) can offer further intelligence relating to the dynamics and nature of the crime. SHU FRG has developed MALDI MS protocols enabling significant contributions in this area[12–17,22] followed by an Australian research group that also contributed to the demonstration of these capabilities.[15,17] MALDI MSP has enabled the detection of blood from fresh and very old specimens (a 9-year-old palm print on a ceramic tile developed using Acid Black 1[13] and a 37-year-old ninhydrin developed fingermark on fabric[7]). While Kamanna *et al.*[17] devised a method to differentiate human and non-human haemoglobin, the method developed by the SHU FRG (within studies supported by the Defence Science and Technology Laboratory, Dstl, in the UK) included additional blood protein targets and was pre-validated with a set of 56 mixed blind samples through a combined bottom-up proteomics approach, enabling the determination of blood provenance, down to animal species[16,21] and also in mixture with other biofluids.[22] These capabilities have been confirmed through MALDI MSI in both fresh[12,13,18] and 4-year-old developed marks.[13] Compatibility of MALDI MS with a presumptive blood test (Kastle–Meyer) has also been demonstrated in casework where the presence of blood was suspected (and then confirmed) on a fragment of a vehicle's smashed window.[12] Regarding compatibility, it has been recently demonstrated, in a collaboration with the Israel Police, that DNA typing is possible following MALDI MS imaging analysis of blood marks enhanced with Acid Black 1.[20] While MALDI MSP and MSI can confirm the presence of blood and specifically human blood, DNA profiling can provide a definitive match with the blood donor. This capability is significant for the integration of the technology/method in an operational environment. The compatibility of molecular fingerprinting, including the confirmatory test for blood, is described more comprehensively in Section 7.4. It is this compatibility that has significantly contributed to the TRL of molecular fingerprinting.

7.3.7 Chemical Intelligence – MALDI MS Application for the Detection and Mapping of Explosives

Explosives are another category of exogenous substances, the detection and mapping of which could significantly contribute to contextual information. However, as for pharmaceuticals, this is a poorly published application. Kaplan *et al.*[57] published a report showing a spot MALDI MS image in the negative mode of both trinitrotoluene (TNT) at m/z 226.0 and of the RDX ion fragment at *m*/*z* 102.1 using α-cyano-4-hydroxycinnamic acid (α-CHCA) as a matrix. In 2015, the authors of this document had invested some efforts in both the detection and mapping of explosives particularly on the components of Semtex, namely, TNT, pentaerythitol tetranitrate (PETN) and nitramide RDX. Detection of PETN has never been possible whereas low intensity negative ion mode MS and MS/MS spectra were obtained for RDX and TNT, using 9-aminoacridine as the matrix, with limits of detection in the imaging mode of 20 ng and 5 ng, respectively (unpublished, data not shown). Very recently, Musah's group published a further report on the application of MALDI MSI to the touch chemistry of explosives,[63] covering TNT, tetryl and picric acid. Here, both profiling and imaging data were shown with these three explosives mapped using 9-aminoacridine as the MALDI matrix, in high spatial resolution onto the fingermark ridge pattern. While this application is interesting and much more robust in the research conducted by Musah's *et al.*, it remains in its infancy and the authors of this chapter speculate that the volatility and the chemical structure of explosives are the reason why this application has not been further developed; however they do not exclude the possibility that the use of dopants might make this application possible in the future.

7.4 Compatibility with Crime Scene Investigation (CSI) and Forensic Examination Laboratory (FEL) Processes

For the operational use of molecular fingerprinting *via* MALDI MS based approaches, any developed capability must be compatible with the prior application of FET and, ideally, it should also not prevent the further downstream processing of the evidence for the extraction of additional information. For this reason, work has been undertaken in the past 15 years starting from the very beginning of this research at SHU, eventually demonstrating compatibility with a number of FETs, namely, powders,[9,38,40,43] vacuum metal deposition (VMD),[40,54] cyanoacrylate fuming (CAF),[38,40,42,54] powder suspensions (PS),[42] acid dyes,[13,16,20] ninhydrin,[7,12,20,43] indandione (Fischer *et al.*, in preparation), as well as with sequential processing routes[20,43,62] which the Home Office's Fingermark Visualisation Manual (FVM) advises deploying in a specific order, to maximise the likelihood of suspect identification by increasing the amount of biometric information visualised (Figure 7.3).

In other instances, the ability to utilise a MALDI MS-based approach in the middle of a sequential workflow also facilitates the recovery of additional intelligence. In the example reported in Section 7.3.6,[20] the possibility to successfully perform DNA typing after (i) enhancing a blood mark by Acid Black 1, and (ii) using MALDI MSI and MSP to confirm the presence of human blood, was of considerable value to the Israel Police (co-authors of the study); these end users subsequently suggested a modification to their workflow

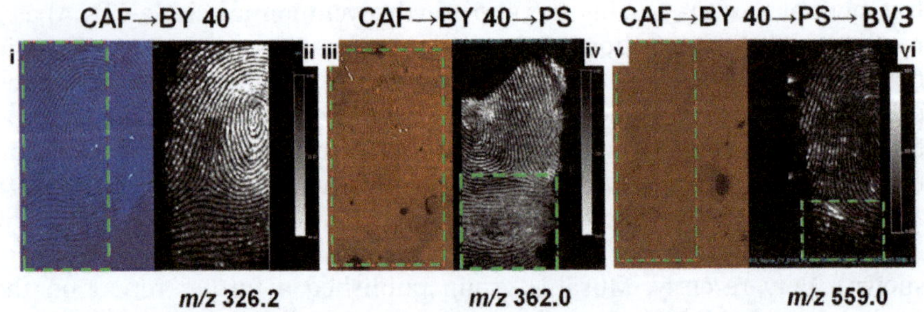

Figure 7.3 Application of MALDI MSI following sequential processing of a latent fingermark on brown tape. Panels i, iii and v indicate optical images of the mark after CAF → BY 40, CAF → BY 40 → PS and CAF → BY 40 → PS → BV3, respectively. Panels ii, iv and vi indicate the corresponding MALDI MS images showing the most ridge flowing/detail. Here, it can be seen that not only does MALDI MS yield the reconstruction of the ridge pattern after the application of multiple FETs but in some cases also yields additional details. Reproduced from ref. 43 with permission from Elsevier Copyright 2021. Abbreviations: CAF = cyanoacrylate fuming; BY40 = basic yellow 40; PS = powder suspension; and BV3 = basic violet 3.

integrating MALDI MS-based approaches when blood marks are involved in casework.[20]

It is this pursued and achieved compatibility with operational workflows and processes that has eventually permitted the introduction of a MALDI MS-based approach into the FVM (2014 edition) as a category C technique[64] and it has subsequently been promoted to category B in the 2023 edition[65] (Section 7.5). The introduction as Cat. C in 2014 has undoubtedly pushed the deployment of MALDI MS-based approaches into operational casework (Section 7.5). Altogether, these achievements have contributed to increase the technology readiness level as already reported by Bradshaw and Francese[45] and covered in Section 7.5.

7.5 Operational Character of the Technique and Integration into Casework

The FVM sets out operational recommendations for formulations, protocols and methods to the major crime units of all 43 police forces in the UK. In addition, the European Network of Forensic Science Institutes (ENFSI) Fingerprint Working Group has adopted the FVM as the key reference document across all European forensic institutes, forming the basis of the ENFSI Fingerprints Best Practice Manual. The FVM has sold >350 copies to agencies across the globe in 37 countries including EU, New Zealand, Australia, US, Scandinavia, South Africa, Israel and Hong Kong. In 2014, MALDI MSP/I was included as a category C technique, indicating "*a process at a developmental stage exhibiting potential [...] for occasional operational use [...] when category A processes have been exhausted.*" This is a five-point scale

(A–E), where categories A–C require particularly rigorous and extensive validation. At the time, the editor commented that "*a category C listing is an achievement in itself as the majority of new methods fail before this point*" (Dr Helen Bandey, personal communication). Given the advancement of molecular fingerprinting by MALDI MSP/MSI and the further validation that took place, the updated version of the FVM published in January 2023[65] features the promotion of this technique to category B, indicating an "*established process [...] likely to offer benefits [...] for occasional operational use [...] and when all Cat. A options have been exhausted*". The editor of the new FVM version commented "*category B processes are generally more established and therefore more likely to be used on casework. [...] It is a lot harder for processes to be included as category A or B due to the additional evidence required to establish its use and effectiveness. [...] It is worth emphasising the importance of such advancements [...] where chemical information within fingermarks, for the first time, has the potential to add significant value to supporting major and more complex crime investigation* (Dr Helen Bandey, personal communication). Building on the framework established by the FVM, a guideline document has been prepared in collaboration between Francese's group and the Forensic Capability Network (FCN) aimed at providing a guidance to police and crime commissioners on the correct exploitation of this technique for fingermark and blood analysis. The document has been placed in the FCN Knowledge Hub (https://fcn.lets-go.live/knowledge-hub). This is an online platform run by the Police Digital Service, hosting, among others, protected documents and is intended to facilitate dissemination of capabilities, ideas and opportunities to promote greater collaboration. The guidelines to use MALDI MSP/MSI in forensic casework involving fingermarks and blood evidence will be featured in the Knowledge Hub Purple, which is primarily for police forces and related organisations. The guidelines document aims to spread awareness on the capability of molecular fingerprinting via MALDI MS-based approaches, as well as to guide the end user to make an informed request of the type of intelligence sought after.

Given the fast pace of the developments and validation studies, it may not be surprising that molecular fingerprinting has been trialled in both pseudo-operational and operational conditions/casework. Altogether, MALDI has been deployed operationally across a total of 8 police/forensic provider casework scenarios (West Midland Police, West Yorkshire Police (WYP), Danish Police, Forensicon BV) (2013–2016) and a number of pseudo-operational trials with WYP (2011–2016) where major crimes were being investigated (murder, attempted murder, harassment, drug dealing, and alleged sexual assault).

MALDI MS capabilities have been subjected to testing since 2013 in a collaboration with WYP, UK. At that time Francese's group was able to attend crime scenes with CSI to (i) understand the constraints under which fingermarks are developed, captured and stored, in order to adapt MALDI MS approaches to those protocols and (ii) to analyse real crime scene marks (although not those of primary evidential value). Although the results of these analyses were never published and the instrument's capabilities at the time were not competitive, this work contributed considerably to the development of protocols for correct handling of evidence, maintaining the chain of custody, transporting and preserving it, in preparation for MALDI MSI analysis. From a technical standpoint, this pseudo-operational work enabled the investigation of the impact of CSI practice on MALDI MS analysis. In particular, it was possible to verify that, for fresh marks, the use of an acetate backing sheet, to which a fingermark lift is adhered for transport and storage, does not prevent successful MALDI MSI when peeling off the lift (Bradshaw, R., Denison, N. and Francese, S., unpublished); specifically, for a mark enhanced with aluminium powder that has been split in half, there is no difference in the clarity of the MALDI MS image of the ridge detail between the primary half lift and the half lift that was placed on the acetate backing sheet and then peeled off. Furthermore, the hypothesis that secondary lifts contained enough material for MALDI MSI analysis was tested.[66] It was observed that upon re-enhancement of the mark, a secondary lift was still viable. While the use of secondary lifts for MALDI MSI enables complete preservation of the primary lift evidence, this practice is not recommended if it can be avoided. Indeed, success in providing biometric as well as chemical information strongly depends on the instrumental sensitivity and on the quantity of the starting material which, for very old or depleted marks, may be limited. All of these observations were captured by Bradshaw *et al.* in 2016 and 2017.[42,66]

Another example of a pseudo-operational trial in which MALDI MS was trialled is represented by the European Network of Forensic Science Institute collaborative exercise organised by the fingerprint working group in 2020. The exercise was advised by the established advisory group (AG).

In this exercise, a compostable bag was selected as the surface of deposition. Due to the relative novelty of this material (bio-based polymers decomposing into natural products, such as carbon dioxide and water when composted), this has not been widely tested, and the AG believed that it could represent a viable challenge for a collaborative exercise. The task, undertaken by 40 state police laboratories across Europe, consisted of using, based on experience, what they believed to be the most effective process or sequence of processes to find and visualise latent fingermarks. While this study was published without Sheffield Hallam University (SHU) Fingermark Research Group (FRG)[67] results, Francese's FRG group did receive the sample and did perform the relevant analyses, as shown in Figure 7.4.

One of the limitations of MALDI MSI is the size of the sample that must fit the sample holder which varies slightly in dimensions according to different instrument manufacturers. The whole specimen provided could not fit the sample holder and one option would have been to cut the item into appropriate squares, analyse each one of them and subsequently stitch the images, if a fingermark was eventually present across two or more of the squares analysed. The other more limiting issue is that given the possibility that marks could be placed on either of the two sides of the item provided,

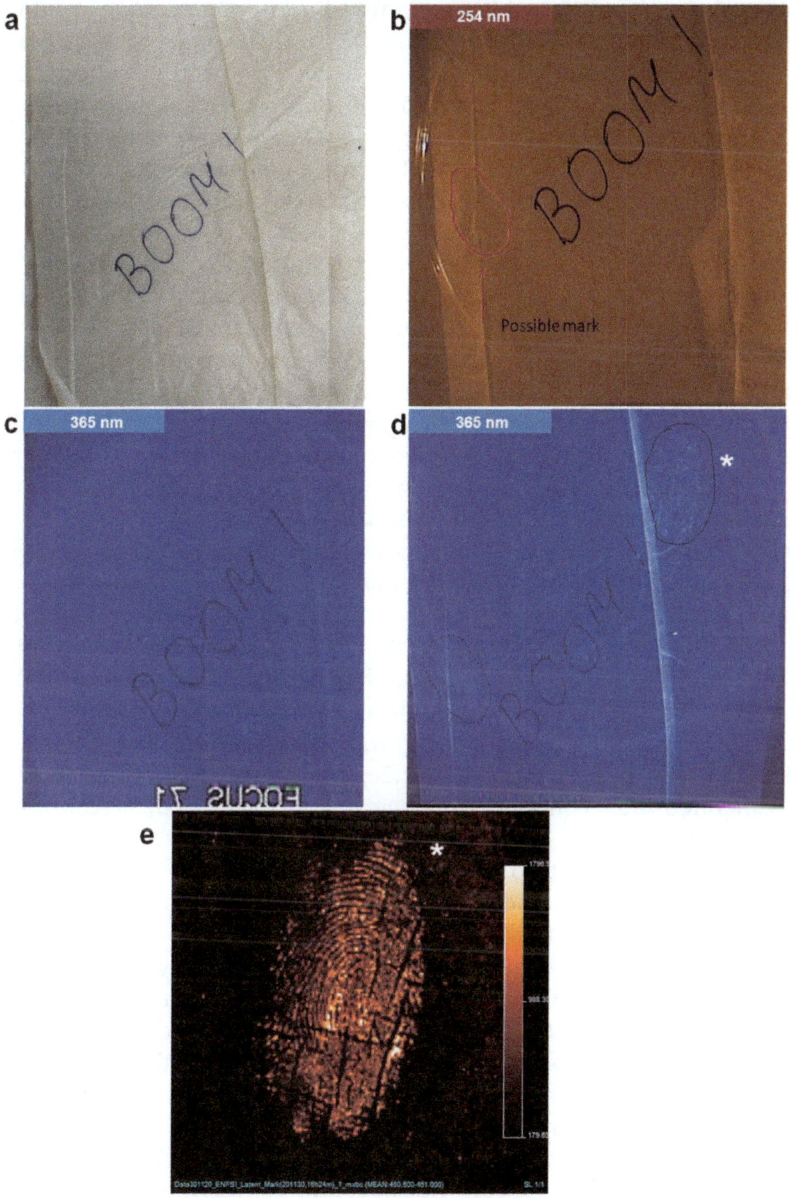

Figure 7.4 MALDI MSI analysis of fingermarks on a compostable bag as part of the ENFSI collaborative exercise 2020. (a) Optical image of the specimen; (b) image of the item at 254 nm possibly revealing a mark on the left-hand side of the letter 'B'; (c) image of the item at 365 nm revealing no marks; (d) image of the item at 365 nm after powdering with (α-CHCA) MALDI matrix revealing two marks (circled) at either side of the word 'BOOM!'; (e) MALDI MSI of the mark labelled with (*) in panel d after spraying a solution of α-CHCA revealing a full fingermark ridge pattern.

once the matrix-untreated side (randomly chosen) was stuck on the double-sided conductive tape and adhered to the MALDI target plate, a possible fingermark present on that side would have been irreparably damaged by the adhesive and physical mounting procedure. In order to simply assess the ability of MALDI MSI to visualise marks on a potentially challenging surface of deposition, the AG advised to analyse the side with the word "BOOM!".

In the absence of any of the classic forensic equipment used by crime laboratories, the "BOOM!" side was initially inspected by angling white light at it but no marks could be located (Figure 7.4a). Subsequently, the item was inspected using the Visual Spectral Comparator (VSC4CX) by Foster and Freeman, under UV light at 254 nm and a possible mark on the side of the letter 'B' was visualised (Figure 7.4b). The VSC was then tuned to 365 nm but no marks were observed (Figure 7.4c). The item was then powdered using α-CHCA according to step 1 of the dry-wet method.[31,68] Subsequent visualisation of the item at 365 nm enabled stronger observation of the mark on the left-hand side of the letter 'B' and a potential additional mark on the right-hand side of the exclamation mark (Figure 7.4d).

Due to lab time restrictions during the COVID period, it was decided to only focus on the latter mark, which was excised and sprayed with a matrix solution (in a modification of the dry-wet method[39,68]) prior to MALDI MSI analysis on a Qstar Elite MALDI qTOF system (Applied Biosystems, UK) in the m/z range 100–1000 Th and at a spatial resolution of 150×150 μm. Although today's MALDI instrumentation is capable of much higher spatial resolution (5–10 μm), a fingermark was visualised and a representative image at m/z 460.8 is given in Figure 7.2e. It was later disclosed that the mark was generated after contaminating the fingertip with hand cream and removing excess.

MALDI MSP/MSI has also been used in casework in the UK and in Europe by both police and forensic providers, although for operational reasons it has mostly not been possible to publish the results. In notable examples published by Bradshaw *et al.*,[38] marks were analysed using MALDI MSP/MSI after visualisation by CSI and/or crime laboratories using the relevant FET, under full operational conditions. While these were difficult marks, for which no additional/complementary biometric information to those provided by FET could be provided by MALDI MSI, chemical intelligence could be recovered, particularly for the harassment case; here, a mark was enhanced from the interior side of a window frame using carbon black powder, lifted, placed on an acetate backing sheet and delivered to the FRG at SHU. In its enhanced state, the fingermark appeared to be smudged with no useful ridge detail and MALDI MSI did not provide the desired biometric information as it yielded a similar molecular image. Indeed, if molecules have migrated away from the ridges, MALDI MSI will provide images of that 'smudged' distribution exactly and will be unable to contribute to biometric intelligence. However, the introduction of ion mobility spectrometry and transfer fragmentation analysis (MALDI IMS–MS/MS) on the Synapt G2 HDMS MALDI qTOF permitted the detection and confirmation of the presence of cocaine and several metabolites (benzoylecgonine (BZE), ecgonine methyl ester (EME), and norcocaine (NC)) within the fingermark. Among these metabolites, cocethylene (a metabolite formed when cocaine is mixed with alcohol) was also detected, indicating not only cocaine consumption, but also contextual alcohol consumption. Unbeknown to the FRG analyst, the suspect had been apprehended and tested positive for cocaine. However, he denied alcohol consumption and he could not be tested to prove/disprove his statement. However, just before the court hearing, the defendant made a confession that included alcohol consumption. The detection of cocethylene by MALDI IMS–MS/MS in this context is of great importance for a number of reasons: (i) it demonstrated the feasibility of using MALDI MS-based methods under full operational conditions, following CSI routine protocols for enhancement, recovery and storage of the fingermark evidence; (ii) it also shows robustness of the technique as this positive analysis was undertaken 28 days after the recovery of the mark and (iii) it demonstrates that, when unable to provide additional biometric information, it can still complement investigations by providing contextual information and, in this case, around the state of mind of the individual (alcohol potentiates the effects of cocaine) while committing the crime.

In 2019, another notable example of operational casework involving fingermarks was handed to SHU (commercial group). This was a murder case investigated by overseas police. As there is still confidentiality around it, only very general details can be disclosed here. The victim had disappeared from home in 2016 and the body was found in a lake 5 months later. A tape wrapped around the victim's neck was recovered and the sticky side was enhanced with carbon-based powder suspension. This process revealed the potential presence of a very low quality mark, which would be classified as grade 1 on the scale proposed by Sears *et al.*[69] The tape, submerged in water for 5 months and enhanced using FETs was analysed by SHU in January 2018, that is, around 18 months from the presumed death of the victim. Although the MALDI MSI analysis was conducted on the lowest performance instrument at the time (Qstar Elite MALDI qTOF system (Applied Biosystems, UK)) and at a low spatial resolution (150×150 μm), it was possible to provide ridge flow in a few areas (grade 2) in addition to ridge flow on the 'matrix etched' mark revealed on the tape after the application of the matrix and MALDI MSI. Data loss was evident in those non-flat areas of the tape (the tape was severely creased), causing laser de-focussing. Although the biometric information was not enough to apprehend the offender, the MALDI MSI outcome was hugely successful compared to the practitioners' process, considering the age and the environmental conditions the mark was exposed to, demonstrating once again operational feasibility.

7.6 Technology Readiness Level

In order to assign a TRL to the different applications described here and elsewhere,[45] in a way that better fits the nature of this technique, an

Table 7.1 Technology readiness level for MALDI–MS-based approaches applied to molecular fingerprinting-based applications developed thus far.

Application	TRL	TRL justification
Visualisation of ridge detail	9	Publication of data from live casework
Detection/mapping of blood	7–8	Pseudo-operational trials (and pre-validation), analysis of evidence under pseudo-operational conditions (7) – operational trials on evidence obtained from casework (8)
Detection/mapping of drugs and metabolites	7–8	Pseudo-operational trials, analysis of evidence under pseudo-operational conditions (7) – operational trials on evidence obtained from casework (8)
Detection/mapping of condom lubricants	7	Operational trials on evidence obtained from casework (8)
Determination of sex	6	Critical mass studies showing clear potential for operational use
Detection/mapping of pharmaceuticals	2	Published papers reporting a technology or data relevant to the area of application being considered
Detection/mapping of explosives	2	Published papers reporting a technology or data relevant to the area of application being considered

adaptation of the FVM TRL scale[64] has been selected as previously chosen by Bradshaw and Francese.[45] On the basis of Sections 7.3–7.5 and on the TRL reference scale proposed by Bradshaw and Francese,[45] Table 7.1 proposes the assignment of TRL to the various molecular fingerprinting MALDI MS-based applications cumulatively developed in different laboratories across the world.[45]

The assignment of the TRL has taken into account whether or not a certain capability has been demonstrated in laboratory settings *versus* an operational context and aims to inform crime commissioners in their decision-making process and consideration of the use of MALDI MS-based approaches.

7.7 Conclusions

MALDI MS-based approaches for molecular fingerprinting have certainly been on a steadily rising implementation trajectory for the past 15 years. Although 15 years may seem a long time, Academia, as the stakeholder undertaking this research, by its very nature and roles, cannot be as fast as industry in the development and deployment of research products/services.

However, Academia does, typically, have very advanced and cutting-edge instrumentation that will not generally be available in forensic laboratories. The achievements discussed in this chapter and the deployment in some operational casework scenarios have greatly benefitted from a very active end-user involvement and they have, to all effects, co-developed the application of this technology to molecular fingerprinting. The authors can safely state that this technology is on track for future promotion to a category A process in the FVM, provided that it meets the appropriate level of validation. However, there are some considerations to be made around implementation and sustainability. As this chapter is being written, a Home Office funded project (STAR scheme) in collaboration with Dorset Police and Yorkshire and Humber Regional Scientific Support Services (YHRSSS) is due to start in July 2023, aimed at testing the technology exclusively in an operational environment through the examination of marks from live, cold and archived police casework. Should MALDI MSI confirm the benefit of this application in providing additional/complimentary biometric information, the question arises as to who will take responsibility for its structured deployment in major crimes, including the possibility to offer expert testimony in a court environment. Indeed, while in the UK at least, it is possible to testify from a non-accredited lab/process, the defence or prosecution will aim to undermine the credibility of the evidence presented. The authors of this chapter see three different models of deployment, namely, delivery by (i) a university spin off, (ii) a forensic provider or (iii) a national, government funded centre of excellence for forensic science. The latter seems unlikely as it would resemble the Forensic Science Service (FSS) which was decommissioned in 2012. However, already existing government laboratories, such as the Ministry of Defence (MOD), Dstl in the UK, could potentially take on this extra capability. While a university spin off is not impossible, and the SHU commercial group does offer molecular fingerprinting service, it does take a very convincing revenue-based business case, in order to make such a considerable investment. A forensic provider, who is particularly engaged in assessing this type of evidence and heavily invested in the quest for additional intelligence from fingermarks, may potentially be interested, although, as for Academia, a convincing projected revenue is still a decisive factor. Consequently, projects such as the previously mentioned Home Office STAR project are incredibly important to disseminate awareness, inform crime commissioners and thus increase the number of instances in which MALDI MS-based capabilities are sought after. An additional step in this direction is facilitated by the demonstration of MALDI MS versatility as a platform for multi-intelligence recovery for several evidence types. This is why the FRG has expanded MALDI MS capabilities to blood and biofluid analysis as well as to hair (in collaboration with the Maastricht Multi-Modal Molecular Imaging Institute (M4I), Netherlands) and recently started two new research lines applying MALDI MS to microbial forensics and questioned documents. Although sustainability models need to be discussed and identified, the power of the technology could be very much increased by the creation of a database of molecular intelligence (drugs, medications, metabolites, condom lubricants, biofluids markers, explosives, *etc.*) linked to the probabilistic recovery of the likely substance being encountered through matching reference MALDI

MS and MS/MS spectra in the same manner as the National Institute of Standards and Technology (NIST) library does for gas chromatography (GC) electron ionisation (EI) MS data. The creation and dynamic expansion of this database would somewhat facilitate the policing (untargeted analysis) request of "what forensic molecular clues are present in that crime scene mark?" Indeed, while it is not possible to individually and manually examine the thousands of ion signals detected by MALDI MS, the increased mass accuracy and mass resolution achievable with high-end instrumentation could facilitate the export of a mass list to be checked against a reference database. MS/MS analysis could then be used to confirm the ion(s) of interest either *via* MALDI MS/MS, or if extraction of the mark is permitted, by liquid chromatography-tandem mass spectrometry (LC–MS/MS).

In terms of hardware, technological developments would be welcome to limit the impact on non-flat surfaces on the imaging data recovered. These developments are in progress at SHU as this chapter is being written; they constitute important developments to increase the number of surfaces analysed and maximise the amount of biometric information obtained, which would have been extremely beneficial in the 2016 murder case described in Section 7.6.

Overall, it is hoped that this chapter has demonstrated that the label 'emerging' is entirely fitting of the to date capabilities of MALDI MS based approaches in molecular fingerprinting.

Acknowledgements

Dr Katie Kennedy is gratefully acknowledged for acquiring the fingermark MALDI MSI data shown in Figure 7.1a and for helping with the data collection in Figure 7.3.

References

1. A. M. Knowles, *J. Phys. E: Sci. Instrum.*, 1978, **11**, 713.
2. G. M. Mong, C. E. Petersen and T. R. W. Clauss, *Advanced Fingerprint Analysis Project Fingerprint Constituents*, Pacific Northwest National Laboratory, Richland, WA, 1999, p. 48.
3. R. S. Ramotowski, *Advances in Fingerprint Technology*, CRC Press, Boca Raton, London, New York, Washington DC, 2nd edn, 2001, p. 63.
4. R. S. Croxton, M. G. Baron, D. Butler, T. Kent and V. G. Sears, *Forensic Sci. Int.*, 2010, **199**, 93.
5. M. J. Bailey, N. J. Bright, S. Francese, L. S. Ferguson, S. Hinder, S. Jickells, B. J. Jones, B. N. Jones, S. G. Kazarian, J. J. Ojeda, R. P. Webb, R. Wolstenholme and S. Bleay, *Anal. Chem.*, 2012, **84**(20), 8514.
6. K. G. Asano, C. K. Bayne, K. M. Horsman and M. V. Buchanan, *J. Forensic Sci.*, 2002, **47**, 4.
7. S. Francese, *Aust. J. Forensic Sci.*, 2019, **51**(6), 623.
8. S. Francese, R. Bradshaw and N. Denison, *Analyst*, 2017, **142**, 2518.

9. R. Wolstenholme, R. Bradshaw, M. R. Clench and S. Francese, *Rapid Commun. Mass Spectrom.*, 2009, **23**, 3031.
10. S. Francese, R. Bradshaw, L. S. Ferguson, R. Wolstenholme, S. Bleay and M. R. Clench, *Analyst*, 2013, **138**, 4215.
11. N. J. Bright, R. P. Webb, K. J. Kirkby, T. R. Willson, D. J. Driscoll, S. M. Reddy, N. I. Ward, M. J. Bailey and S. Bleay, *Forensic Sci. Int.*, 2013, **230**, 81.
12. R. Bradshaw, S. Bleay, M. R. Clench and S. Francese, *Sci. Justice*, 2014, **54**, 110.
13. E. Patel, P. Cicatiello, L. Deininger, M. R. Clench, G. Marino, P. Giardina, G. Langenburg, A. West, P. Marshall, V. Sears and S. Francese, *Analyst*, 2015, **141**, 191.
14. L. Deininger, E. Patel, M. R. Clench, V. Sears, C. Sammon and S. Francese, *Proteomics*, 2016, **16**, 1707.
15. S. Kamanna, J. Henry, N. Voelcker, A. Linacre and K. Kirkbride, *Int. J. Mass Spectrom.*, 2016, **397–398**, 18.
16. K. Kennedy, C. Heaton, G. Langenburg, L. Cole, T. Clark, M. R. Clench, V. Sears, M. Sealey, R. McColm and S. Francese, *Sci. Rep.*, 2020, **10**, 17087.
17. S. Kamanna, J. Henry, N. Voelcker, A. Linacre and K.P. Kirkbride, *Rapid Commun. Mass Spectrom.*, 2017, **31**, 1927.
18. M. Witt, K. Kennedy, C. Heaton, G. Langenburg and S. Francese, *Poster presented in part at the 69th Annual American Society of Mass Spectrometry Conference*, Philadelphia, November, 2021 (last accessed 12/04/2023).
19. C. Heaton, M. Witt, L. Cole, J. Eyre, S. Tazzyman, R. McColm and S. Francese, *Analyst*, 2021, **146**, 4290.
20. K. Kennedy, R. Bengiat, C. Heaton, Y. Herman, C. Oz, M. L. Elad, L. Cole and S. Francese, *Forensic Sci. Int.*, 2021, **323**, 110774.
21. K. Kennedy, L. Cole, M. Witt, M. Sealey and S. Francese, *Molecules*, 2022, **27**, 2039.
22. K. Kennedy, C. Gannicliffe, L. M. Cole, M. Sealey and S. Francese, *Sci. Justice*, 2022, **62**, 602.
23. T. Porta, C. Grivet, T. Kraemer, E. Varesio and G. Hopfgartner, *Anal. Chem.*, 2011, **83**, 4266.
24. E. Beasley, S. Francese and T. Bassindale, *Anal. Chem.*, 2016, **88**, 10328.
25. B. Flinders, E. Beasley, R. Verlaan, E. Cuypers, S. Francese, T. Bassindale, M. Clench and R. M. A. Heeren, *J. Am. Soc. Mass Spectrom.*, 2017, **28**, 2462.
26. B. Flinders, T. Bassindale and R. M. A. Heeren, Emerging Technologies for the Analysis of Forensic Traces, in *Advanced Sciences and Technologies for Security Applications*, ed. S. Francese, Springer, Cham, 2019.
27. B. Flinders, E. Cuypers, H. Zeijlemaker, J. Tytgat and R. M. A. Heeren, *Drug Test. Anal.*, 2015, **7**, 859.
28. M. Karas, D. Bachmann and F. Hillenkamp, *Anal. Chem.*, 1985, **57**, 2935.
29. K. Root, R. Frey, D. Hilvert and R. Zenobi, *Helv. Chim. Acta*, 2017, **100**, 1700166.
30. HM5 and HM5 Pearl: Fifth Generation High-Mass MALDI Retrofit System, https://covalx.com/pdf/CovalXBrochure-HighMassMALDI.pdf (accessed 6 April 2023).

31. H. Ehring, M. Karas and F. Hillenkamp, *Org. Mass Spectrom.*, 1992, **27**, 472.
32. M. Karas, M. Glückmann and J. Schäfer, *J. Mass Spectrom.*, 2000, **35**, 1.
33. T. W. Jaskolla and M. Karas, *J. Am. Soc. Mass Spectrom.*, 2011, **22**, 976.
34. B. Spengler, M. Hubert and R. Kaufmann, *Presented in part at the 42nd Annual Conference on Mass Spectrometry and Allied Topics Chicago*, Illinois, May, 1994.
35. R. M. Caprioli, T. B. Farmer and J. Gile, *Anal. Chem.*, 1997, **69**, 4751.
36. C. Russo, C. Heaton, L. Flint, O. Voloaca, S. Haywood-Small, M. R. Clench, S. Francese and L. M. Cole, *J. Spectral Imaging*, 2020, **9**, a13.
37. K. Kennedy, A. Black, L. Cole and S. Francese, *HTX Imaging Application Note*, 2021 (accessed 12 April 2023).
38. S. Francese, R. Bradshaw and N. Denison, *Analyst*, 2017, **142**, 2518.
39. L. Ferguson, R. Bradshaw, R. Wolstenholme, M. R. Clench and S. Francese, *Anal. Chem.*, 2011, **83**, 5585.
40. R. Bradshaw, S. Bleay, R. Wolstenholme, M. R. Clench and S. Francese, *Forensic Sci. Int.*, 2013, **232**, 111.
41. K. Scotcher and R. Bradshaw, *Sci. Rep.*, 2018, **8**, 8765.
42. R. Bradshaw, N. Denison and S. Francese, *Analyst*, 2017, **142**, 1581.
43. R. Bradshaw, G. Wilson, N. Denison and S. Francese, *Forensic Sci. Int.*, 2021, **319**, 110643.
44. R. Bradshaw, W. Rao, R. Wolstenholme, M. R. Clench, S. Bleay and S. Francese, *Forensic Sci. Int.*, 2012, **222**, 318.
45. R. Bradshaw and S. Francese, *MALDI Mass Spectrometry Imaging: From Fundamentals to Spatial Omics*, The Royal Society of Chemistry, 1st edn, 2021.
46. E. Claude, E. Bardin and Z. Takats, *Waters Application Note*, 2017 (accessed 5 April 2023).
47. I. Mekkaoui Alaou and J. Halamek, *J. Forensic Sci.*, 2019, **64**, 1495.
48. Z. Zhou and R. N. Zare, *Anal. Chem.*, 2017, **89**, 1369.
49. L. S. Ferguson, F. Wulfert, R. Wolstenholme, J. M. Fonville, M. R. Clench, V. A. Carolan and S. Francese, *Analyst*, 2012, **137**, 4686.
50. C. Heaton, C. Bury, E. Patel, R. Bradshaw, F. Wulfert, R. M. Heeren, L. Marchant, N. Denison, R. McColm and S. Francese, *Forensic Chem.*, 2020, **20**, 1.
51. C. Russo, L. Wyld, A. Da Costa, C. Bury, C. Heaton, L. M. Cole and S. Francese, *Sci Rep.*, 2023, **13**, 1868.
52. C. Heaton, M. Witt, L. Cole, J. Eyre, S. Tazzyman, R. McColm and S. Francese, *Analyst*, 2021, **146**, 4290.
53. M. J. Bailey, R. Bradshaw, S. Francese, T. L. Salter, C. Costa, M. Ismail, R. Webb, I. Bosman, K. Wolff and M. de Puit, *Analyst*, 2015, **140**, 6254.
54. G. Groeneveld, M. dePuit, S. Bleay, R. Bradshaw and S. Francese, *Sci. Rep.*, 2015, **5**, 11716.
55. A. Skriba and V. Havlicek, *Eur. J. Mass Spectrom.*, 2018, **24**, 124.
56. L. Sundar and F. Rowell, *Analyst*, 2014, **139**, 633.
57. K. Kaplan-Sandquist, M. A. LeBeau and M. L. Miller, *Forensic Sci. Int.*, 2014, **235**, 68.

58. C. Costa, C. Frampas, K. A. Longman, V. Palitsin, M. Ismail, P. Sears, R. Nilforooshan and M. J. Bailey, *Rapid Commun. Mass Spectrom.*, 2019, **14**, 8553.
59. A. de Ronde, B. Kokshoorn, C.J. de Poot and M. de Puit, *Forensic Sci. Int.*, 2019, **302**, 109904.
60. A. de Ronde and C. J. de Poot, Friction Ridge Impressions: Activity Level Propositions, in *Houck, Encyclopedia of Forensic Sciences*, ed. M. Max, Elsevier, 3rd edn, 2023, pp. 774–783.
61. R. Bradshaw, R. Wolstenholme, L. S. Ferguson, C. Sammon, K. Mader, E. Claude, R. D. Blackledge, M. R. Clench and S. Francese, *Analyst*, 2013, **138**, 2546.
62. R. Bradshaw, R. Wolstenholme, R. D. Blackledge, M. R. Clench, L. S. Ferguson and S. Francese, *Rapid Commun. Mass Spectrom.*, 2011, **25**, 415.
63. C. M. Longo and R.A. Musah, *Forensic Chem.*, 2020, **20**, 100269.
64. H. Bandey, V. Bowman, S. Bleay, R. Downham and V. H. Sears, *Fingermark Visualisation Manual*, CAST (Home Office), Sandridge, UK, 1st edn, 2014.
65. H. Bandey, *Fingermark Visualisation Manual*, Dstl (Home Office), Porton Down, UK, 2nd edn, 2023.
66. R. Bradshaw, N. Denison and S. Francese, *Anal. Methods*, 2016, **8**, 6795.
67. F. Zampa, M. Hilgert and A. Mattei, *Forensic Sci. Int.*, 2022, **335**, 111276.
68. L. S. Ferguson, S. Creasey, R. Wolstenholme, M. R. Clench and S. Francese, *J. Mass Spectrom.*, 2013, **48**, 677.
69. V. G. Sears, S. M. Bleay, H. L. Bandey and V. J. Bowman, *Sci. Justice*, 2012, **52**, 145.

CHAPTER 8

Emerging Technologies: Use of Secondary Ion Mass Spectrometry for the Analysis of Forensic Evidence

CATIA D. S. COSTA[a] AND MELANIE J. BAILEY*[b]

[a] University of Surrey Ion Beam Centre, Guildford, Surrey, GU2 7XH, UK; [b] School of Chemistry and Chemical Engineering, University of Surrey, Guildford GU2 7XH, UK
*Email: m.bailey@surrey.ac.uk

Abstract

Secondary ion mass spectrometry (SIMS) is a technique that can be used to provide high resolution images of elements and molecules in 3D, and it has been widely used for material characterisation, particularly of inorganic materials. Recent developments in SIMS instrumentation are now enabling the analysis of organic materials, and there is, therefore, considerable scope for exploitation in forensic science. In this chapter, we describe the principles of operation of SIMS and outline the progress that has been made towards its application in forensic science.

8.1 Introduction to SIMS

Secondary ion mass spectrometry (SIMS) is an analytical technique for surface analysis that can probe the elemental and molecular composition of a

New Developments in Mass Spectrometry No. 14
Applications of Mass Spectrometry for the Provision of Forensic Intelligence: State-of-the-art and Perspectives
Edited by Simona Francese and Stephen Bleay
© The Royal Society of Chemistry 2024
Published by the Royal Society of Chemistry, www.rsc.org

solid sample at the surface or near the surface. The technique relies on a primary ion beam to sputter (a process through which particles are ejected from a surface through bombardment with energetic particles[1,2]) positive, negative and neutral species from the surface of a sample (normally the first 1–2 nm), which are then collected, passed through a mass analyser, and detected.[3–6]

SIMS is an established analytical technique and has found some key application areas in electronics manufacturing, materials science, geochemistry and cosmochemistry. Most of these areas rely on the ability to measure and identify elemental species and isotopes, but recent developments in ion sources (namely, cluster beams) have seen an increase in applications in the biological[7–18] and forensic areas.[19–24]

The popularity of SIMS stems from its ability to not only detect elements and molecules (for example, inorganic and organic species such as fragments of lipids, metabolites, and drugs), which is an unusual trait for a mass spectrometry technique, but also due to its sensitivity (down to sub parts per billion), excellent lateral and depth resolution (below 1 μm) and minimal sample preparation requirements.[4–6,25]

8.2 Instrumentation

The standard SIMS equipment comprises a primary ion source, an analysis chamber, a mass analyser and a detector.[4,5,25]

8.2.1 Primary Ion Sources

Primary ion beams are accelerated to energies between 1–50 keV and the choice of ion species depends on the type of experiment to be performed and on the type of sample. Primary ion beams can be monoatomic, polyatomic or clusters (*e.g.*, Bi_n^+, Ga^+, Au_n^+, C_{60}^+, Ar_n^+ and $(H_2O)_n^+$), each offering different advantages in terms of imaging resolution and levels of molecular fragmentation. The primary ion source can be one of the following three types:[6,8,9,25]

- Liquid metal ion sources (LMISs): This type of source is used to produce beams of gold (Au), gallium (Ga), and bismuth (Bi), for example. These are bright (intense) ion sources, providing stable operation, spot sizes down to sub-micron sizes and high secondary ion yields. These features make these ion sources preferred for imaging experiments. The high energy per atom projectile originating from this source can cause high levels of damage, and thus the static limit (which will be explained later in this chapter) must be observed.[26]
- Buckminsterfullerene (C_{60}) polyatomic ion sources: This type of source is widely used in SIMS for imaging experiments. It produces higher yields of intact molecular species and causes less damage in comparison to LMIS and therefore has been more widely used for the analysis of biological samples.[26]

- Gas cluster sources: The use of large clusters has been found to reduce molecular fragmentation, increase secondary ion yield and reduce damage. Commonly found gas cluster sources include water (H_2O), argon (Ar) and oxygen (O_2). The arrival of water cluster ion sources has led to enhancements in ion signals compared to an argon beam, mostly due to the formation of proton adducts, making the technique more akin to other commercially available imaging mass spectrometry techniques like desorption electrospray ionisation (DESI), for example.[6,27–30]

8.2.2 Analysis Chamber

The analysis chamber, where the sample is housed during the measurements, is maintained under ultra-high vacuum conditions (10^{-10} mbar) so that primary and secondary ions do not collide with atmospheric particles, causing energy loss. The sample chamber often contains a sample stage which enables manipulation of the sample in x, y, z directions along with rotation capabilities. The analysis chamber also contains a secondary ion extractor, responsible for extracting the sputtered ions into the mass analyser prior to detection.[4–6,25]

Many of the samples analysed by SIMS will be insulators and these samples are bombarded with positive ion beams. This causes the surface potential to rise due to the increase in positive charge and generation of secondary electrons. Besides modifying the sample to increase its conductivity (*e.g.*, depositing a layer of conductive material or placing a metallic grid in close proximity to the sample), a gentle flow of electrons can be directed towards the sample using an electron flood gun to help dissipate the accumulation of positive charges. Striking a balance in these situations is not trivial and charge compensation through electron bombardment can contribute to sample degradation and even contribute to ion emission.[4–6,25]

8.2.3 Detector

There are three classical types of SIMS instrument, and each employ a different type of mass analyser. Each type presents different benefits and limitations in terms of mass resolving power, sensitivity and duty cycle.[4–6,25,31]

- *Time-of-flight secondary ion mass spectrometry (ToF-SIMS):* This is probably the most common type of instrumentation where the sputtered ions are extracted into a flight tube, accelerated, and allowed to travel along a defined path to the detector. The velocity of the ion is inversely proportional to its mass and therefore, ions with different masses will have different flight times (heavier ions are slower than lighter ones). Increasing the length of the flight tube will increase the mass resolution of the mass spectrometer. ToF-type instruments have the ability to detect all ions simultaneously (for a given polarity); however they rely on the packet of secondary ions to be pulsed (either the primary ion beam is pulsed or the secondary ion beam, depending on the instrument manufacturer).[31–35]

- *Magnetic sector SIMS:* This type of SIMS instruments use electrostatic and magnetic sector analysers for velocity and mass analysis of the secondary ions. Magnetic fields are used to deflect the ion beam, whereby lighter ions suffer a greater deflection than heavier ones, creating different ion paths and ultimately distinctive ion beams. Magnetic sector SIMS instruments rely on continuous beams and high primary beam currents and, therefore, have found limited applications in areas where samples are susceptible to beam damage.[34,35]
- *Quadrupole SIMS:* These are less commonly used versions of the instrument due to the limited mass resolution inherent in quadrupole mass analysers. Quadrupoles use a resonating electric field so that only selected ions with specific masses maintain stable trajectories through the quadrupoles. Much like magnetic sectors, these are often used with high primary ion beam currents.[31,34,35]

New advances in mass spectrometry have catapulted the technique to new horizons with the advent of the OrbiSIMS instrument. This dual analyser detector comprises the commonly found ToF component in addition to a Orbitrap Q Exactive™ HF mass analyser. The ToF analyser provides fast scanning speeds and this can be used to rapidly image and find areas of interest. In contrast, the Orbitrap is slower, but offers high mass resolution, which can be used to confirm peak assignment in the ToF images.[36,37] This has, for example, enabled untargeted metabolomic analysis at the single cell level.[36]

8.3 Modes of Operation

8.3.1 Static and Dynamic SIMS

Static SIMS refers to the analysis mode where the purpose is to encourage desorption and ionisation of the topmost layer of the samples only. To preserve the sensitivity of the measurements to the top layer, the number of primary ions delivered to the sample is carefully controlled. The ion dose used in static SIMS experiments should be kept within the static limit of less than 10^{13} ions per cm^2. In dynamic SIMS, a high dose of primary ions is used, which deliberately removes material from the sample to enable depth profiling.[4–6,25]

8.3.2 Depth Profiling

SIMS depth profiling is used to obtain compositional information about a sample as a function of depth, utilising a layer-by-layer approach. A depth profile can be obtained by alternating between slowly eroding on the sample surface and performing surface analysis, to obtain a plot of the secondary ion intensity as a function of the primary ion dose delivered to the sample. Beams such as C_{60}^+ and argon clusters are often used to sputter the sample at depth. The depth resolution (smallest depth that can be resolved) depends on primary ion beam (species and energy) as well as on the uniformity and size of the area being analysed.[4–6,25]

8.3.3 Imaging

Using SIMS in the imaging mode will allow the generation of ion maps, portraying the spatial distribution of molecules and elements over the surface of the sample. The primary ion beam can be focussed to a sub-micron dimension and scanned across the surface of the sample, producing a spectrum per pixel. The primary beam can be scanned over relatively small areas of the samples (typically up to 500 × 500 μm), but if larger area mapping is necessary, the sample can be moved under the beam to the adjacent areas to build a mosaic scan of a larger section. Post-acquisition, selection of individual or a combination of mass-to-charge (*m*/*z*) peaks from the total combined spectrum will generate ion distribution maps (Figure 8.1).[4–6,25]

8.4 Data Analysis

Like most mass spectrometry imaging approaches, SIMS experiments can result in datasets which are complex, big and difficult to interpret. Many publications in the literature have addressed some of the issues associated

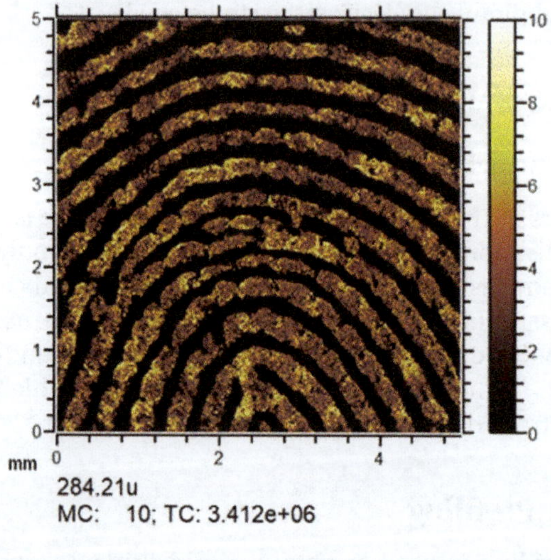

Figure 8.1 Example 2D ion image of *m*/*z* 284.21 taken from a fingerprint deposited on a silicon wafer and analysed using a 25 keV Bi_3^+ primary ion source on a ToF-SIMS instrument.

with SIMS data processing (mass calibration, peak intensity repeatability, peak identification, *etc.*) as well as how conditions applied before, during and after data acquisition can affect the quality of the data obtained using SIMS. Parameters that can affect the quality of the data include, but are not limited to, sample preparation and substrate choice, primary ion choice and instrumental parameters, damage induced by the analysis and data processing parameters.[38–40]

8.4.1 Data Formats

The different modes of operation of SIMS result in different data outputs – mass spectrum, depth profile and imaging – which present their own individual challenges.

A mass spectrum, generated for both static and dynamic SIMS modes, is a plot of the intensity (plotted on the ordinate), (plotted on the ordinate) *versus* the corresponding mass-to-charge (m/z) ratio (plotted on the abscissa) of the detected secondary ions.[4,41] The mass spectrum will contain all the detected secondary ions detected for a given polarity (data can only be acquired in one polarity at a time – positive or negative ion) and can be used to identify the elemental and molecular composition of the sample through identification of the peaks in the spectrum. Peak identification can be rather complicated with SIMS due to the high levels of fragmentation observed for some of the primary ion beams commonly used, which can lead to peak overlaps. Additionally, peak identification can be difficult when a mass spectrometer with low mass resolution is used for the analysis.[42] Low mass resolution can lead to interferences from species that have similar m/z ratios which cannot be resolved.

Depth profiles are obtained when "dynamic SIMS" conditions are applied to selectively remove layers of the samples. The depth of material probed depends on the sputtering time, sputtering conditions and on the matrix of the sample. A depth profile will consist of a plot of intensities (or relative abundance) of the ions of interest (plotted on the ordinate) *versus* the sputtering time (or depth; plotted on the abscissa). The relative abundance as a function of depth is not a direct instrumental output but can be derived from data processing.[4,43]

Imaging (2D or 3D) SIMS reveals the spatial (or volumetric, in the case of 3D SIMS) distribution of secondary ions over the analysis area, see Figure 8.1. The secondary ion distribution is represented using a false colour scale, which is indicative of the intensity of that m/z peak over the areas analysed. Variations in topography and/or fluctuations in instrumental parameters can lead to erroneous intensity distributions in SIMS imaging experiments. Often, researchers normalise the ion maps to the total ion count map or to a known matrix peak to account for these instabilities. 3D images can be obtained by overlaying standard 2-dimensional maps collected as a function of sputtering time, in a similar fashion to depth profiling. These data sets must be carefully processed as distortions can happen during the measurements.[4,6]

8.4.2 Mass Calibration

Similar to many other mass spectrometry techniques, mass calibration is essential to ensure accurate peak identification. Taking ToF-SIMS as an example, mass calibration should be performed on each individual spectrum as changes in sample matrix and topography can lead to different arrival times of the sputtered secondary ions, therefore affecting their measured mass-to-charge (m/z) ratios. According to a Versailles project on Advanced Materials and Standards (VAMAS) Interlaboratory Study, there are no widely accepted and implemented protocols for ToF-SIMS mass calibration and laboratories simply rely on local procedures. It is acknowledged that a mass accuracy (difference between the measured m/z and the theoretical m/z of an ion) of less than 10 ppm is necessary for peak identification.[38–40,44] Many factors can contribute to poorer mass accuracy, but Gilmore *et al.* highlighted 5 factors that can help in improving the mass accuracy of ToF-SIMS spectral identification:[38–40,44]

(1) Optimisation of the instrumental parameters.
(2) Using ions that have low degradation or fragmentation from the parent ion.
(3) Not including atomic species in the calibration list for molecular peak identification.
(4) Not including hydrogen in the final calibration.
(5) Opting for a calibration mass range that is 55% higher than the peak that is to be assigned.

Of course, some of these requirements will require a level of SIMS expertise that may not be available in novice SIMS users, presenting a slight disadvantage of ToF-SIMS for operational deployment, which will be discussed in Section 8.6.

8.4.3 Multivariate Analysis (MVA)

Software packages available from SIMS instrument manufacturers offer limited options for multivariate analysis (MVA) solutions. MVA ecompasses a range of statistical methods for the analysis of datasets containing more than one variable/observation (m/z peaks and respective intensities, in the case of SIMS). These data analysis approaches reduce the size of the dataset, allowing for correlations to be made using readily available algorithms. There are several software tools available for MVA of SIMS data (*e.g.*, simsMVA, PLS_Toolbox and MIA_Toolbox, NBtoolbox and

MCR-ALS toolbox) but some of these are targeted towards specific areas of research and cannot handle all types of samples routinely analysed using SIMS.[45,46]

8.5 Limitations and Constraints

SIMS is regarded as one of the most sensitive surface analysis techniques available. However, like every other analytical technique, it is not suitable for all types of samples and analysis. It also presents some limitations, which will be covered in this section.

8.5.1 Sample Damage and Modification

One of the main issues associated with SIMS analysis is its capacity to cause damage to the sample. The SIMS analysis process – *i.e.*, sputtering – relies on the removal layers of material. This consequentially may lead to measurable changes in the sample, especially when performing dynamic SIMS where the static limit is deliberately surpassed. Changes and damage to the sample may preclude further measurements on the same sample.[4,47] Additionally, ion beam mixing has been found to occur during depth profiling, whereby irradiation of a sample with the primary ion beam results in intermixing of species at the interface of different layers, which results in incorrect depth profiles.[48]

8.5.2 High Vacuum

Given the keV energy ranges of the primary ion beams, it is essential that SIMS analysis is performed under ultra-high vacuum conditions as explained in Section 8.2. Consequently, samples that are vacuum labile (*e.g.*, volatile) are not suitable for SIMS analysis.[49]

8.5.3 Matrix Effects

Matrix effects are well known in SIMS analysis and represent one of the major hurdles for quantitative SIMS measurements. It is not a SIMS-unique issue as all mass spectrometry techniques suffer from similar issues. Matrix effects take place when the ionisation of a given analyte is influenced by other components of the sample, manifesting as either ion suppression or ion enhancement. These effects are present in both dynamic and static SIMS, but a few approaches can be employed to mitigate and overcome matrix effects.[4,50–56] Examples include:

- *Laser post-ionisation:*[51] Also called secondary neutral mass spectrometry, this relies on using a laser to ionise the neutral species released by the sputtering process. This separates the sputtering and ionisation processes and has been found to enhance the sensitivity of the analysis.

- *Reactive primary ion beams:*[53,57] Polyatomic ion beams offer many advantages over atomic ion beams as explained in Section 8.2.1. Reports in the literature have shown how the employment of reactive ion beams (*e.g.*, methane doped argon beam or water doped argon beam) improved the magnitude of matrix effects.
- *Sample modification:*[55,56,58,59] The sample can be modified (*e.g.*, addition of a thin layer of metal such as gold or silver) to modify the ionisation process. This is commonly referred to as metal-assisted SIMS (MetA-SIMS) and has been shown to reduce matrix effects and consequentially increase sensitivity. Other organic matrices (*e.g.*, trehalose/sucrose mix) have been used for this purpose with similar effects and increased linearity in quantitation.

8.6 Application of SIMS in Forensics

8.6.1 Nuclear Forensics

Nuclear forensics is a specialised branch of forensic science pertaining the analysis of nuclear and radioactive materials. Analysis of such materials is of great importance for nuclear safety and anti-terrorism, and it aims to determine any processing that has been applied to the seized material and deduce its intended use.[19]

Mass spectrometry provides fundamental intelligence regarding the isotopic composition of a sample, offering information about the sample's provenance, age, and processing record. SIMS on its own can provide information on the isotope ratios and elemental composition with great sensitivity (in the ppb range). It can also, when used in the imaging mode, provide information on the spatial distribution of such analytes with unparalleled resolution (which can be as good as 50 nm laterally and monolayer in depth) In Figure 8.2, a uranium certified reference material (CRM) particle is shown.[19,20,24]

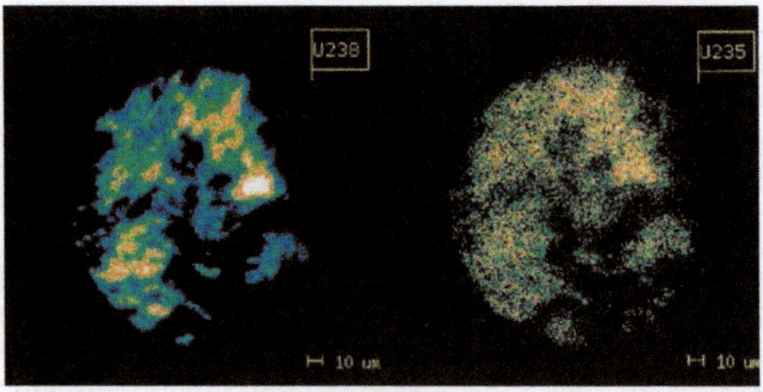

Figure 8.2 SIMS ion map for ^{238}U and ^{235}U in a uranium certified reference material particle. Reproduced from ref. 118 with permission from Elsevier, Copyright 1998.

The matrix effects described above in this chapter can have a detrimental effect on the semi-quantitative potential of SIMS when compared to other elemental techniques such as inductively coupled plasma-mass spectrometry (ICP–MS) as found by a study looking at the $^{235}U/^{238}U$ isotopic ratio. Here the authors found that the ratio measured using SIMS did not match with that found through ICP–MS, presumably due to the heterogeneity of the soil particles analysed.[20,23,60,61]

SIMS has been used in nuclear forensics for several decades. Its excellent spatial resolution has found key application in the determination of heterogeneity of samples previously thought to be homogeneous (*e.g.*, uranium fuel pellets). This characterisation provides information on the conditions of manufacturing of the fuel pellets, helping in the tracing of its origin.[21,22]

8.6.2 Ink Analysis

Examination of questioned documents (*e.g.*, legal proceedings, checks, and forged documents) is an integral part of some forensic investigations. A forensic document examiner will look at handwriting comparison, identifying text deletions, indentations and determining the authenticity of documents, for example. One of the key areas is the examination and identification of inks (ballpoint pen ink, fountain pen ink, inkjet printer ink, laser printer toners, *etc.*) and paper in questioned documents. In such documents, it is necessary to ascertain whether the entry, especially if handwritten, was executed using a single pen and at a single point in time. Forensic scientists prefer non-destructive methods of analysis in order to preserve the exhibit for further testing if necessary, and methods include infrared absorption, luminescence spectral analysis, hyperspectral analysis and optical microscopy. However, it is not always possible to do so and, in some instances, destructive methods such as thin layer chromatography (TLC) and liquid chromatography (LC) are employed.[62–64] Strictly speaking, SIMS is a destructive technique as it relies on the sputtering of material from the sample. However, as explained earlier in this chapter, maintaining the dose below the *static limit* will minimise the damage inflicted on the sample, and in many cases, there is no visible damage to the sample following SIMS analysis.

SIMS can provide more than just chemical identification of the inks used in the document. For example, recent reports in the literature have explored the determination of the order of deposition of inks, which is mostly inaccessible to optical methods (*e.g.*, optical microscopy).[62,64]

The use of SIMS for the analysis of inks dates back to as early as 1994,[63] when authors explored twenty-one pen inks and sixteen printed paper exhibits to determine the colourants and composition of each ink. In this work, the use of static SIMS resulted in visibly non-destructive analysis (no burns on the papers), allowing further investigations to take place if necessary. The authors acknowledged that the size of the exhibit is limited by the size of the analysis chamber/sample stage, but careful removal of selected fibres or portions of the document can easily overcome such problems. Other reports in the literature have explored the identification of ink compositions and how SIMS can be used after TLC separation.[64,65] One other advantage of using SIMS for ink analysis is the ability to monitor ink ageing and associated changes over time. Goacher *et al.* described how SIMS analysis of different ballpoint pen inks at different times points post-deposition (4 to 671 days) yielded spectral differences, which could be used to distinguish marks made using the same pen at different time-points, with the greatest likelihood of success when the marks were deposited one month apart or longer. The best results were obtained when analysis took place 3–6 months after deposition, after which differentiation became difficult.[66] While these initial results show promise, more work is needed to validate their use for different types of pen, paper and storage conditions.

In addition to allowing identification of ink components, SIMS has also been applied for the analysis of multiple overlapping layers of ink on paper.[67] Different coloured inks were individually analysed so that the chemical fingerprint of each individual ink could be determined and then used to distinguish between the overlapping signals and eventually determine the one used last. The overlap order of black pen inks was also studied with some success. Exploration of different primary ion sources, ion polarity and advanced data processing methods (MVA, Section 8.4.3) did not help determine the correct deposition order, highlighting some of the challenges associated with using SIMS for this purpose.[68]

A variant of SIMS, which uses MeV ion beams (instead of keV energies), has also been applied to determine the deposition order of ballpoint pen inks.[69] When compared to optical methods, MeV-SIMS was found to perform better in determining the deposition order when the ink was oil-based (rather than water-based). Importantly, MeV SIMS was found to perform well in the cases where optical methods had failed. The use of statistical methods on the MeV-SIMS spectral data was found to further improve the discriminating power of this technique.[70] One of the limitations of MeV-SIMS is the inability to resolve intersections of inkjet printer ink due to its adherence to the paper and penetration, especially in cases when this ink was applied last. The combination of MeV-SIMS and an elemental mapping technique known as particle induced X-ray emission (PIXE) was found to complement the mass spectral data by probing deeper into the sample and enabling determination of the deposition order in the presence of inkjet printer ink for most scenarios tested.[71]

One final key application area is the determination of the overlapping order of inks and fingerprints.[72–74] Bright *et al.* described how the deposition order of fingerprints and ballpoint pen inks could be determined by performing line scans across the intersection and calculating the normalised standard deviation across the line. Bailey *et al.* applied MeV-SIMS to probe the deposition order of fingerprints (doped with moisturiser) and ink (marker pen) based on the relative intensity of fingerprint and ink signals over the measurement time (depth profiling) and resulting ion maps.[75]

Montalto *et al.* explored the intersection of fingerprints and laser printed ink and found that components found in paper and fingerprints (*e.g.*, sodium and $C_3H_5^+$ ions) were mostly absent from laser printed ink. Mapping of these ions enabled the visualisation of fingerprint ridge details, but only when the fingerprint was above the ink. The methodology was applied to twenty-one blind samples and yielded a 100% success rate. Fingerprint ageing (up to 28 days) did not affect the determination of the deposition order.[74] Follow-on reports from the same group explored a range of variables including paper, ink type, ink style (solid, text, draft), ink colours, fingerprint donors, fingerprint depletion series and fingerprint development process (ninhydrin, indandione and iodine fuming). The success rate of the process depended on several factors including the amount of ions (sodium and potassium), the amount of ink on the surface and the type of ink. The age of the laser printer ink did not affect the success rate. Ballpoint pen ink results depended greatly on the type of ink and how it was deposited on the surface – solvent-based inks that were lightly applied were found to sit on the surface, precluding determination of the deposition order. The process relies on the fingerprint residues situated on top of the ink, which is only possible if the ink layer acts as an impermeable barrier into the paper. Therefore, inkjet inks are incompatible with this methodology. Prior fingerprint development using ninhydrin was found to compromise the results, but indandione and iodine fuming produced successful determinations.[73]

8.6.3 Fingerprints

In a police investigation, fingerprints can play a vital role in the development of a criminal case and can be used to narrow down a list of suspects due to their individuality. However, detection and recovery of fingerprints depends on various factors, such as the type of fingerprints (visible, plastic and latent), the circumstances of deposition and the surface on which they are deposited. Fingerprint visualisation techniques are summarised in the finFingermark Visualisation Manual (FVM),[76] published by the UK's Home Office, which describes a range of techniques (chemical, physical and optical techniques) available to UK police forces to locate and enhance latent fingerprints. Despite improvements in fingerprint enhancement techniques, the process of matching to a suspect relies on having a sufficient proportion of the fingerprint available (and not smudged) and on database matching, which is not always available. Therefore, research groups internationally have been exploring the possibility of using chemical mapping techniques to image fingerprints. These efforts were recently reviewed by Bleay *et al.*[77] In addition to providing images of the fingerprint ridges, chemical mapping techniques can provide additional information about the circumstances of fingerprint deposition. Detection of endogenous and exogenous fingerprint components can provide information on donors' age, sex, medical history, use of illicit substances and even whether a suspect has handled explosives or discharged a firearm recently.

SIMS is included in the FVM as a category C process, designating it a potential technique for fingerprint recovery to be considered by law enforcement agencies, if other approaches fail. The high spatial resolution of SIMS is well-suited for accurately mapping the fingerprint ridge detail down to the third level detail (*i.e.*, sweat pores)[78,79] (Figure 8.3).

In some instances, TOF-SIMS has shown greater sensitivity than conventional developers used by forensic laboratories.[80,81] Continuous fingerprint ridge detail was observed using SIMS on aluminium foil after cyanoacrylate (CA) fuming in a region where the CA only partially developed the ridge. SIMS also enabled the visualisation of continuous ridge detail on a hand grenade handle (a simulated exhibit provided by Israel police) where conventional developers used by that police force (CA and crystal violet, vacuum metal deposition and small particle reagent) struggled previously. Finally, fingerprint patterns were developed on glass after samples were either buried in soil or submerged in sea water and where conventional developers failed.[80] In an expanded study, Charlton *et al.* explored the sensitivity of TOF-SIMS against FVM category A processes on three different substrates (paper, stainless steel and polyethylene) and using fingerprint depletion series (consecutive deposition of fingerprints to reduce the amount of material deposited each time). Overall, SIMS was able to image fingerprints further down the depletion series compared to conventional developers and in cases where both techniques were successful, the quality of information (improved ridge quality and visualisation of sweat pores) was higher with SIMS. Furthermore, successive SIMS imaging provided additional enhancement in every sample tested compared to conventional developers.[81]

SIMS has also been used to visualise fingerprint patterns on challenging surfaces such as banknotes, metallic surfaces and even on brass bullet casings.[82–84] The latter shows an important advancement in SIMS imaging experiments, whereby a rotation stage was installed in the analysis chamber, allowing for imaging of fingerprints in a curved sample such as a bullet

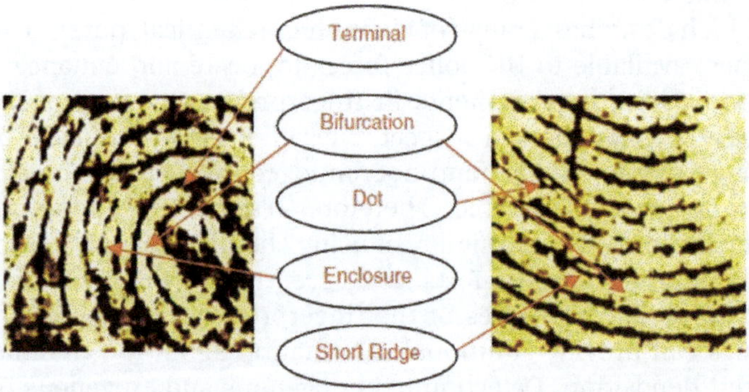

Figure 8.3 ToF-SIMS image of fingerprint ridge details showing the minutiae. Reproduced from ref. 78 with permission from John Wiley & Sons, Copyright 2010.

casing. SIMS requires the sample to be as flat as possible (even slight surface roughness can contribute to variations in secondary ion yield), which precludes analysis of certain types of evidence.[84]

Other than enabling visualisation of ridge details, SIMS can also provide information on the chemical composition of a fingerprint sample.[85] Identification of the chemicals in fingerprint residues can be used to assess fingerprint ageing[86] and the presence of external contaminants (often referred to as exogenous substances)[87] such as drugs of abuse, either by dermal contact[88,89] or ingestion[89,90] and gunshot residue,[91] for example. Costa *et al.* described how the distribution of cocaine and benzoylecgonine (BZE) (a metabolite, but also an impurity in street cocaine) could be used to distinguish between cocaine administration and dermal contact with the substances using a series of mass spectrometry imaging techniques, including ToF-SIMS. The sensitivity of TOF-SIMS towards cocaine and BZE in the fingerprints of drug users using bismuth (Bi_3^+) as the primary ion source was poor. However, employment of water clusters enabled the detection of BZE in the fingerprints of a participant who testified ingesting cocaine, showing the enhanced sensitivity when using cluster ion sources. The detection of BZE in this samples was consistent with the results obtained using DESI and MALDI.[89]

8.6.4 Hair Analysis

Here, we briefly describe some of the key application areas for hair analysis using SIMS.

8.6.4.1 Drug Testing

SIMS can be used to detect drugs and their metabolites in hair samples, providing information about drug use over an extended period. Hair as a drug testing matrix presents a few issues relating to the distinction between active users (*e.g.*, a person who ingests a drug) and contaminated users (*e.g.*, a person whose hair contains a drug through unintentional exposure).[92] SIMS and other mass spectrometry imaging techniques have been employed to further explore the usefulness of hair as a drug testing matrix and understand some of the issues surrounding sample preparation (*e.g.*, the decontamination process).[93–97] One of most debated issues in hair sample preparation is whether the decontamination solvents used to remove external contaminants, hair products and sebum, cause the ingress of the drugs into the hair. If this is the case, detection of a substance in hair is not solely indicative of ingestion. SIMS imaging can assist in assessing whether decontamination protocols cause ingress and can also distinguish between different scenarios of exposure (soaking, powder exposure or ingestion).[93,97] SIMS, more specifically nanoSIMS, has also been used to image arsenic (As) distribution in hair. The high spatial resolution of the technique allowed for quantitative location of As across the medulla, cortex and cuticle, enabling distinction between ingestion and surface contamination.[98]

8.6.4.2 *Identification*

Although SIMS analysis of hair has not been directly used to identify the donor, SIMS analysis can provide some basic information regarding a person based on a hair sample. For example, SIMS has been employed to study the interaction of hair dyes with fine hair structures and whether hair bleaching was performed, which could be used to identify whether a person has dyed hair.[99,100] Finally, a study employing MeV-SIMS has looked at the detection of lithium (Li) in hair samples. Detection of such substance may be indicative of lithium ingestion in therapeutic settings, helping narrow down a list of suspects. However, care must be taken as the substance may also be nutritionally ingested and/or presence in hair care products.[101]

8.6.5 Gunshot Residue

Gunshot residue (GSR) describes all the material that is expelled from the muzzle of a firearm following the discharge of a bullet and comprises both inorganic and organic components originating from the primer, propeller and other firearm components. After a shooting incident, GSR will deposit on any surface around the discharge point, including the hands of the shooter. However, detection of GSR in a person's hands, body or clothing is not an indication that the person has fired the weapon. GSR contamination may arise through handling of the firearm pre- or post-shooting, handling of objects/surfaces contaminated with GSR or being in the proximity of the firearm at the time of the shooting.[102–104]

Current practices for GSR analysis employ scanning electron microscopy – energy dispersive X-ray spectrometry (SEM-EDX) and involve detecting the presence of elements such as lead (Pb), barium (Ba) and antimony (Sb) as well as characteristic morphology and presence of other trace elements.[103–106]

SIMS has shown great potential for the analysis of GSR as described by many publications over the years.[103,105,107,108] However, the main advantage of SIMS is its ability to provide information on both the organic and inorganic composition of GSR.[103,109] Coumbaros *et al.*[107] described how SIMS provided enhanced elemental sensitivity compared to SEM-EDX while enabling detection of other elemental species not achievable by SEM-EDX and organic components related to GSR. Mahoney *et al.*[109] employed SIMS to distinguish different types of gunpowder (smokeless powders and black powder samples) using principal component analysis (PCA). Distinction was made based on additive composition, distribution of said additives within the gunpowder and presence of unique contaminants.

8.6.6 Provenance Determination

It is well established that the chemical composition of a sample can be used to provide source level (provenance) information, and SIMS can be an attractive tool for chemical characterisation of solid samples as sample preparation is not required. Examples of SIMS for this type of analysis in forensics include glass,[110] lipstick[111] and automobile paint analysis.[112] SIMS has been applied as a tool to study and observe counterfeiting products. Counterfeit refers to a fake or imitation product or currency that is designed to look like the real item in order to deceive people into believing that it is genuine. Counterfeit products can include anything from clothing and accessories to electronics, pharmaceuticals, and currency. Counterfeiting is illegal and can have serious consequences for those involved in the production, distribution, or sale of counterfeit goods. Examples include analysis of porcelain art items and determining whether they are counterfeit or genuine based on their composition,[113] analysis of counterfeit gold coins made to look like ancient artefacts,[114] analysis of suspected counterfeit electronic goods[115] and analysis of pharmaceuticals for anti-counterfeiting measures.[116]

8.7 Conclusions

From the applications described above, it is clear that SIMS is a versatile analytical tool with many applications in forensics. Despite its many applications, SIMS is still finding its place in routine forensic casework. The acquisition cost of a SIMS system ranges from US \$200k to \$2M and so SIMS systems are not commonly found in forensic or toxicology laboratories. As such, its use in these scenarios is rather limited to academic and research settings. However, the spatial information that can be retrieved from SIMS imaging experiments is complementary to bulk hair analysis techniques like liquid chromatography–mass spectrometry (LC–MS) and may provide extra information regarding surface contamination, penetration and surface reactions.[117]

References

1. *Sputtering by Particle Bombardment*, ed. R. Behrisch and W. Eckstein, 1st edn, Springer Berlin, Heidelberg, 2007.
2. P. van der Heide, *Secondary Ion Mass Spectrometry*, 2014, pp. 44–143.
3. P. van der Heide, *Secondary Ion Mass Spectrometry*, 2014, pp. 1–19.
4. P. van der Heide, *Secondary Ion Mass Spectrometry: An Introduction to Principles and Practices*, John Wiley & Sons, Incorporated, New York, United States, 2014.
5. J. C. Vickerman and D. Briggs, *ToF-SIMS: Materials Analysis by Mass Spectrometry*, SurfaceSpectra, 2013.
6. A. Alnajeebi, PhD thesis, The University of Manchester, 2017.
7. P. Massonnet and R. M. A. Heeren, *J. Anal. At. Spectrom.*, 2019, **34**, 2217–2228.
8. P. Agüi-Gonzalez, S. Jähne and N. T. N. Phan, *J. Anal. At. Spectrom.*, 2019, **34**, 1355–1368.

9. F. Gyngard and M. L. Steinhauser, *J. Anal. At. Spectrom.*, 2019, **34**, 1534–1545.
10. S. Chandra, *J. Anal. At. Spectrom.*, 2019, **34**, 1998–2003.
11. M. Dowlatshahi Pour, L. Ren, E. Jennische, S. Lange, A. G. Ewing and P. Malmberg, *J. Anal. At. Spectrom.*, 2019, **34**, 1581–1587.
12. H. Tian, P. Rajbhandari, J. Tarolli, A. M. Decker, T. V. Neelakantan, T. Angerer, F. Zandkarimi, J. Daniels, H. Remotti, G. Frache, N. Winograd and B. R. Stockwell, *bioRxiv*, 2022, DOI: 10.1101/2022.09.26.508878.
13. H. Tian, S. Sheraz Née Rabbani, J. C. Vickerman and N. Winograd, *Anal. Chem.*, 2021, **93**, 7808–7814.
14. L. J. Sparvero, H. Tian, A. A. Amoscato, W.-Y. Sun, T. S. Anthonymuthu, Y. Y. Tyurina, O. Kapralov, S. Javadov, R.-R. He, S. C. Watkins, N. Winograd, V. E. Kagan and H. Bayır, *Angew. Chem., Int. Ed.*, 2021, **60**, 11784–11788.
15. K. Dimovska Nilsson, A. Karagianni, I. Kaya, M. Henricsson and J. S. Fletcher, *Anal. Bioanal. Chem.*, 2021, **413**, 4181–4194.
16. S. Sheraz, H. Tian, J. C. Vickerman, P. Blenkinsopp, N. Winograd and P. Cumpson, *Anal. Chem.*, 2019, **91**, 9058–9068.
17. S. Sämfors, M. Ståhlman, M. Klevstig, J. Borén and J. S. Fletcher, *Int. J. Mass Spectrom.*, 2019, **437**, 77–86.
18. D. Touboul, S. Roy, D. P. Germain, P. Chaminade, A. Brunelle and O. Laprévote, *Int. J. Mass Spectrom.*, 2007, **260**, 158–165.
19. Z. Varga, M. Wallenius, M. Krachler, N. Rauff-Nisthar, L. Fongaro, A. Knott, A. Nicholl and K. Mayer, *TrAC, Trends Anal. Chem.*, 2022, **146**, 116503.
20. M. D. Straub, J. Arnold, J. Fessenden and J. L. Kiplinger, *Anal. Chem.*, 2021, **93**, 3–22.
21. S. V. Jovanovic, P. K. Weber, A. J. Pidduck, A. M. Gaffney, P. Girard, F. Pointurier, M. Hedberg, A. J. Simons, V. Stebelkov, T. Kell, K. Knight, T. Parsons-Davis, M. Kristo, R. W. Williams, K. C. Treinen, N. J. Montgomery, J. King, A. Wickenden, D. Knight, A.-L. Fauré, A. Hubert, N. Albert, M.-C. Vincent, M. Wallenius, I. A. Elantyev, K. D. Zhizhin, J. M. Schwantes, O. Marsden and F. Taylor, *J. Radioanal. Nucl. Chem.*, 2020, **326**, 1853–1866.
22. R. Kips, P. K. Weber, M. J. Kristo, B. Jacobsen and E. C. Ramon, *Anal. Chem.*, 2019, **91**, 11598–11605.
23. F. E. Stanley, A. M. Stalcup and H. B. Spitz, *J. Radioanal. Nucl. Chem.*, 2013, **295**, 1385–1393.
24. K. Mayer, M. Wallenius and I. Ray, *Analyst*, 2005, **130**, 433–441.
25. R. N. S. Sodhi, *Analyst*, 2004, **129**, 483–487.
26. M. Dubey, J. Brison, D. W. Grainger and D. G. Castner, *Surf. Interface Anal.*, 2011, **43**, 261–264.
27. C. M. Mahoney, *Cluster Secondary Ion Mass Spectrometry: Principles and Applications*, Wiley, 2013.
28. N. Winograd, *Anal. Chem.*, 2005, **77**, 142A–149A.

29. K. Hiraoka, in *Fundamentals of Mass Spectrometry*, ed. K. Hiraoka, Springer, New York, New York, NY, 2013, pp. 199–230.
30. I. Berrueta Razo, S. Sheraz, A. Henderson, N. P. Lockyer and J. C. Vickerman, *Rapid Commun. Mass Spectrom.*, 2015, **29**, 1851–1862.
31. G. Kaklamanos, E. Aprea and G. Theodoridis, in *Encyclopedia of Food and Health*, eds. B. Caballero, P. M. Finglas and F. Toldrá, Academic Press, Oxford, 2016, pp. 661–668.
32. M. Guilhaus, in *Encyclopedia of Analytical Science*, ed. P. Worsfold, A. Townshend and C. Poole, Elsevier, Oxford, 2nd edn, 2005, pp. 412–423.
33. C. Pareige, W. Lefebvre-Ulrikson, F. Vurpillot and X. Sauvage, in *Atom Probe Tomography*, eds. W. Lefebvre-Ulrikson, F. Vurpillot and X. Sauvage, Academic Press, 2016, pp. 123–154.
34. O. De Castro, J.-N. Audinot, H. Q. Hoang, C. Coulbary, O. Bouton, R. Barrahma, A. Ost, C. Stoffels, C. Jiao, M. Dutka, M. Geryk and T. Wirtz, *Anal. Chem.*, 2022, **94**, 10754–10763.
35. M. M. Houck and J. A. Siegel, in *Fundamentals of Forensic Science*, ed. M. M. Houck and J. A. Siegel, Academic Press, San Diego, 2nd edn, 2010, pp. 99–122.
36. M. K. Passarelli, A. Pirkl, R. Moellers, D. Grinfeld, F. Kollmer, R. Havelund, C. F. Newman, P. S. Marshall, H. Arlinghaus, M. R. Alexander, A. West, S. Horning, E. Niehuis, A. Makarov, C. T. Dollery and I. S. Gilmore, *Nat. Methods*, 2017, **14**, 1175–1183.
37. L. Matjacic, M. P. Seah, G. F. Trindade, A. Pirkl, R. Havelund, J.-L. Vorng, E. Niehuis and I. S. Gilmore, *Surf. Interface Anal.*, 2022, **54**, 331–340.
38. I. S. Gilmore and M. P. Seah, *Surf. Interface Anal.*, 2000, **29**, 624–637.
39. I. S. Gilmore, M. P. Seah and F. M. Green, *Surf. Interface Anal.*, 2005, **37**, 651–672.
40. I. S. Gilmore, F. M. Green and M. P. Seah, *Surf. Interface Anal.*, 2007, **39**, 817–825.
41. J. L. Koenig, in *Spectroscopy of Polymers*, ed. J. L. Koenig, Elsevier Science, New York, 2nd edn, 1999, pp. 441–480.
42. T. Miyayama and S.-I. Iida, *J. Vac. Sci. Technol., B: Microelectron. Nanometer Struct.*, 2018, **36**, 03F126.
43. C. W. Magee and R. E. Honig, *Surf. Interface Anal.*, 1982, **4**, 35–41.
44. F. M. Green, I. S. Gilmore and M. P. Seah, *J. Am. Soc. Mass Spectrom.*, 2006, **17**, 514–523.
45. G. F. Trindade, PhD thesis, Universirty of Surrey, 2018.
46. G. F. Trindade, M.-L. Abel and J. F. Watts, *Chemom. Intell. Lab. Syst.*, 2018, **182**, 180–187.
47. A. V. Walker, *Anal. Chem.*, 2008, **80**, 8865–8870.
48. M. Saggio, C. Montandon, A. Bourenkov, L. Frey and P. Pichler, *Radiat. Eff. Defects Solids*, 1997, **141**, 37–52.
49. J. S. Fletcher, *Analyst*, 2009, **134**, 2204–2215.
50. M. P. Seah and A. G. Shard, *Appl. Surf. Sci.*, 2018, **439**, 605–611.

51. L. Breuer, N. J. Popczun, A. Wucher and N. Winograd, *J. Phys. Chem. C*, 2017, **121**, 19705–19715.
52. A. Priebe, T. Xie, G. Bürki, L. Pethö and J. Michler, *J. Anal. At. Spectrom.*, 2020, **35**, 1156–1166.
53. M. Lagator, I. Berrueta Razo, T. Royle and N. P. Lockyer, *Surf. Interface Anal.*, 2022, **54**, 349–355.
54. A. M. Alnajeebi, J. C. Vickerman and N. P. Lockyer, *Rapid Commun. Mass Spectrom.*, 2018, **32**, 1962–1970.
55. L. A. McDonnell, R. M. A. Heeren, R. P. J. de Lange and I. W. Fletcher, *J. Am. Soc. Mass Spectrom.*, 2006, **17**, 1195–1202.
56. J. J. D. Fitzgerald, P. Kunnath and A. V. Walker, *Anal. Chem.*, 2010, **82**, 4413–4419.
57. H. Tian, A. Wucher and N. Winograd, *J. Am. Soc. Mass Spectrom.*, 2016, **27**, 2014–2024.
58. A. Delcorte, *Appl. Surf. Sci.*, 2006, **252**, 6582–6587.
59. M. Inoue and A. Murase, *Surf. Interface Anal.*, 2005, **37**, 1111–1114.
60. P. Rubio Montero and A. Martı?n Sánchez, *J. Environ. Radioact.*, 2001, **55**, 157–165.
61. C. Gascó, M. P. Antón, A. Espinosa, A. Aragón, A. Alvárez, N. Navarro and E. García-Toraño, *J. Radioanal. Nucl. Chem.*, 1997, **222**, 81–86.
62. W. L. Leaver, in *The Forensic Laboratory Handbook: Procedures and Practice*, ed. A. Mozayani and C. Noziglia, Humana Press, Totowa, NJ, 2006, pp. 223–248.
63. S. J. Pachuta and J. S. Staral, *Anal. Chem.*, 1994, **66**, 276–284.
64. J. Coumbaros, K. P. Kirkbride, G. Klass and W. Skinner, *Forensic Sci. Int.*, 2009, **193**, 42–46.
65. J. A. Denman, W. M. Skinner, K. P. Kirkbride and I. M. Kempson, *Appl. Surf. Sci.*, 2010, **256**, 2155–2163.
66. R. E. Goacher, *Forensic Chem.*, 2021, **22**, 100298.
67. A. He, D. Karpuzov and S. Xu, *Surf. Interface Anal.*, 2006, **38**, 854–858.
68. R. E. Goacher, L. G. DiFonzo and K. C. Lesko, *Anal. Chem.*, 2017, **89**, 759–766.
69. M. C. Malloy, I. Bogdanović Radović, Z. Siketić and M. Jakšić, *Forensic Chem.*, 2018, **7**, 75–80.
70. M. Barac, A. Filko, Z. Siketić, M. Brajković, A. Ledić and I. B. Radović, *Forensic Sci. Int.*, 2022, **331**, 111136.
71. K. L. Moore, M. Barac, M. Brajković, M. J. Bailey, Z. Siketić and I. Bogdanović Radović, *Anal. Chem.*, 2019, **91**, 12997–13005.
72. N. J. Bright, R. P. Webb, S. Bleay, S. Hinder, N. I. Ward, J. F. Watts, K. J. Kirkby and M. J. Bailey, *Anal. Chem.*, 2012, **84**, 4083–4087.
73. N. Attard-Montalto, J. J. Ojeda, A. Reynolds, M. Ismail, M. Bailey, L. Doodkorte, M. de Puit and B. J. Jones, *Analyst*, 2014, **139**, 4641–4653.
74. N. Attard Montalto, J. J. Ojeda and B. J. Jones, *Sci. Justice*, 2013, **53**, 2–7.

75. M. J. Bailey, B. N. Jones, S. Hinder, J. Watts, S. Bleay and R. P. Webb, *Nucl. Instrum. Methods Phys. Res., Sect. B*, 2010, **268**, 1929–1932.
76. *Fingermark Visualisation Manual*, ed. H. Bandey, Home Office & Dstl, 2nd edn, 2022, ISBN: 978-1-3999-0976-1.
77. S. M. Bleay, M. J. Bailey, R. S. Croxton and S. Francese, *WIREs Forensic Sci.*, 2021, **3**, e1403.
78. M. I. Szynkowska, K. Czerski, J. Grams, T. Paryjczak and A. Parczewski, *Imaging Sci. J.*, 2007, **55**, 180–187.
79. S. J. Hinder and J. F. Watts, *Surf. Interface Anal.*, 2010, **42**, 826–829.
80. M. J. Bailey, M. Ismail, S. Bleay, N. Bright, M. L. Elad, Y. Cohen, B. Geller, D. Everson, C. Costa, R. P. Webb, J. F. Watts and M. de Puit, *Analyst*, 2013, **138**, 6246–6250.
81. D. Charlton, C. Costa, G. F. Trindade, S. Hinder, J. F. Watts and M. J. Bailey, *Sci. Justice*, 2023, **63**, 9–18.
82. W.-J. Li, L.-H. Sun, W. You, L.-X. Wang, Y.-B. Zhao and Z.-P. Li, *Chin. J. Anal. Chem.*, 2020, **48**, 1511–1518.
83. T. D. Thandauthapani, A. J. Reeve, A. S. Long, I. J. Turner and J. S. Sharp, *Sci. Justice*, 2018, **58**, 405–414.
84. C. J. Lee, D. J. Scurr, L. Jiang, A. Kenton, S. R. T. Beebe and J. S. Sharp, *Analyst*, 2021, **146**, 7563–7572.
85. M. J. Bailey, N. J. Bright, R. S. Croxton, S. Francese, L. S. Ferguson, S. Hinder, S. Jickells, B. J. Jones, B. N. Jones, S. G. Kazarian, J. J. Ojeda, R. P. Webb, R. Wolstenholme and S. Bleay, *Anal. Chem.*, 2012, **84**, 8514–8523.
86. S. Muramoto and E. Sisco, *Anal. Chem.*, 2015, **87**, 8035–8038.
87. M. I. Szynkowska, K. Czerski, J. Rogowski, T. Paryjczak and A. Parczewski, *Surf. Interface Anal.*, 2010, **42**, 393–397.
88. M. I. Szynkowska, K. Czerski, J. Rogowski, T. Paryjczak, A. Parczewski, C. Costa, M. Jang, J. de Jesus, R. T. Steven, C. J. Nikula, E. Elia, J. Bunch, A. T. Bellew, J. F. Watts, S. Hinder and M. J. Bailey, *Analyst*, 2021, **146**, 4010–4021.
89. C. Costa, M. Jang, J. de Jesus, R. T. Steven, C. J. Nikula, E. Elia, J. Bunch, A. T. Bellew, J. F. Watts, S. Hinder and M. J. Bailey, *Analyst*, 2021, **146**, 4010–4021.
90. M. J. Bailey, R. Bradshaw, S. Francese, T. L. Salter, C. Costa, M. Ismail, R. P. Webb, I. Bosman, K. Wolff and M. de Puit, *Analyst*, 2015, **140**, 6254–6259.
91. M. I. Szynkowska, A. Parczewski, K. Szajdak and J. Rogowski, *Surf. Interface Anal.*, 2013, **45**, 596–600.
92. R. E. Joseph Jr., K. M. Höld, D. G. Wilkins, D. E. Rollins and E. J. Cone, *J. Anal. Toxicol.*, 1999, **23**, 396–408.
93. E. Cuypers, B. Flinders, C. M. Boone, I. J. Bosman, K. J. Lusthof, A. C. Van Asten, J. Tytgat and R. M. A. Heeren, *Anal. Chem.*, 2016, **88**, 3091–3097.
94. R. Erne, L. Bernard, A. E. Steuer, M. R. Baumgartner and T. Kraemer, *Anal. Chem.*, 2019, **91**, 4132–4139.

95. L. Jeromel, N. Ogrinc, Z. Siketić, P. Vavpetič, Z. Rupnik, K. Bučar, B. Jenčič, M. Kelemen, M. Vencelj, K. Vogel-Mikuš, J. Kovač, R. M. A. Heeren, B. Flinders, E. Cuypers, Ž. Barba and P. Pelicon, *PLoS One*, 2022, **17**, e0263338.
96. B. Flinders, E. Cuypers, H. Zeijlemaker, J. Tytgat and R. M. A. Heeren, *Drug Test. Anal.*, 2015, **7**, 859–865.
97. R. Erne, L. Bernhard, M. Kawecki, M. R. Baumgartner and T. Kraemer, *Analyst*, 2020, **145**, 4906–4919.
98. J. N. Audinot, S. Schneider, M. Yegles, P. Hallegot, R. Wennig and H. N. Migeon, *Appl. Surf. Sci.*, 2004, **231–232**, 490–496.
99. T. Kojima, H. Yamada, T. Yamamoto, Y. Matsushita and K. Fukushima, *Colloids Surf., B*, 2013, **106**, 140–144.
100. T. Kojima, H. Yamada, M. Isobe, T. Yamamoto, M. Takeuchi, D. Aoki, Y. Matsushita and K. Fukushima, *Sking Res. Technol.*, 2014, **20**, 416–421.
101. K. J. Cloete, B. Jenčič, Ž. Šmit, M. Kelemen, K. Mkentane and P. Pelicon, *Anal. Methods*, 2017, **9**, 5249–5253.
102. M. M. Houck, *Encyclopedia of Forensic Sciences*, Elsevier Science, 2022.
103. O. Dalby, D. Butler and J. W. Birkett, *J. Forensic Sci.*, 2010, **55**, 924–943.
104. F. S. Romolo, in *Emerging Technologies for the Analysis of Forensic Traces*, ed. S. Francese, Springer International Publishing, Cham, 2019, pp. 183–202.
105. P. Shrivastava, V. K. Jain and S. Nagpal, *Egypt. J. Forensic Sci.*, 2021, **11**, 11.
106. M. J. Bailey, M. de Puit and F. S. Romolo, *Annu. Rev. Anal. Chem.*, 2022, **15**, 173–196.
107. J. Coumbaros, K. P. Kirkbride, G. Klass and W. Skinner, *Forensic Sci. Int.*, 2001, **119**, 72–81.
108. P. Collins, J. Coumbaros, G. Horsley, B. Lynch, K. P. Kirkbride, W. Skinner and G. Klass, *J. Forensic Sci.*, 2003, **48**, 538–553.
109. C. M. Mahoney, G. Gillen and A. J. Fahey, *Forensic Sci. Int.*, 2006, **158**, 39–51.
110. J. Coumbaros, J. Denman, K. P. Kirkbride, G. S. Walker and W. Skinner, *J. Forensic Sci.*, 2008, **53**, 312–320.
111. J. Lee and Y. Lee, *J. Anal. Chem.*, 2021, **76**, 854–867.
112. Y. Lee, S. Han, J.-H. Yoon, Y.-M. Kim, S.-K. Shon and S.-W. Park, *Anal. Sci.*, 2001, **17**, 757–761.
113. D. Wainstein, V. Vakhrushev, A. Kovalev, E. Konovalov, A. Volkov and M. Shipova, Preprints.org, 2023, DOI: 10.20944/preprints202302.0106.v1.
114. F. Reiff, M. Bartels, M. Gastel and H. Ortner, *Fresenius' J. Anal. Chem.*, 2001, **371**, 1146–1153.
115. M. Simard-Normandin, C. Banks, N. Havercroft, P. Clark and E. Tallarek, ISTFA 2019, 2019.
116. A. K. Deisingh, *Analyst*, 2005, **130**, 271–279.
117. G. Gillen, S. Roberson, C. Ng and M. Stranick, *Scanning*, 1999, **21**, 173–181.
118. G. Tamborini, M. Betti, V. Forcina, T. Hiernaut, B. Giovanonne and L. Koch, *Spectrochim. Acta B*, 1998, **53**, 1289–1302.

CHAPTER 9

Transferable Mass Spectrometry Methods – Sustainable Forensic Intelligence: Mass Spectrometry-based Methods for Monitoring Fentanyl-related Substances in Wastewater-based Epidemiology†

LEON P. BARRON,*[a,b,c] DERRYN GRANT,[a,b] TIM MARCZYLO[a,b,c,d] AND TIMOTHY W. GANT[a,b,d]

[a] MRC Centre for Environment and Health, Environmental Research Group, Faculty of Medicine, Imperial College London, 86 Wood Lane, W12 0BZ, UK; [b] NIHR Health Protection Research Unit in Environmental Exposures and Health, Imperial College London, 86 Wood Lane, W12 0BZ, UK; [c] NIHR Health Protection Research Unit in Chemical and Radiation Threats

†A supplementary table is available as electronic supplementary information (ESI) at https://doi.org/10.1039/9781837671922-00205.

New Developments in Mass Spectrometry No. 14
Applications of Mass Spectrometry for the Provision of Forensic Intelligence: State-of-the-art and Perspectives
Edited by Simona Francese and Stephen Bleay
© The Royal Society of Chemistry 2024
Published by the Royal Society of Chemistry, www.rsc.org

and Hazards, Imperial College London, 86 Wood Lane, W12 0BZ, UK;
[d] UK Health Security Agency, Harwell Science Campus, Oxon, OX11 0RQ, UK
*Email: Leon.barron@imperial.ac.uk

Abstract

An overview of the latest technological advancements and knowledge in wastewater-based epidemiology (WBE) for monitoring illicit drug usage, using mass spectrometry-based analytical methods, is critically discussed with a particular focus on fentanyl-related substances. The higher potency of many such substances in comparison to other opioids, and thus the amount used, has posed significant challenges to their detection and reliable monitoring in wastewater matrices at the low nanogram per litre concentration level. Sample preparation approaches for multi-analyte qualitative screening and quantitative monitoring are discussed, with a particular emphasis on scalability for large national and international surveillance programmes. Following analytical measurement, this chapter also overviews the back-calculation process to determine community wastewater loadings and consumption, as well as some of the key constraints and limitations of the WBE methodology, especially regarding interpretation and ethics. Finally, examples of key WBE studies are discussed for fentanyl-related substances globally and in context with other opioids and illicit drugs. The use of WBE to measure the consumption of such substances represents a useful new and sustainable way to gather intelligence to guide interventions and actions to protect public health.

9.1 Introduction

Wastewater based epidemiology (WBE) is a recently developed methodology to monitor various markers of community health, including consumption and exposure to a variety of biological and chemical substances in near real-time at the population level.[1–3] It is based on the fundamental premise that excretion of human biomarkers into the sewershed represents a pooled community urine and faeces sample, which can then be monitored. Mass spectrometry (MS) underpins many WBE studies, and it has rapidly become the preferred analytical detector for WBE of chemical residues, especially when coupled to various modes of liquid chromatography (LC) or gas chromatography (GC) to add even more selectivity for structurally similar compounds. It offers excellent selectivity by identifying analytes based on their mass-to-charge (m/z) ratio and/or *via* various modes of molecular fragmentation to give further structural information and selectivity. MS importantly provides sufficiently high sensitivity, allowing the detection and quantification of trace chemical residues at the parts-per-billion to parts-per-trillion concentration levels typically encountered in complex wastewater extracts.[4] Additionally, MS detection frequency is rapid, with the capability to analyse multiple samples in a short timeframe,[5] as well as large numbers of analytes simultaneously which are frequently present together in wastewater,[6,7] making it particularly suitable for high-throughput applications for large-scale WBE programs at national or international levels.[8,9] More recently, the ability to perform non-targeted and suspect screening analysis of wastewater with high resolution accurate mass spectrometry (HRMS) has made MS very useful for rapid tentative identification of new substances, often without the need for sample re-analysis.[10,11]

WBE has several benefits including, primarily, its ability to act as a very practical early-warning system, which can be deployed at scale.[12,13] It also offers more objectivity. For example, WBE is not influenced by individual participant bias or lack of knowledge of exact exposure sources or consumption patterns for certain substances. Wastewater from the community is normally combined and collected at a municipal wastewater treatment plant (WWTP), from which biomarkers can be measured, usually on a daily resolution. Such biomarkers have been used for the estimation of exposure to pharmaceuticals,[14–16] illicit drugs,[17,18] pesticides[19–22] and several other contaminants of emerging concern.[23,24] Further methods can be used to estimate the prevalence of infectious disease markers[25,26] and of non-communicable disease.[27] Wastewater analysis has also been successfully used to monitor threats to public safety (*e.g.*, clandestine laboratory synthesis of drugs[28] or explosives[29]), to understand the development of anti-microbial resistance,[30,31] and as a basis for environmental chemical risk assessment,[6,32] among other applications. The most recent example of WBE in action in a public health crisis was during the SARS-CoV-2 pandemic from 2019–2023 to determine where and when outbreaks were likely to occur.[33–36] Importantly, infrastructure for sample collection and analysis at scale for routine WBE was rapidly developed and many countries now have national level programs for infectious disease monitoring. This is a particularly exciting period for WBE, not least as the same samples used for infectious diseases can be used for chemical monitoring too.[24] One of the best examples where WBE had been under development for more than a decade prior to the pandemic was for the monitoring of illicit drug use within communities. Some illicit drug biomarkers, or drug target residues (DTRs) in this context, have been more widely studied than others including those for large-scale use substances like cocaine, (meth)amphetamines and heroin. A long-standing, coordinated, multi-country WBE monitoring programme running annually in Europe is co-led by the Sewage Analysis Core Group (SCORE) and the European Monitoring Centre for Drugs and Drug Addiction (EMCDDA), encompassing 118 cities, 132 WWTPs and 41 laboratories as of 2022. Formed in 2010, its focus has been mainly on a small selection of stimulant-type compounds including cocaine (*via* the benzoylecgonine metabolite as the DTR), amphetamine, methamphetamine, and 3,4-methylenedioxymethamphetamine (MDMA). Some data for other compounds also exist including ketamine, heroin (*via* measurement of 6-monoacetylmorphine, 6-MAM, as the DTR), and cannabis (*via* Δ^9-carboxy tetrahydrocannabinol, THC-COOH, as the DTR). Inter-laboratory assessment exercises have been critical and are performed every year to ensure data quality.[37] This research network has also incorporated WBE for alcohol use, including co-consumption with illicit drugs,[38] as well as the use of novel psychoactive substances (NPS).[39,40] Despite the existing power of WBE approaches for a suite of illicit drugs, data on the development of capability and estimates of consumption for new and emerging opioids on the same scale are lacking, particularly for highly potent synthetic opioids such as fentanyl and its analogues, among others. However, some key examples do exist and the approaches, methods and trends will form the primary focus of this chapter, especially given the dominance of opioid-related deaths in overdose cases in the EU,[41] the UK[42] and the USA.[43] Particular focus is placed on the requirements and capabilities of MS-based methods for fentanyl-related substance monitoring using WBE.

9.2 From Opium to Fentanyl

An opiate is a naturally occurring chemical that can be isolated from poppies, including substances such as morphine, codeine and thebaine. Although descriptions of opium use have been found in ancient Greek, Roman, Indian, Egyptian, and Chinese texts, its use became more widespread during the Middle Ages, when it was identified to be addictive.[44] Refined morphine isolation and medical usage began in the 18th and 19th centuries.[45] The analgesic properties of meperidine were later discovered in 1939[46] and in the 1940s, Janssen theorised that the piperidine rings present in both meperidine and morphine were fundamental to their analgesic effects. He later synthesised new molecules which had faster on-set effects and fewer side-effects.[47,48] Among these was fentanyl, developed in 1960 (Figure 9.1).[49] Its structure has been adapted extensively since then, primarily through modification(s) of one or more of four main features, *i.e.*, the piperidine ring, the anilinophenyl ring, the 2-phenylethyl substituent and a carboxyamide linked to the anilino-nitrogen. In terms of monitoring, many fentanyl substances fall under broad-scope control measures based on their core-structure (Table 9.1), but some differences do occur between countries. For example, the US Drug Enforcement Agency (DEA) currently enforces controls over several fentanyl-type compounds, with 43 controlled under schedule I and eight under schedule II. In the United Kingdom, most fentanyl derivatives are currently controlled under the Misuse of Drugs Act 1971 and the Misuse of Drugs Regulations 2001 as class A, schedule 2 substances, but only two substances (3-methylfentanyl and alpha-methylfentanyl) are categorised as schedule 1, meaning that they are considered to have no therapeutic value and therefore cannot be lawfully possessed.

9.3 Fentanyl on the Market and the US Opioid Crisis

Fentanyl was immediately available as an intravenous anaesthetic in the USA after its discovery in 1960. The US Center for Disease Control and Prevention

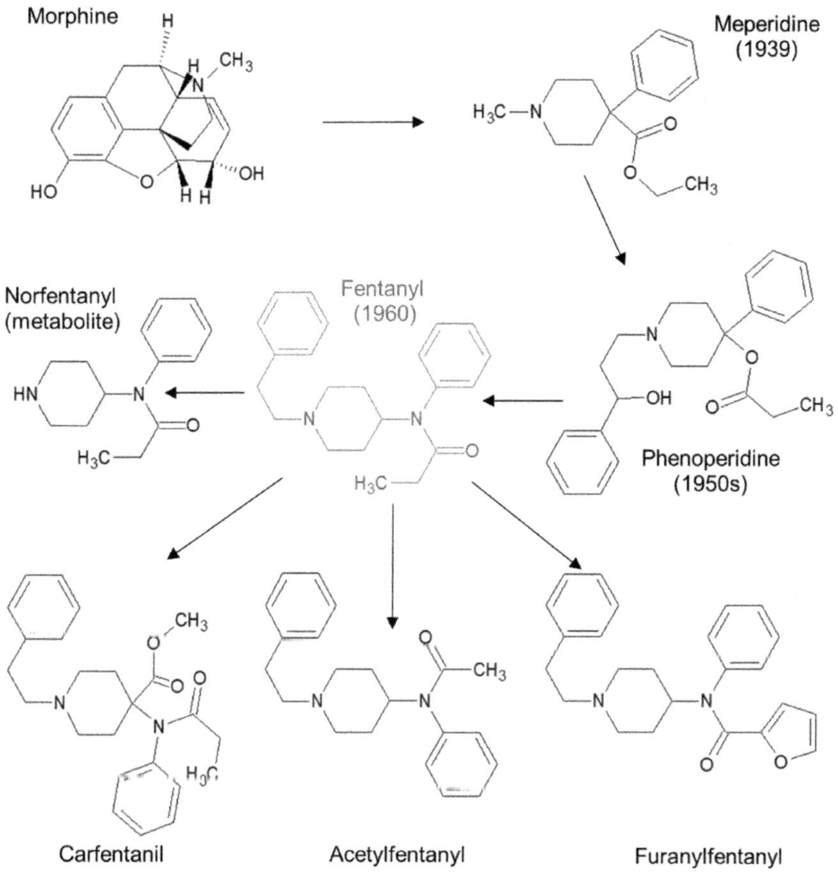

Figure 9.1 The development of fentanyl from morphine, meperidine and phenoperidine. Also shown is the metabolite norfentanyl, used as a DTR in WBE for fentanyl consumption as well as three frequently detected fentanyl-related substances in post-mortem toxicology investigations and in wastewaters.

reported that drug overdose deaths have quadrupled between 1999 and 2020, resulting in 564 000 overdose deaths from opioids alone.[50,51] It continues to be one of the worst-case examples, globally, affecting such a large population. In 2020, 75% of overdose deaths involved opioids in the USA with more than 56 000 deaths from synthetic opioids specifically (other than methadone). This accounted for 82% of all opioid-related deaths there, reflecting a 13-fold increase in deaths recorded from synthetic opioids in the USA over the last decade. Primarily, this has been attributed to fentanyl, but more recently other analogues have emerged, such as acetylfentanyl, furanylfentanyl and carfentanil, as well as other synthetic opioids such as U-47700.[52] Consumption has been both intentional (either *via* known

Table 9.1 List of scheduled fentanyl-related substances controlled by the US Drug Enforcement Agency and UK Home Office as well as selected physicochemical properties and relative potency to fentanyl.

Compound name	Monisotopic mass (Da)	Molecular formula	log *P* (ACD labs)	US DEA schedule[b]	UK class and schedule[c]	UN control status schedule[d]	No. identified metabolites or groups[e]	Relative potency[e] (to fentanyl)
2′-Fluoro *ortho*-fluorofentanyl	372.20132	$C_{22}H_{26}F_2N_2O$	3.86	I	A2	—	—	—
3-Methylfentanyl	350.23581	$C_{23}H_{30}N_2O$	4.13	I	A1	I, IV	5	20 : 1 (+); 0.2 : 1 (−) for *cis* isomer ~1 : 1 for *trans* isomer
3-Methylthiofentanyl	356.19223	$C_{21}H_{28}N_2OS$	4.09	I	A2	I, IV	—	—
4-Anilino-*N*-phenethyl-4-piperidine (ANPP)	280.19395	$C_{19}H_{24}N_2$	4.07	II	A2	—	—	—
4-Fluoroisobutyrylfentanyl	368.22639	$C_{23}H_{29}FN_2O$	4.49	I	A2	—	7	—
4′-Methyl acetyl fentanyl	336.22016	$C_{22}H_{28}N_2O$	4.09	I	A2	—	—	—
Acetyl fentanyl	322.20450	$C_{21}H_{26}N_2O$	3.75	I	A2	—	5	0.3 : 1
Acetyl-alpha-methylfentanyl	336.22015	$C_{22}H_{28}N_2O$	3.91	I	A2	I, IV	—	—
Acrylfentanyl	334.20450	$C_{22}H_{26}N_2O$	3.76	I	A2	—	—	~0.75
Alfentanyl	416.25360	$C_{21}H_{32}N_6O_3$	2.05	II	A2	I	1	~0.3 : 1
Alpha-methylfentanyl	350.23581	$C_{23}H_{30}N_2O$	4.10	I	A1	I, IV	3	~1 : 1
Alpha-methylthiofentanyl	356.19223	$C_{21}H_{28}N_2OS$	4.04	I	A2	I, IV	—	—
Benzfentanyl	322.20450	$C_{21}H_{26}N_2O$	3.34	I	A2	—	—	—
Beta-hydroxy-3-methylfentanyl	366.23071	$C_{23}H_{30}N_2O_2$	3.24	I	A2	I, IV	—	—
Beta-hydroxyfentanyl	352.21509	$C_{22}H_{28}N_2O_2$	3.07	I	A2	I, IV	—	—
Beta-hydroxythiofentanyl	358.17151	$C_{20}H_{26}N_2O_2S$	2.78	I	A2	—	—	—
Beta-methyl fentanyl	350.23581	$C_{23}H_{30}N_2O$	4.10	I	A2	—	—	—
Beta′-phenyl fentanyl	412.25146	$C_{28}H_{32}N_2O$	5.49	I	A2	—	—	—
Butyryl fentanyl	350.23581	$C_{23}H_{30}N_2O$	4.39	I	A2	—	7	0.03–0.13 : 1
Carfentanil	394.22565	$C_{24}H_{30}N_2O_3$	3.50	II	A2	Not scheduled	6	30–100 : 1
Crotonyl fentanyl	348.22015	$C_{23}H_{28}N_2O$	3.92	I	A2	—	—	—
Cyclopentyl fentanyl	376.25146	$C_{25}H_{32}N_2O$	4.99	I	A2	—	—	—
Cyclopropyl fentanyl	348.22016	$C_{23}H_{28}N_2O$	4.04	I	A2	—	4	3 : 1
Fentanyl	336.22016	$C_{22}H_{28}N_2O$	4.09	II	A2	I	—	1 : 1
Fentanyl carbamate	352.21508	$C_{22}H_{28}N_2O_2$	4.31	I	A2	—	—	—
Furanyl fentanyl	374.19943	$C_{24}H_{26}N_2O_2$	4.12	I	A2	—	6	7 : 1

Isobutyryl fentanyl	350.23581	$C_{23}H_{30}N_2O$	4.10	I	A2	—	—	0.13 : 1
Lofentanyl	408.24129	$C_{25}H_{32}N_2O_3$	4.29	—	A2	—	—	—
Merthoxyacetylfentanyl	352.21508	$C_{22}H_{28}N_2O_2$	3.29	I	A2	—	5	0.3 : 1
Norfentanyl	232.15756	$C_{14}H_{20}N_2O$	1.72	II	A2	—	—	—
Ocfentanyl	370.20566	$C_{22}H_{27}FN_2O_2$	3.21	I	A2	—	3	2.5 : 1
ortho-Fluoroacryl fentanyl	352.19509	$C_{22}H_{25}FN_2O$	3.82	I	A2	—	—	—
ortho-Fluorobutyryl fentanyl	368.22639	$C_{23}H_{29}FN_2O$	4.18	I	A2	—	—	—
ortho-Fluorofentanyl	354.21074	$C_{22}H_{27}FN_2O$	4.02	I	A2	—	1	—
ortho-Fluoroisobutyryl fentanyl	368.22639	$C_{23}H_{29}FN_2O$	4.18	I	A2	—	—	—
ortho-Methyl acetylfentanyl	336.22016	$C_{22}H_{28}N_2O$	4.09	I	A2	—	—	—
ortho-Methyl-methoxyacetyl fentanyl	366.23073	$C_{23}H_{30}N_2O2$	4.02	I	A2	—	—	—
para-Chloroisobutyryl fentanyl	384.19684	$C_{23}H_{29}ClN_2O$	5.05	I	A2	—	—	—
para-Fluorobutyrylfentanyl	368.22639	$C_{23}H_{29}FN_2O$	4.45	I	A2	—	—	—
para-Fluorofentanyl	354.21074	$C_{22}H_{27}FN_2O$	3.87	I	A2	I, IV	—	—
para-Fluorofuranyl fentanyl	392.19001	$C_{24}H_{25}FN_2O_2$	4.13	I	A2	—	—	—
para-Methoxybutyryl fentanyl	380.24638	$C_{24}H_{32}N_2O_2$	4.34	I	A2	—	—	—
para-Methyl fentanyl	350.23581	$C_{23}H_{30}N_2O$	4.68	I	A2	—	—	—
Phenyl fentanyl	384.22016	$C_{26}H_{28}N_2O$	5.16	I	A2	—	—	—
Remifentanyl	376.19982	$C_{20}H_{28}N_2O_5$	1.85	II	A2	I	1	~1 : 1
Sufentanil	386.20280	$C_{22}H_{30}N_2O_2S$	3.78	II	A2	I	4	~10 : 1
Tetrahydrofuranyl fentanyl	378.23073	$C_{24}H_{30}N_2O_2$	3.31	I	A2	—	4	~0.2 : 1
Thenylfentanyl[a]	328.16093	$C_{19}H_{24}N_2OS$	3.42	I[a]	A2	—	—	—
Thiafentanyl	416.17698	$C_{22}H_{28}N_2O_4S$	3.12	II	A2	—	—	—
Thiofentanyl	342.17658	$C_{20}H_{26}N_2OS$	3.79	I	A2	I, IV	—	—
Thiofuranylfentanyl	390.17658	$C_{24}H_{26}N_2OS$	4.60	I	A2	—	—	—
Valeryl fentanyl	364.25147	$C_{24}H_{32}N_2O$	4.84	I	A2	—	—	—

[a] Expired 11/29/86.
[b] DEA "Orange Book" https://www.deadiversion.usdoj.gov/schedules/orangebook/orangebook.pdf.
[c] UK Home Office Misuse of Drugs Regulations https://www.legislation.gov.uk/uksi/2001/3998/pdfs/uksi_20013998_en.pdf (broader control of all fentanyl-related substances is included under these regulations).
[d] Single convention on narcotic drugs, 1961 (as amended by the 1972 protocol), United Nations https://www.unodc.org/pdf/convention_1961_en.pdf.
[e] According to ref. 50.

misuse or as prescription pain relief) and unintentional (*e.g.*, through inadvertent consumption through its addition to other illicit drugs such as heroin, cocaine or counterfeit medicines). The opioid epidemic in the US can be split into three phases (Figure 9.2): increase in prescription opioids primarily for pain relief in the 1990s, leading to increased overdose deaths since at least 1999; a second phase involving a rise in heroin-related deaths beginning in 2010 as a result of opioid dependency, leading to heroin use and its cheaper availability; and the third phase is characterised by the rise in synthetic opioid-related deaths, primarily fentanyl, since 2013. There is now some suggestion that the drug epidemic may be entering a fourth phase as a result of polysubstance misuse.[53]

Quantifying how the US opioid crisis affects or reflects that of the wider world is difficult. Wang and colleagues recently examined the roles of both China and India as the two largest suppliers of fentanyl globally, including illicit manufacture.[54] They concluded that global legislation and regulation of fentanyl-type compounds remains insufficient, especially in countries without a domestic fentanyl crisis, where systemic corruption exists and/or where international cooperation is limited, and this enabled new fentanyl substances and precursors to consistently emerge. In particular, the US Department of State has reported that the vast majority of illicit US fentanyl and fentanyl analogue supply was trafficked from China since 2013 through Canada and Mexico[55] until stricter controls were implemented in China in 2019, resulting in a decrease in illicit shipments of fentanyl-related substances to the USA. The World Health Organisation (WHO) reported in 2021 that 70% of drug-related deaths worldwide were related to opioids.[56] WHO also attributed the rising number of overdoses in many countries to increased use of opioids in pain management. The Global Burden of Diseases, Injuries and Risk Factors Study estimated that 40.5 million people globally were dependent on opioids in 2017.[57,58] More generally, opioid dependence is currently the third most important substance-use disorder globally behind alcohol and tobacco.[59] In 2019, Drummer reviewed the global research describing fatalities involving 15 synthetic opioids, including acetylfentanyl, acrylfentanyl, 3-methylfentanyl, alpha-methylfentanyl, fentanyl, furanylfentanyl, butyrylfentanyl, ocfentanil and sufentanil.[60] Detection was very challenging in post-mortem samples and broad analytical sensitivity, including for unknowns, was required down to the low $ng\,mL^{-1}$ concentration level particularly in cases where the circumstances indicate the potential use of other substances, aiming to prevent misinterpretation of the primary toxicant. In the UK specifically, Edmund *et al.* found fentanyl or its analogues in 40 of 84 suspected fentanyl deaths following post-mortem analysis of biological samples using LC–HRMS.[61] Post-mortem blood concentrations for fentanyl and its more potent analogue carfentanil ranged from 0.21–107 $ng\,mL^{-1}$ (median: 2.66 $ng\,mL^{-1}$) and 0.03–1.66 $ng\,mL^{-1}$ (median: 0.24 $ng\,mL^{-1}$), respectively. With all of this in mind, monitoring the exposure of such drugs using novel approaches like WBE therefore offers a distinct and unique advantage.

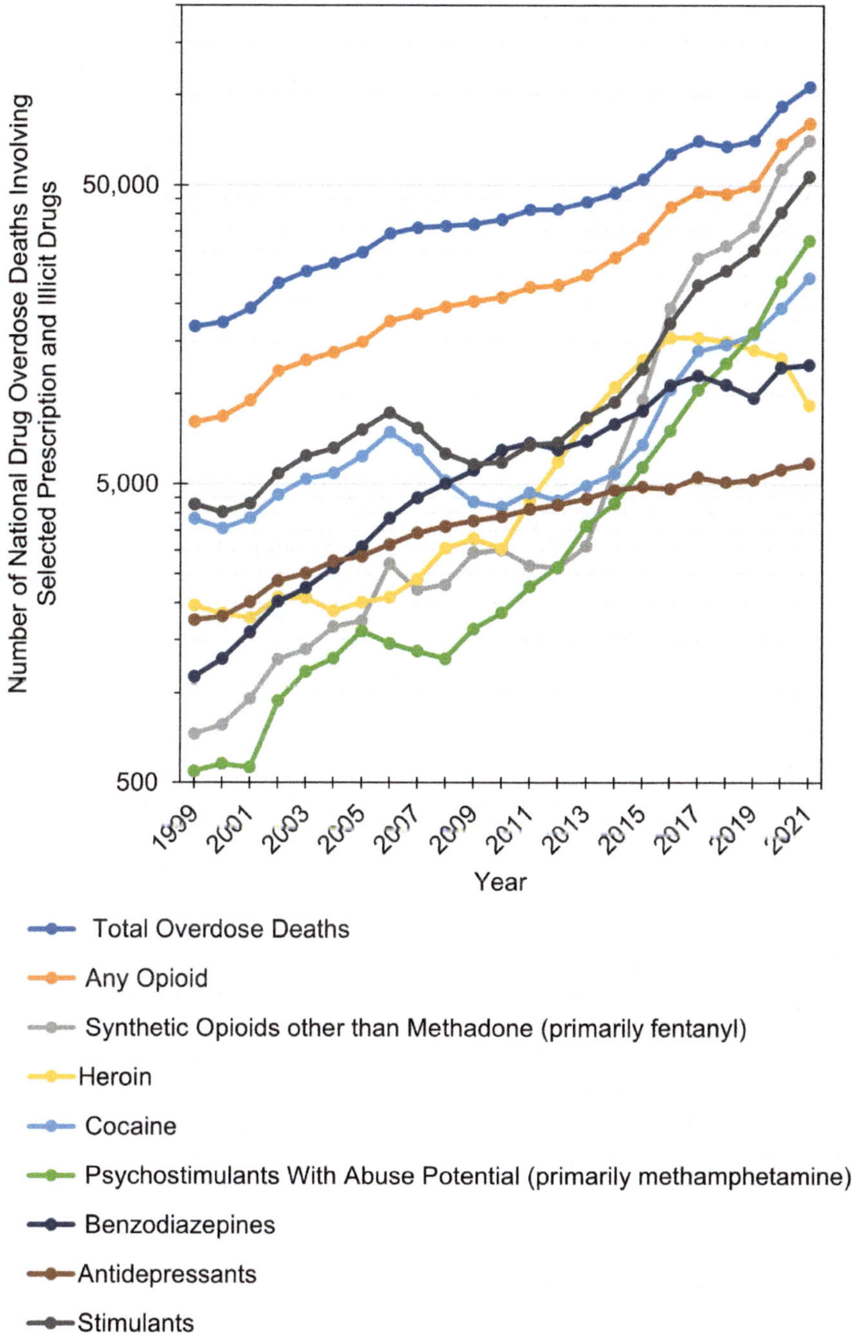

Figure 9.2 The number of overdose deaths from 1999 to 2021 in the USA and the related drug types. Source: Death Rate Maps & Graphs (2022) Centers for Disease Control and Prevention. Centers for Disease Control and Prevention. Data available at https://www.cdc.gov/drugoverdose/deaths/index.html (accessed: March 25, 2023).

9.4 Pharmacokinetics

Fentanyl-related substances are administered intranasally, intravenously, intramuscularly and transdermally. They are available both as a prescribed medication and illicitly. Prescribed fentanyl medications typically come in several forms including patches, tablets, lozenges, injectable solutions and nasal sprays. For deep sedation, fentanyl is usually administered at 1–2 $\mu g\,kg^{-1}$.[62] Illicit fentanyls also come in several different forms including powders on blotting paper, in eyedroppers or nasal sprays, or counterfeit medicine pills. In addition, other illicit drugs, including other opioids, can be contaminated or cut with fentanyls.[63] Given these multiple administration routes and such uncertainties relating to contamination, WBE estimates of the number of doses per 1000 people per day in a catchment can be challenging and varies in the literature (*e.g.*, defined daily doses of between 0.2–0.6 mg have been used for fentanyl, see the supplementary table available as ESI[†]).

Fentanyls are relatively lipophilic (Table 9.1) and they bind to human opioid receptors in the brain governing pain and emotion. They have several effects including analgesia, euphoria, confusion, hallucinations, delirium, drowsiness, respiratory depression, nausea, and death, among others. While other metabolic pathways may exist,[64] fentanyl itself is thought to be primarily metabolised in the liver by CYP3A4 and 3A5 isoenzymes, resulting in predominantly the *N*-dealkylated metabolite, norfentanyl, which is inactive and often the biomarker used for WBE of fentanyl consumption (its other 14 minor metabolites account for <1% of the administered dose[65]). Fentanyl has a large apparent volume of distribution (3.5–8 $L\,kg^{-1}$), also likely due to its lipophilicity, but clearance is relatively rapid (30–72 $L\,hour^{-1}$).[66] In rats, median effective and lethal doses (ED_{50} and LD_{50}) were determined to be 0.0061 $mg\,kg^{-1}$ and 62 $mg\,kg^{-1}$, respectively.[67] However, as little as 2 mg is estimated to be lethal in humans.[68,69] The terminal elimination half-life of fentanyl following intravenous administration varies from 2.3–14.2 h.[70] In 2013, Lötsch and co-workers reviewed the pharmacokinetics of non-intravenous forms of fentanyl administration and found that the maximum plasma concentrations were <3 $ng\,mL^{-1}$ for oral transmucosal, intranasal and transdermal formulations of varying dosages from 10.5 to 1600 μg.[71] An elimination half-life ($t_{1/2}$) of as little as 3–4 h was found for some nasal sprays and up to 25 h for sub-lingual tablets ($t_{1/2}$ tested across several dose levels and formats). In general, less than 10% of the administered fentanyl dose was found to be excreted as the parent molecule[72,73] while >85% was recoverable mostly in the form of metabolites in urine and faeces after 72 h.[74]

The relative potencies of other fentanyl-related substances exhibit a range of variations (Table 9.1). For example, carfentanil has an estimated 30 to 100-fold higher potency than fentanyl itself.[50] In 1976, van Bever *et al.* summarised the relative potencies of several opioids in rats with carfentanil exhibiting the highest potency at 10 031-fold that of morphine (fentanyl had 292-fold the potency of morphine).[75] This potency has a knock-on effect, necessitating much smaller doses, making the detection of the administered compound or its metabolites much more challenging, especially in wastewater samples where significant dilution occurs. Many other fentanyl compounds are, primarily, metabolised in a similar way to fentanyl[50] to generate *N*-dealkylated products. Some analogues generate an extensive suite of metabolites such as butyrylfentanyl and 4-fluoroisobutyrylfentanyl. As WBE operates at a population level, the development and application of analytical laboratory methods to separate and detect all of these substances potentially occurring together at low concentrations in a sample pose a major challenge, even before considering the sample matrix and its varying complexity. This is where mass spectrometry is particularly suitable.

9.5 Estimating Community Consumption of Illicit Drugs Using WBE

9.5.1 Selection of a Suitable Drug Target Residue Biomarker

The first step in performing quantitative WBE is selecting a suitable DTR, which should have several preferred characteristics.[76,77] Arguably, the most critical characteristics among these include specificity to the ingested parent drug, slow in-sewer biotransformation kinetics, the extent of metabolism, low sorption to suspended solids, and the ability to be reliably identified and measured at low to sub $ng\,L^{-1}$ concentrations, ideally using either LC–MS/MS and/or LC–HRMS. In reality, some selected DTRs are less than ideal as no better alternative may exist, and therefore these require additional considerations to mitigate uncertainty. For example, trained personnel may be required to treat or prepare samples at WWTPs directly, which may involve additional materials, reagents, training and storage. A number of different approaches have been used, including filtration, chilling/freezing and addition of chemical reagents such as sodium metabisulphite or hydrochloric acid to minimise microbiological transformation and maintain stability of some DTRs. An extensive study by Lin and colleagues showed the effect of various in-sample instability mitigation approaches, including for several opioids, but each approach showed different levels of success.[78] With respect to opioids, 6-MAM showed poor stability undergoing almost complete transformation with no sample pre-treatment under relatively hot climate conditions over a period of just one week. As a result, heroin consumption estimations using WBE are therefore often more challenging. When feasible, fortification with suitable stable-isotope internal standards can account for losses when added to samples as soon as possible after sampling. Interestingly, in-sample stability for norfentanyl was good, regardless of sample treatment type, thus indicating its viability as a relatively good DTR, although acidification and freezing at −20 °C were still recommended. However, the stability of fentanyl substances has also not always been consistent.

For example, Krisman-Matasic *et al.* showed that fentanyl, norfentanyl and sufentanil in spiked wastewater stored at −20 °C for 100 days exhibited a statistical decrease in concentration of >25% after 7 days.[79] Baker *et al.* also observed losses over relatively short periods of time between 0 (12 h at 2 °C) and 14% (24 h at 19 °C).[77] Therefore, stability should be investigated in the matrix of the intended sites under study and the number of freeze-thaw cycles should also be considered carefully in all cases when (re)analysing frozen samples as this may increase DTR instability.[80] With a predicted log *P* of 1.72 (ACD Labs), norfentanyl is more polar than most other related analogues and precursors and previous studies have shown that it also displays moderate-to-low sorption to organic matter.[81] However, for the *N*-dealkylated metabolites of other fentanyls, some challenges exist, which make specificity of WBE estimations difficult. For example, carfentanil and remifentanil both metabolise into norcarfentanil,[82] making differentiation difficult without the presence of other metabolites and/or the intact parent compound. This is also the case for alfentanil and sufentanil.

9.5.2 Sampling Approaches

The general workflow for sampling, analysis and data analysis to determine population consumption of illicit drugs is shown in Figure 9.3. Samples are

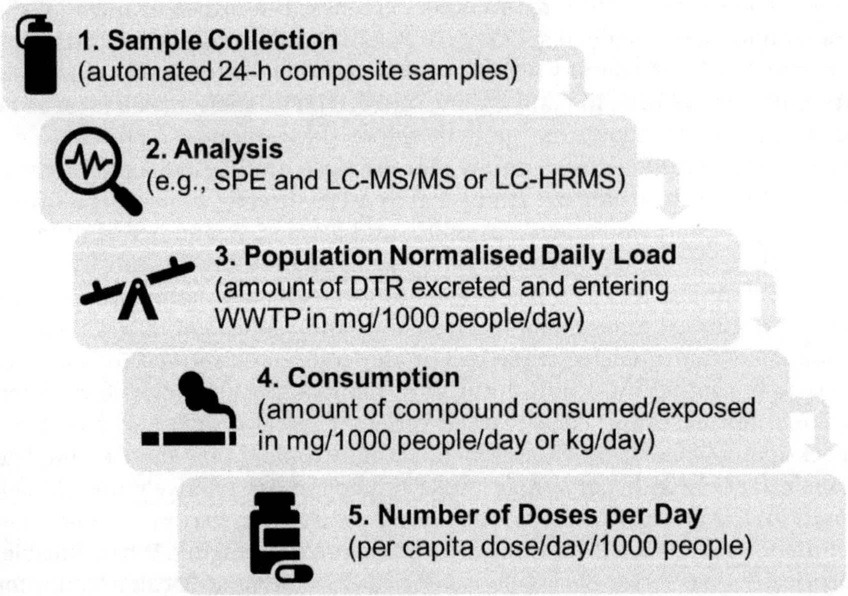

Figure 9.3 General analytical workflow for wastewater-based epidemiology of illicit drugs. SPE: solid-phase extraction; LC–MS/MS and LC–HRMS: liquid chromatography–tandem mass spectrometry and liquid chromatography–high resolution accurate mass spectrometry, DTR: drug target residue; and WWTP: wastewater treatment plant.

generally taken after the fine-screen using automated composite samplers, manual grab sampling or autonomous passive samplers. For the most reliable quantitative estimation of community consumption, automated composite sampling is preferred, and samples can be taken in a time- or flow-integrated manner, usually over a 24 hour period. Therefore, any such sample is generally assumed to represent consumption in the preceding 24 hours. Normally, campaigns aim to cover each day of the week to assess any recreational usage trends at weekends. However, representativeness depends on a number of variables including the drug elimination rate, toilet location with respect to the residence of the individual (these can, for example, lie inside/outside the catchment served by the WWTP due to population movement), sewer transit time to the WWTP, and intra-day sampling interval.[83] The frequency of replicate sampling campaigns is also important for inter/intra year representativeness and arguably this is still lacking in WBE for illicit drugs generally. For example, while a few studies in this review cover an impressively large geographical area and estimated population, sampling campaigns last mostly for seven consecutive days with repeat campaigns often separated by a few months at a time. This introduces the chance of missed instances of short-term usage changes, in some cases linked to specific events such as national holidays,[84,85] sporting competitions[86] or large festivals.[87,88] Although simpler, grab sampling suffers from significant uncertainty and lacks generalisability, as it represents a single 'snapshot' of DTR at the exact time of sampling. Multiple grab samples can be pooled manually, but achieving this consistently and on a large scale needs significant time commitment.

A less time-consuming and labour-intensive strategy is to use passive sampling in which devices containing a suitable sorbent material are placed *in situ* in the WWTP influent flow for a period of 1–2 weeks.[89–92] This approach offers the advantage of a time-weighted average concentration of DTR and often offers improved method sensitivity as the DTR is accumulated by the sorbent. As a result, passive sampling offers an excellent means to undertake suspect screening and/or non-targeted analysis for new compounds. This latter use in WBE still in its nascent stages and is mainly utilised for environmental chemical risk assessment.[93–95] The main drawback of passive sampling is that it requires calibration to determine the DTR uptake rate (R_s) to extrapolate measurements to concentrations in water. Recent approaches have attempted machine-learning based prediction of R_s based on quantitative structure–activity relationships.[96] Many R_s data are also derived in the laboratory and may not be field-representative.[97] In contrast, *in situ* calibration can work well, but relies on occurrence of compounds in wastewater and can be heavily dependent on site-specific variation in conditions governing uptake such as flow rate, temperature and pH.[98,99] In terms of scaling up to national and international levels, this approach represents a practical alternative not just for drug monitoring, but also for biological targets, such as infectious diseases like SARS-CoV-2.[100] More recently, low-cost 3D-printed passive samplers have emerged, including for opioids, which not only aid in scaling up for WBE but also help in environmental risk assessment.[101–103] Rodayan *et al.* used polar organic chemical integrative samplers (POCISs) configured with a mixed-mode cation exchange sorbent to determine fentanyl in wastewater effluent, river water and drinking water.[104]

9.5.3 Sample Preparation

With the relatively high potency of fentanyl and its analogues, samples have generally been prepared by isolation of trace DTRs from the sample matrix followed by pre-concentration to enable instrumental analysis. In most published works (see the supplementary table available as ESI[†]), sample preparation for fentanyls has generally been performed using offline solid-phase extraction (SPE). Several methods have employed hydrophilic–lipophilic balanced (HLB) materials, which are broadly applicable sorbents for soluble organics comprising ethyl- or divinyl-benzene with either hydroxylation of the phenyl rings or co-polymerision with polar monomers (such as vinylpyrrolidone, for example). These have been used for many other WBE and environmental applications, including pharmaceuticals and illicit drugs.[81,105,106] Although HLB-based SPE works best for basic fentanyls at alkaline pH, samples have generally not been buffered and are assumed to be in the region of pH 6.5–8.5. Alternatively, mixed-mode cation exchangers have been used which employ the same base HLB material with modified carboxylic or sulphonic acid functionalities. In this format, samples are generally acidified to promote ionisation. O'Rourke *et al.* evaluated the use of HLB as well as weak and strong mixed-mode cation exchangers for SPE of 12 fentanyl analogues among a total of 40 NPS,[107] with higher recoveries recorded for most fentanyls using mixed-mode strong cation exchangers (MCX). Salgueiro-González *et al.* also investigated recoveries on both HLB and MCX cartridges for 62 NPS including 20 fentanyl-related substances and performance was better using MCX.[108] Despite this, relatively high recoveries in the region of 80–120% on average have been reported for fentanyl-type compounds using either sorbent chemistry, enabling methods to be quantitative.

Sample volumes have ranged from 50–300 mL with the majority of positive detections of fentanyl substances reported using 50 or 100 mL in general. Several performance and practical limitations exist which can prevent the effective use of larger volumes and therefore require careful consideration. Sorbents will have some affinity for the matrix which can result in interference both during the extraction process and in signal enhancement/suppression during LC–MS/MS measurement.[84] Furthermore, using smaller volumes can be more practical for larger WBE programs as it facilitates collection, transportation and storage of larger numbers of frozen samples for replicate or later analysis, potentially reducing the reagent use. To further boost sensitivity and to ensure compatibility with LC–MS analysis, SPE extracts have generally been reduced to dryness under nitrogen gas and reconstituted in solvents representing the starting LC mobile phase condition, with overall concentration factors ranging from 50 to 1500-fold to achieve sufficient sensitivity (the median concentration factor for most works in the supplementary table available as ESI[†] was 250-fold).

9.5.4 Instrumental Analysis

For all WBE studies of fentanyls and opioids, the instrumental technique was gradient reversed-phase LC–MS/MS or LC–HRMS. Level 1 identification of drugs in wastewater has been generally performed according to the Schymanski framework and based on retention time, MS and MS/MS spectral matching with a reference material.[109] For triple quadrupole MS analysers, acquisition is generally conducted in multiple reaction monitoring mode (MRM) using at least two transitions, where possible, with one of these (usually the most intense signal) used for quantification purposes while the other(s) serve as qualifiers to confirm identity. In addition, MRM ion transition ratios are also taken into consideration for added assurance (as an example, Figure 9.4 shows MRM transitions for 14 fentanyl related substances in wastewater from O'Rourke *et al.*[107]). For HRMS in the targeted mode under data-dependent acquisition (DDA), accurate *m*/*z* measurement is now readily achievable in the sub 5 ppm range including for time-of-flight instruments (TOF-MS) with 40 000 full-width-half-maximum (FWHM) resolution.[110,111] For Orbitrap mass analysers, resolution is even further enhanced to over 100 000 FWHM, making elemental analysis more reliable for compounds with very similar *m*/*z*.[112] The real advantage of HRMS analysis is the ability to perform suspect screening to identify new substances using the data-independent acquisition mode (DIA) or sequential window acquisition of all theoretical mass spectra (SWATH-MS).[111,113,114] However, mass spectral identification of some fentanyl compounds can be challenging, as many are structurally similar and some are positional isomers or enantiomers (Table 9.1). Therefore, a greater reliance on adequate chromatographic resolution exists to differentiate such compounds. To put this in perspective, in 2019, the US Department of Homeland Security estimated that 3024 fentanyl-related derivatives could be synthesised,[115] but very few reference materials exist for reliable identification or quantification. This is particularly true for stable isotope-labelled internal standard (SIL-IS) compounds. In some cases, suspect screening workflows have used retention time or retention index prediction using machine learning, among other approaches, to shortlist potentially viable candidates.[10,116–118]

In comparison to human biological fluids,[119,120] very few WBE methods have included a large number of fentanyl compounds with one of the most comprehensive being that by Salgueiro-Gonzalez *et al.*[108] The matrix effect for most compounds was found to lie within ±40% (total variance was ±80%) with half of all fentanyl-related compounds suffering from ion suppression. For LC, the stationary phase choice has mostly been octadecylsilica. However, phenyl, biphenyl and pentafluorophenyl media have also

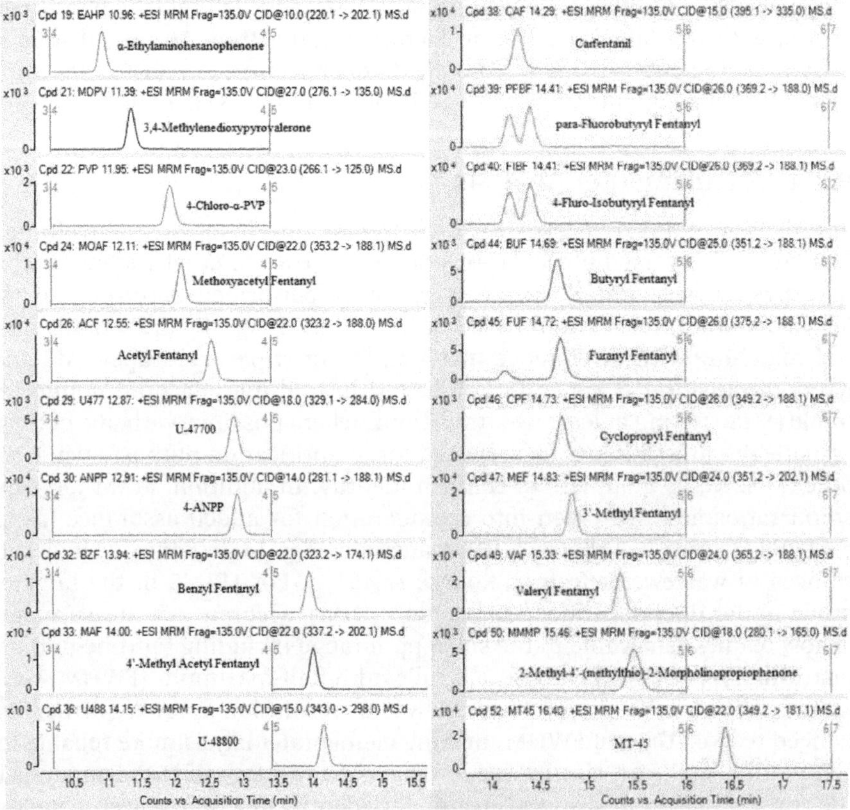

Figure 9.4 Selected reaction monitoring chromatograms of multiple drugs including 14 fentanyl-related substances a spiked wastewater extract (50 or 100 ng). Reproduced from ref. 107 with permission from American Chemical Society, Copyright 2023.

been used and have been shown to have higher capacity and selectivity for mid-to-high-polarity drugs and metabolites, enabling the execution of rapid analysis.[8,85,121] Generally, the mass analyser of choice for quantitative analysis has been a triple quadrupole with electrospray ionisation in the positive ion mode. When combined with SPE, lower limits of quantitation (LOQ) have generally been achieved at the pg L^{-1} level. One of the most sensitive methods by Bade *et al.* used a biphenyl LC column and a triple quadrupole MS to separate several opioid NPS including fentanyls and together with a 1000-fold SPE concentration factor, it achieved quantification limits between 50 and 100 pg L^{-1} for three fentanyl opioids in influent wastewater (butyrylfentanyl, furanyl fentanyl and valeryl fentanyl).[122,123]

Quantification has generally been performed using matrix-matched calibration either using an external pooled wastewater matrix or *via* standard addition. Although SIL-ISs are generally recommended for more reliable quantification, very few SIL-ISs exist and closely eluting SIL-IS analogues are often selected which can be subjected to some unknown degree of matrix interference uncertainty between samples.[107] When compounds are present in the external calibration matrix used for matching, background subtraction is also performed.

9.5.5 Drug Consumption Estimation

Following the determination of analyte concentration in wastewater, eqn (9.1) is used to calculate the population normalised daily load (PNL), which is the amount of DTR excreted by the community and entering the WWTP per day (mg per 1000 people per day).

$$\begin{aligned} \text{PNL} = [\text{DTR}] \times \text{flow} \times \left(\frac{1000}{\text{PE}}\right) \times \left(\frac{100}{100 + \text{stability}_{\text{DTR}}}\right) \\ \times \left(\frac{100}{100 - \text{sorption}}\right) \times \left(\frac{100}{100 - \text{exfiltration}}\right) \end{aligned} \tag{9.1}$$

This calculation is performed by multiplying the measured DTR concentration, the recorded influent flow in litres per day as well as the population equivalent (PE), and a set of correction factors governing the stability and biosolid sorption behaviour of the DTR in the sewer network and any losses of influent flow prior to the WWTP (Exfiltration) due to misconnections, damaged pipes or combined sewer overflows, for example (all correction factors expressed as %). PE is calculated based on the organic biodegradable load having a five-day biochemical oxygen demand (BOD) of 60 g of oxygen per person per day.[124] In some cases, population count from census data is used instead of PE.

Following this, the estimate of consumption can be calculated using eqn (9.2).

$$\text{Consumption} = (\text{PNL} \times \text{correction factor}) - \text{OS} \tag{9.2}$$

where consumption is the mg of parent drug consumed per 1000 people per day, OS represents other sources of drug consumption (*e.g.*, medical use) and the correction factor describes the correction for excretion of the metabolised drug and the ratio of the molecular weight (MW) of ingested parent drug (Par) to that of the DTR, as eqn (9.3).

$$\text{Correction factor} = \left(\frac{100}{100 - \text{excretion}}\right) \times \left(\frac{\text{MW}_{\text{Par}}}{\text{MW}_{\text{DTR}}}\right) \tag{9.3}$$

In addition to reporting consumption in milligrams per 1000 people per day, it is possible to express the total mass of substance consumed per day.[8] When sufficient spatio-temporal monitoring programs are performed, this is a very valuable method to potentially estimate the size of an illicit drug market. In some cases, authors have also expressed WBE estimates as the number of doses in a catchment in a day, especially for stimulants.[9]

9.5.6 Considerations for Interpretation of WBE Estimates

There are several sources of uncertainty in WBE which have been the focus of several works.[125–129] These include the issues with representative spatio-temporal sampling discussed above, the inability to reliably calculate the population in an area[130] including the movement of people,[131–133] multiple drug administration routes, metabolism and excretion rates for different drugs,[81,134] the amount and frequency of wastewater influent loss before arriving at the WWTP,[32] partitioning to organic matter and stability of the DTR in a specific catchment,[135] a reliable estimation of OS[136,137] and uncertainties arising from the laboratory analysis.[138] In many cases, researchers have chosen to compare WBE data across sites using PNL data to minimise uncertainty. In addition, the exact purity of the ingested drug cannot be determined, nor can data be used to estimate the price of materials (aside from being able to estimate the approximate value of a market from WBE estimations of use[139]). Identification of individuals also cannot be performed, nor the exact number of people contributing to PNL, or how many doses each person consumes in a day. This draws out a further important issue: the ethical considerations for the implementation of WBE. It is generally assumed, as a non-individualised measurement of population exposure and consumption, that ethical approval is not required due to the impracticality (and sometimes impossibility) of gathering informed consent on such a scale. Nonetheless, where WBE studies could impact or stigmatise residents, vulnerable communities, local or regional authorities within a catchment, and particularly those on a more localised scale, ethical guidance and approval should be sought.[140] The EMCDDA, for example, has issued ethical guidelines for those undertaking WBE studies, which broadly consider many such scenarios.[141] In particular, WBE at specific sites such as workplaces, schools, prisons and hospitals is discussed. Although WBE can benefit participants, the use of WBE data to enforce or implement punitive measures in prison settings poses a risk to the rehabilitation and reintegration process (in addition to staff or visitors also likely being included in WBE measurements). For schools, workplaces and hospitals, although all very different scenarios, some similar issues arise around the risk of stigmatisation and potential damage to the institutional reputation, including economic harm and implementation of strict or unfair work or study conditions.

9.6 Application of WBE to Fentanyl Monitoring Internationally

The number of papers reporting the occurrence of and the ability to monitor fentanyl-related compounds is much smaller than reported works on other illicit drugs. Methods are generally applied to analyse wastewater for a small number of fentanyls together with other drugs in a single analytical run. In many cases, the paucity of occurrence could be simply due to low usage despite adequate method sensitivity.[81] Although this is not a systematic review of the literature, there are several papers worthy of detailed discussion which will be outlined in the following sections.

9.6.1 Europe

Of all analytical methods developed, that by Salgueiro-González *et al.* included the most fentanyl-related compounds ($n = 20$) for an Italian nationwide study. Across all 33 locations, only two fentanyl-related substances were detected, fentanyl and norfentanyl, and in only 9 and 11 WWTPs, respectively. PNLs were similar across all these cities and between <0.050 mg per 1000 people day (Trani) to 0.25 mg per 1000 people per day (Cagliari). No statistical difference was observed between weekend or weekday use and this was considered likely due to its predominant use as a prescription medication. Therefore, at least in Italy, significant population usage and exposure seem to be currently low. Elsewhere in Europe, occurrence data for fentanyls in wastewater also seem to be low.[79,81,142,143]

9.6.2 The Americas

The US successfully implemented the national wastewater surveillance system in response to the SARS-CoV-2 pandemic, but prior WBE applications have been sparse, despite an ongoing opioid crisis. There has also been reluctance to capitalise on this new infrastructure to adopt WBE on a similarly coordinated level for illicit drugs due to potential community stigmatisation and mistrust between public health and law enforcement officials about how substance misuse is tackled, including the use of WBE data.[144–146] However, some notable examples do exist for the USA. Bade and co-workers recently conducted an international wastewater screening comparison of eight countries for >200 NPS during the 2019/2020 New Year period. They found only one related substance, acetyl fentanyl, and only in samples from the USA.[147] In 2018, Gushgari *et al.* reported the first case of fentanyl consumption on a university campus in the Southwestern USA with norfentanyl and fentanyl detected in 7 and 5% of the samples, respectively.[148] This team later published work on municipal WWTPs in two cities in the Southwestern USA showing relatively similar PNLs 10 ± 0.2 and 9 ± 2.7 mg per 1000 people per day for fentanyl, but slightly differing PNLs for norfentanyl at 18 ± 7 and 47 ± 18 mg per 1000 people per day.[149] The population of both cities was also different at 130 000 and 45 000, respectively. Overall, there was a detection frequency of 62% ($n = 45$ samples) and 100% ($n = 21$ samples) across both sites for fentanyl and norfentanyl, with concentrations measured up to 198 ng L^{-1}. In 2019, O'Rourke *et al.* reported concentrations in wastewater in four communities in South Illinois, USA.[107]

Of 13 fentanyl-related substances studied, three were quantifiable in wastewater (carfentanil, furanyl fentanyl and methoxyacetyl fentanyl at <2.5 ng L^{-1} on average at each WWTP). Carfentanil was quantifiable in all samples at two WWTPs and PNLs were subsequently determined to be 0.51 ± 0.03 and 0.41 ± 0.04 mg per 1000 people per day. Skees and colleagues from the same group also performed a prior WBE assessment at one WWTP in Tennessee, USA.[150] Fentanyl was found to be 1.60 ± 0.1 ng L^{-1} in influent wastewater and a consumption of 1.27 ± 0.10 mg per 1000 people per day was estimated. Interestingly, effluent concentrations were much higher at 19 ± 26 ng L^{-1}. In the neighbouring state of Kentucky, Croft *et al.*[151] determined higher PNLs by comparison at 1.32 and 0.86 mg per 1000 people per day in two urban WWTPs and 0.03 mg per 1000 people per day in a rural WWTP. This team also estimated that fentanyl was consumed at a rate of 134 and 169 mg per 1000 people per day in the urban areas while in a much lower consumption rural catchment, it was estimated to be 31 mg per 1000 people per day. A study of six WWTPs in New York around major public holidays by Centazzo *et al.* detected fentanyl only once out of 48 samples, *i.e.*, the day after the 4th July celebrations with a concentration of 5.9 ng L^{-1} in wastewater (expressed as 4.3 ng mg^{-1} of creatinine).[152] As grab samples were taken in this case, PNLs or consumption values could not be reliably calculated for comparison. In Canada, Rodayan *et al.* detected fentanyl in 100% of samples of effluent wastewater, river water and even in drinking water at low ng L^{-1} concentration levels.[104] No data were gathered on influent for a WBE assessment of fentanyl consumption. Although the WWTP discharge point was 19 km upstream of the drinking water treatment plant (and with fentanyl concentrations in two grab samples ranging between 331 ± 36 and 478 ± 47 ng L^{-1}), high concentrations in the treated drinking water were observed (~11–12 ng L^{-1} on average), indicating a potentially wider source of exposure in that catchment. Canada's wastewater survey (CWS) has been ongoing since March 2019 at five major WWTPs.[153] In comparison to the USA, PNLs of between ~4 and 12 mg per 1000 people per day were reported on average during the SARS-CoV-2 pandemic in 2019 (with the highest consumption in Vancouver) with a reduction in maximum PNLs of <6 mg per 1000 people per day in 2020. Nevertheless, this indicates that relatively high fentanyl consumption also exists there. In contrast, WBE was applied by Cruz-Cruz in 15 Mexican cities in 2017–2018. Both fentanyl and norfentanyl were included in the analytical method, but neither were detected with any significant frequency with the exception of fentanyl detected at one site in two out of seven 24 hours composite daily samples (Culiacan) and norfentanyl in one sample at another site (Veracruz).

9.6.3 Asia

Two research papers published by Du *et al.* focused on drug monitoring in Asia. The first involved sampling at two WWTPs in Kuala Lumpur in 2017 and analysis for 24 DTRs including several opioids.[154] However, fentanyl was not detected. The second study focussed only on tramadol and fentanyl in a nationwide study in China from 2016–2019 across 77 WWTPs with a combined population of almost 60 million people.[155] The largest proportion of all samples was taken in 2016 ($n = 257$) and 2017 ($n = 299$), and fentanyl was present in these samples at low concentrations at or below 3.6 ng L^{-1} (LOD = 0.2 ng L^{-1}) and at a low detection frequency of ≤5%. Subsequently, no fentanyl was detected in a much smaller number of WWTPs in 2018 (63 samples at 9 WWTPs) and 2019 ($n = 35$ samples at 5 WWTPs). This suggested limited domestic use and this observation was in agreement with other surveys of low fentanyl seizures in Asia.

9.6.4 Australasia

WBE monitoring of illicit drugs has been very active in Australia and New Zealand and constitute highly co-ordinated partnerships across academia, industry and government stakeholders. In New Zealand, Chappell *et al.* published a large nationwide study covering 71% of the North and South islands in 2019/20.[156] However, fentanyl was only detected at or below the limit of reporting and PNL or consumption data were not presented. The longest-running and regular national-scale WBE program for illicit drug consumption has been performed in Australia since 2016 and is coordinated by the Australian Criminal Intelligence Commission (ACIC). Results have been published every quarter since March, 2017.[157] Before this, Tscharke *et al.* reported development and application of an analytical method to monitor four stimulants, six opioids (including fentanyl) and 11 NPS.[158] The method was tested on approximately 700 wastewater samples from 4 WWTPs every second month from Dec 2011–Dec 2015 in South Australia. Temporally, moderately increasing trends for the consumption of fentanyl were observed from mid-2014. In contrast to US studies, PNLs were estimated to be lower, ranging from 0.3–0.7 mg per 1000 people per day and corresponding to ∼8–17 doses per day per 1000 people. Very little statistical difference in use was observed between days of the week and only ∼1% increase was observed at the weekend. Following this, the ACIC national programme was launched monitoring 12 substances across 58 WWTPs (57% of the Australian population) and has included 9075 samples to date (as of March 2023).[159] Two types of sampling campaigns have been implemented and each is run on an alternating basis every two months. The first campaign includes up to 22 capital sites across all territories. The second campaign includes both the capital sites and up to 36 regional sites. Seven 24-hour composite samples are taken per site in each campaign. As a relatively low throughput 1000-fold SPE step is still required to determine norfentanyl sensitively and reliably in comparison to direct-injection analysis for most of the other illicit drugs monitored, samples from only two days from each campaign have been analysed for this substance. Nonetheless, the data generated over such a long period of time at this frequency and with such high spatial coverage have generated an excellent overview of fentanyl use in Australia. As can be clearly observed in Figure 9.5,

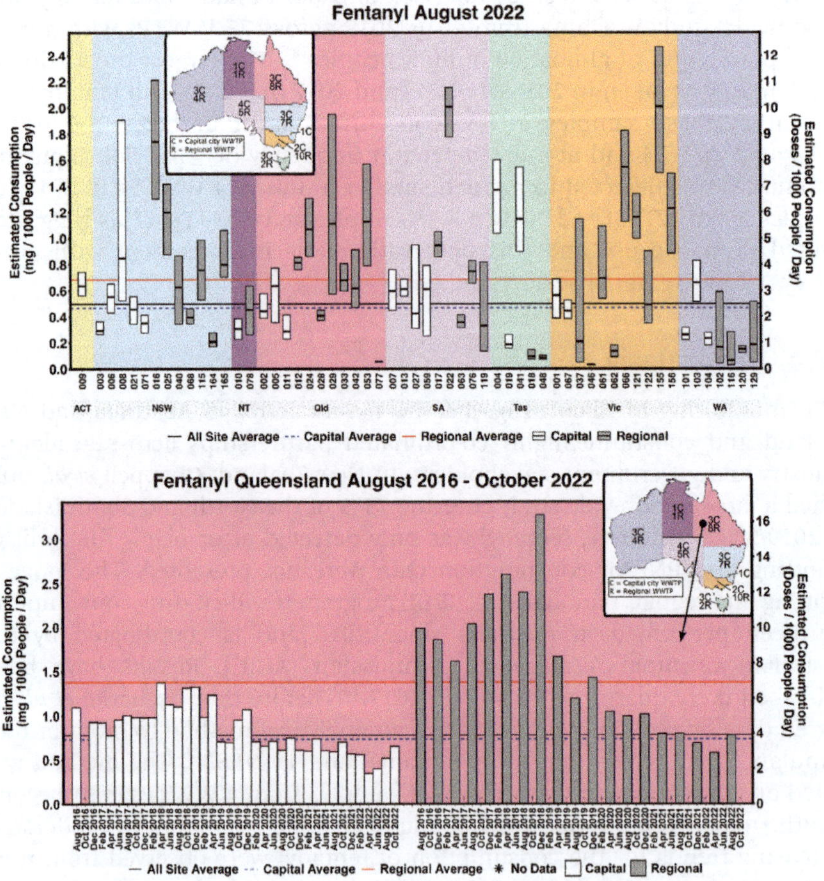

Figure 9.5 National scale WBE of fentanyl as part of the ACIC program in Australia covering 58 WWTPs and 57% of the population. Top: A single month snapshot of fentanyl consumption (left axis) and the number of doses (right axis) across all sites in August 2022. Bottom: Comparison of capital (C, white bars) *versus* rural (R, grey bars) consumption in Queensland from 2016–2022. Inset maps show the number of regional and capital WWTP sites included in each bar plot. Figures produced by and with permission from the University of Queensland and ACIC.

one of the most notable issues is the difference between capital and regional sites with a consumption on average across the full time period (data shown from August 2016–Oct 2022) being roughly double in regional sites. In December 2018, regional average consumption exceeded 3 mg per 1000 people per day, but has continued to decline nationally since then at all sites (capital and regional). By August 2022, differences between both types of sites also narrowed considerably. In its latest report, fentanyl is still detectable with ≥89% and 77% frequency at capital and regional sites, respectively.[139] However, regional sites in South Australia still ranked first nationally for fentanyl consumption, but Tasmania ranked first nationally for capital sites. The value of this coordinated programme over this length of time has allowed ACIC to develop strong academic partnerships to identify critical high-risk markets for illicit materials particularly at regional sites and WBE is being used to aid law enforcement investigations to evaluate the effectiveness of responses and reaction of organised crime activity to particular interventions, including seizures. In addition to fentanyl, it has been feasible to establish correlations of consumption with other substances. Although it is difficult to determine whether this is related to non-prescribed co-consumption or medicinal usage in general, fentanyl consumption tracked well with oxycodone use in some areas, particularly in four territories including New South Wales, Northern Territories, Queensland and Victoria. In contrast, fentanyl and heroin consumption did not track together. Other research by this team has expanded the WBE remit to both opioid and non-opioid analgesics and reported mean defined daily doses (DDDs) to be 0.037 and 0.029 DDD per day per person, respectively, with regional and remote areas indicating a greater burden of pain.[160]

9.7 Conclusion

The sustainable use of a waste material for flexible monitoring of a variety of public health-related indicators is emerging through the development and application of WBE. Fentanyl-related substance WBE monitoring is underpinned by LC–MS based methods for both targeted and untargeted analysis at the low-sub ng L^{-1} concentration level. In general, the elevated potency and lower doses consumed of such substances mean that much higher analytical method sensitivity is needed in comparison to WBE of other illicit drugs, often including pre-concentration factors of several hundred to a thousand-fold achieved using SPE. WBE methods clearly need to evolve to include more fentanyl related analogues, precursors as well as other synthetic opioids and NPSs to keep up with advances in forensic science and human biomonitoring. Nevertheless, several nationwide and international comparison studies have emerged that have successfully demonstrated the application of WBE for fentanyl and some of its related substances. It is clear that WBE has been capable of correctly identifying intercontinental differences in fentanyl-related substance use, indicating that recreational usage has been limited or minimal on a global level. Some encouraging examples also exist that have taken WBE to the next level of continuous high frequency monitoring at the national level. In some cases, these have seen successful engagement with authorities and other stakeholders to translate the findings into impacts within communities to not only identify hotspots of consumption but, also, to enable a unique means to quantify criminal market reactivity to interventions for a variety of substances in near real time. During the SARS-CoV-2 pandemic in 2020, many nations used WBE for monitoring the viral loads in wastewater on a very large scale and the infrastructure investments made during this period offer many opportunities to expand WBE to a meaningful early-warning and public health monitoring mechanism.

Acknowledgements

The authors gratefully acknowledge funding from the National Institute for Health Research (NIHR) Health Protection Research Units in Environmental Exposures and Health, and Chemical and Radiation Threats and Hazards, both partnerships between the UK Health Security Agency and Imperial College London. The views expressed are those of the author(s) and not necessarily those of the NIHR, UK Health Security Agency or the Department of Health and Social Care. Special thanks are extended to Stav Friedman at Imperial College London for draft review and to Ben Tscharke, Jake O'Brien and Kevin Thomas at The University of Queensland for providing data and figures from the ACIC National Wastewater Drug Monitoring Program in Australia.

References

1. P. M. Choi, B. J. Tscharke, E. Donner, J. W. O'Brien, S. C. Grant and S. L. Kaserzon, *et al.*, *TrAC, Trends Anal. Chem.*, 2018, **105**, 453–469.
2. M. Kitajima, W. Ahmed, K. Bibby, A. Carducci, C. P. Gerba and K. A. Hamilton, *et al.*, *Sci. Total Environ.*, 2020, **739**, 139076.
3. M. Lorenzo and Y. Picó, *Curr. Opin. Environ. Sci. Health*, 2019, **9**, 77–84.
4. N. Fontanals, E. Pocurull, F. Borrull and R. M. Marcé, *Curr. Opin. Environ. Sci. Health*, 2019, **9**, 26–33.
5. L. Couchman, D. S. Fisher, K. Subramaniam, S. A. Handley, R. J. Boughtflower and C. M. Benton, *et al.*, *Drug Test. Anal.*, 2018, **10**, 323–329.
6. H. Rapp-Wright, F. Regan, B. White and L. P. Barron, *Sci. Total Environ.*, 2023, **860**, 160379.
7. R. Bade, L. Bijlsma, T. H. Miller, L. P. Barron, J. V. Sancho and F. Hernandez, *Sci. Total Environ.*, 2015, **538**, 934–941.
8. K. T. Ng, H. Rapp-Wright, M. Egli, A. Hartmann, J. C. Steele and J. E. Sosa-Hernández, *et al.*, *J. Hazard. Mater.*, 2020, **398**, 122933.
9. I. González-Mariño, J. A. Baz-Lomba, N. A. Alygizakis, M. J. Andrés-Costa, R. Bade and A. Bannwarth, *et al.*, *Addiction*, 2020, **115**, 109–120.
10. J. Boelrijk, D. van Herwerden, B. Ensing, P. Forré and S. Samanipour, *J. Cheminf.*, 2023, **15**, 28.
11. H. Mohammed Taha, R. Aalizadeh, N. Alygizakis, J. P. Antignac, H. P. H. Arp and R. Bade, *et al.*, *Environ. Sci. Eur.*, 2022, **34**, 104.
12. European Monitoring Centre for D, Drug A, European drug report 2013: trends and developments: Publications Office, 2013.
13. N. Sims and B. Kasprzyk-Hordern, *Environ. Int.*, 2020, **139**, 105689.
14. L. Duan, Y. Zhang, B. Wang, G. Yu, J. Gao and G. Cagnetta, *et al.*, *Water Res.*, 2022, **216**, 118321.
15. A. Galani, N. Alygizakis, R. Aalizadeh, E. Kastritis, M. A. Dimopoulos and N. S. Thomaidis, *Sci. Total Environ.*, 2021, **798**, 149014.

16. L. E. Tomsone, I. Perkons, V. Sukajeva, R. Neilands, K. Kokina and V. Bartkevics, *et al.*, *Water Res.*, 2022, **221**, 118800.
17. J. W. O'Brien and W. D. Hall, *Addiction*, 2019, **114**, 1591–1592.
18. B. Kasprzyk-Hordern, K. Proctor, K. Jagadeesan, L. Lopardo, K. J. O'Daly and R. Standerwick, *et al.*, *Environ. Int.*, 2021, **147**, 106331.
19. N. I. Rousis, E. Zuccato and S. Castiglioni, *Environ. Int.*, 2017, **99**, 213–220.
20. D. A. Devault and S. Karolak, *Environ. Sci. Pollut. Res.*, 2020, **27**, 4695–4702.
21. L. Vera-Herrera, D. Sadutto and Y. Picó, *Molecules*, 2021, **26**, 3688.
22. N. I. Rousis, E. Zuccato and S. Castiglioni, *Sci. Total Environ.*, 2016, **571**, 1349–1357.
23. H. Wang, S. Tang, X. Zhou, R. Gao, Z. Liu and X. Song, *et al.*, *Environ. Res.*, 2022, **204**, 112398.
24. B. Kasprzyk-Hordern, N. Sims, K. Farkas, K. Jagadeesan, K. Proctor and M. J. Wade, *et al.*, *J. Hazard. Mater.*, 2023, **450**, 130989.
25. W. Ahmed, N. Angel, J. Edson, K. Bibby, A. Bivins and J. W. O'Brien, *et al.*, *Sci. Total Environ.*, 2020, **728**, 138764.
26. A. Albastaki, M. Naji, R. Lootah, R. Almeheiri, H. Almulla and I. Almarri, *et al.*, *Sci. Total Environ.*, 2021, **760**, 143350.
27. V. Amin, D. A. Bowes and R. U. Halden, *Sci. Total Environ.*, 2023, **858**, 160103.
28. M. K. Green, A. L. Ciesielski and J. R. Wagner, *Forensic Chem.*, 2020, **19**, 100253.
29. H. Rapp-Wright, G. McEneff, B. Murphy, S. Gamble, R. Morgan and M. Beardah, *et al.*, *J. Hazard. Mater.*, 2017, **329**, 11–21.
30. S. Rodriguez-Mozaz, S. Chamorro, E. Marti, B. Huerta, M. Gros and A. Sànchez-Melsió, *et al.*, *Water Res.*, 2015, **69**, 234–242.
31. J. Wang, L. Chu, L. Wojnárovits and E. Takács, *Sci. Total Environ.*, 2020, **744**, 140997.
32. K. Munro, C. P. B. Martins, M. Loewenthal, S. Comber, D. A. Cowan and L. Pereira, *et al.*, *Sci. Total Environ.*, 2019, **657**, 1099–1111.
33. M. N. F. Shaheen, E. M. Elmahdy and Y. E. Shahein, *J. Water Health*, 2022, **20**, 1212–1222.
34. G. La Rosa, M. Iaconelli, C. Veneri, P. Mancini, G. Bonanno Ferraro and D. Brandtner, *et al.*, *Sci. Total Environ.*, 2022, **837**, 155767.
35. E. Róka, B. Khayer, Z. Kis, L. B. Kovács, E. Schuler and N. Magyar, *et al.*, *Sci. Total Environ.*, 2021, **786**, 147398.
36. M. Rusiñol, I. Zammit, M. Itarte, E. Forés, S. Martínez-Puchol and R. Girones, *et al.*, *Sci. Total Environ.*, 2021, **787**, 147463.
37. A. L. N. van Nuijs, F. Y. Lai, F. Been, M. J. Andres-Costa, L. Barron and J. A. Baz-Lomba, *et al.*, *TrAC, Trends Anal. Chem.*, 2018, **103**, 34–43.
38. Y. Ryu, D. Barceló, L. P. Barron, L. Bijlsma, S. Castiglioni and P. de Voogt, *et al.*, *Sci. Total Environ.*, 2016, **565**, 977–983.
39. R. Bade, L. Bijlsma, J. V. Sancho, J. A. Baz-Lomba, S. Castiglioni and E. Castrignanò, *et al.*, *Chemosphere*, 2017, **168**, 1032–1041.

40. R. Bade, J. M. White, L. Nguyen, B. J. Tscharke, J. F. Mueller and J. W. O'Brien, *et al.*, *Sci. Total Environ.*, 2020, **731**, 139209.
41. European Monitoring Centre for Drugs and Drug Addiction E, Drug-related deaths and mortality in Europe: update from the EMCDDA expert network, Technical report, Publications, Office of the European Union, Luxembourg, 2021.
42. UK Office for National Statistics, Deaths related to drug poisoning in England and Wales: 2021 registrations, 2022.
43. Centers for Disease Control and Prevention, National Centre for Health Statistics, Multiple causes of death 1999–2021 on CDC Wonder Database, 2023.
44. K. Brook, J. Bennett and S. P. Desai, *J. Anesth. Hist.*, 2017, **3**, 50–55.
45. R. B. Raffa, J. V. Pergolizzi Jr, J. A. LeQuang, R. Taylor Jr, N. R. Group and S. Colucci, *et al.*, *J. Clin. Pharm. Ther.*, 2018, **43**, 154–158.
46. K. S. Latta, B. Ginsberg and R. L. Barkin, *Am. J. Ther.*, 2002, **9**, 53–68.
47. P. A. Janssen and N. B. Eddy, *J. Med. Pharm. Chem.*, 1960, **2**, 31–45.
48. P. A. Janssen, A. H. Jageneau, P. J. Demoen, C. Van De Westeringh, A. H. Raeymaekers and M. S. Wouters, *et al.*, *J. Med. Pharm. Chem.*, 1959, **1**, 105–120.
49. T. H. Stanley, T. D. Egan and H. Van Aken, *Anesth. Analg.*, 2008, **106**, 451–462.
50. M. Wilde, S. Pichini, R. Pacifici, A. Tagliabracci, F. P. Busardò and V. Auwärter, *et al.*, *Front. Pharmacol.*, 2019, **10**, 238.
51. The US Centre for Disease Control and Prevention, Understanding the Opioid Overdose Epidemic. Available from: https://www.cdc.gov/opioids/basics/epidemic.html#:~:text=The%20first%20wave%20began%20with,overdose%20deaths%20involving%20heroin4 (accessed 22/05/2023).
52. J. K. O'Donnell, J. Halpin, C. L. Mattson, B. A. Goldberger and R. M. Gladden, *MMWR Morb. Mortal. Wkly Rep.*, 2017, **66**, 1197–1202.
53. D. Ciccarone, *Curr. Opin. Psychiatry*, 2021, **34**, 344–350.
54. C. Wang, N. Lassi, X. Zhang and V. Sharma, *Int. J. Environ. Res. Public Health*, 2022, **19**, 2074.
55. US Department of State, International Narcotics Control Strategy Report, Volume I: Drug and Chemical Control, 2020, https://www.state.gov/2020-international-narcotics-control-strategy-report.
56. Centers for Disease Control and Prevention, Death Rate Maps & Graphs 2022. Available from: https://www.cdc.gov/drugoverdose/deaths/index.html (accessed 25 March 2023).
57. GBD 2017 Disease and Injury Incidence and Prevalence Collaborators, *Lancet*, 2018, **392**, 1789–1858.
58. GBD 2017 Causes of Death Collaborators, *Lancet*, 2018, **392**, 1736–1788.
59. L. Degenhardt, J. Grebely, J. Stone, M. Hickman, P. Vickerman and B. D. L. Marshall, *et al.*, *Lancet*, 2019, **394**, 1560–1579.
60. O. H. Drummer, *Forensic Sci. Res.*, 2019, **4**, 95–110.
61. E. Rab, R. J. Flanagan and S. Hudson, *Forensic Sci. Int.*, 2019, **300**, 13–18.

62. H. S. M. Vahedi, H. Hajebi, E. Vahidi, A. Nejati and M. Saeedi, *World J. Emerg. Med.*, 2019, **10**, 27–32.
63. A. Macmadu, J. J. Carroll, S. E. Hadland, T. C. Green and B. D. L. Marshall, *Addict. Behav.*, 2017, **68**, 35–38.
64. V. C. Ziesenitz, S. K. König, N. S. Mahlke, G. Skopp, W. E. Haefeli and G. Mikus, *J. Clin. Pharmacol.*, 2015, **55**, 708–717.
65. E. J. Kuip, M. L. Zandvliet, S. L. Koolen, R. H. Mathijssen and C. C. van der Rijt, *Br. J. Clin. Pharmacol.*, 2017, **83**, 294–313.
66. P. W. Peng and A. N. Sandler, *Anesthesiology*, 1999, **90**, 576–599.
67. Y. Higashikawa and S. Suzuki, *Forensic Toxicol.*, 2008, **26**, 1–5.
68. F. Schifano, S. Chiappini, J. M. Corkery and A. Guirguis, *Front. Pharmacol.*, 2019, **10**, 46.
69. K. A. Marquardt, R. S. Tharratt and N. A. Musallam, *Ann. Pharmacother.*, 1995, **29**, 969–971.
70. C. S. Reilly, A. J. Wood and M. Wood, *Anaesthesia*, 1985, **40**, 837–843.
71. J. Lötsch, C. Walter, M. J. Parnham, B. G. Oertel and G. Geisslinger, *Clin. Pharmacokinet.*, 2013, **52**, 23–36.
72. D. E. Feierman and J. M. Lasker, *Drug Metab. Dispos.*, 1996, **24**, 932–939.
73. S. Grond, L. Radbruch and K. A. Lehmann, *Clin. Pharmacokinet.*, 2000, **38**, 59–89.
74. D. A. McClain and C. C. Hug Jr, *Clin. Pharm. Ther.*, 1980, **28**, 106–114.
75. W. F. Van Bever, C. J. Niemegeers, K. H. Schellekens and P. A. Janssen, *Arzneimittelforschung*, 1976, **26**, 1548–1551.
76. Z. Gao, P. Li, H. Lin, W. Lin and Y. Ren, *Environ. Sci. Pollut. Res.*, 2023, **30**, 5516–5529.
77. D. R. Baker and B. Kasprzyk-Hordern, *J. Chromatogr. A*, 2011, **1218**, 8036–8059.
78. X. Lin, P. M. Choi, J. Thompson, T. Reeks, R. Verhagen and B. J. Tscharke, *et al.*, *Environ. Sci. Technol.*, 2021, **55**, 7418–7429.
79. I. Krizman-Matasic, P. Kostanjevecki, M. Ahel and S. Terzic, *J. Chromatogr. A*, 2018, **1533**, 102–111.
80. K. B. Palmquist and M. J. Swortwood, *J. Anal. Toxicol.*, 2021, **45**, 870–877.
81. D. R. Baker, L. Barron and B. Kasprzyk-Hordern, *Sci. Total Environ.*, 2014, **487**, 629–641.
82. N. Allibe, N. Fouilhe Sam-Lai, T. Willeman, J-F Jourdil, M. Bartoli and M. Mallaret, *et al.*, *Forensic Toxicol.*, 2019, **37**, 488–495.
83. C. Ort, M. G. Lawrence, J. Reungoat and J. F. Mueller, *Environ. Sci. Technol.*, 2010, **44**, 6289–6296.
84. N. Centazzo, B. M. Frederick, A. Jacox, S. Y. Cheng and M. Concheiro-Guisan, *Forensic Sci. Res.*, 2019, **4**, 152–167.
85. R. Bade, G. Eaglesham, K. M. Shimko and J. Mueller, *Talanta*, 2023, **251**, 123767.
86. D. Gerrity, R. A. Trenholm and S. A. Snyder, *Water Res.*, 2011, **45**, 5399–5411.

87. J. Brett, K. J. Siefried, A. Healey, M. E. Harrod, E. Franklin and M. J. Barratt, *et al.*, *Clin. Toxicol.*, 2022, **60**, 440–445.
88. T. Mackul'ak, P. Brandeburová, A. Grenčíková, I. Bodík, A. V. Staňová and O. Golovko, *et al.*, *Sci. Total Environ.*, 2019, **659**, 326–334.
89. R. Z. Hahn, C. Augusto do Nascimento and R. Linden, *Front. Chem.*, 2021, **9**, 596875.
90. A. Rodayan, S. Afana, P. A. Segura, T. Sultana, C. D. Metcalfe and V. Yargeau, *Environ. Toxicol. Chem.*, 2016, **35**, 843–849.
91. A. G. Shakallis, H. Fallowfield, K. E. Ross and H. Whiley, *Water*, 2022, **14**, 3478.
92. N. Bishop, T. Jones-Lepp, M. Margetts, J. Sykes, D. Alvarez and D. E. Keil, *Sci. Total Environ.*, 2020, **745**, 140697.
93. A. C. Taylor, G. A. Mills, A. Gravell, M. Kerwick and G. R. Fones, *Sci. Total Environ.*, 2021, **787**, 147519.
94. S. Wang, R. Basijokaite, B. L. Murphy, C. A. Kelleher and T. Zeng, *Environ. Sci. Technol.*, 2022, **56**, 16726–16736.
95. A. K. Richardson, M. Chadha, H. Rapp-Wright, G. A. Mills, G. R. Fones and A. Gravell, *et al.*, *Anal. Methods*, 2021, **13**, 595–606.
96. T. H. Miller, J. A. Baz-Lomba, C. Harman, M. J. Reid, S. F. Owen and N. R. Bury, *et al.*, *Environ. Sci. Technol.*, 2016, **50**, 7973–7981.
97. A. Garnier, C. Bancon-Montigny, S. Delpoux, S. Spinelli, M. Avezac and C. Gonzalez, *Talanta*, 2020, **219**, 121316.
98. C. Harman, M. Reid and K. V. Thomas, *Environ. Sci. Technol.*, 2011, **45**, 5676–5682.
99. S. McKay, B. Tscharke, D. Hawker, K. Thompson, J. O'Brien and J. F. Mueller, *et al.*, *Sci. Total Environ.*, 2020, **704**, 135891.
100. C. Schang, N. D. Crosbie, M. Nolan, R. Poon, M. Wang and A. Jex, *et al.*, *Environ. Sci. Technol.*, 2021, **55**, 10432–10441.
101. U. Kalsoom, C. K. Hasan, L. Tedone, C. Desire, F. Li and M. C. Breadmore, *et al.*, *Anal. Chem.*, 2018, **90**, 12081–12089.
102. A. K. Richardson, R. C. Irlam, H. R. Wright, G. A. Mills, G. R. Fones and S. R. Stürzenbaum, *et al.*, *Sci. Total Environ.*, 2022, **839**, 156260.
103. F. Nitti, M. I. G. S. Almeida, R. Morrison, R. W. Cattrall, V. J. Pettigrove and R. A. Coleman, *et al.*, *Microchem. J.*, 2018, **143**, 359–366.
104. A. Rodayan, S. Afana, P. A. Segura, T. Sultana, C. D. Metcalfe and V. Yargeau, *Environ. Toxicol. Chem.*, 2016, **35**, 843–849.
105. G. McEneff, L. Barron, B. Kelleher, B. Paull and B. Quinn, *Sci. Total Environ.*, 2014, **476–477**, 317–326.
106. C. Lacey, G. McMahon, J. Bones, L. Barron, A. Morrissey and J. M. Tobin, *Talanta*, 2008, **75**, 1089–1097.
107. C. E. O'Rourke and B. Subedi, *Environ. Sci. Technol.*, 2020, **54**, 6661–6670.
108. N. Salgueiro-González, E. Zuccato and S. Castiglioni, *Sci. Total Environ.*, 2022, **843**, 156982.
109. E. L. Schymanski, J. Jeon, R. Gulde, K. Fenner, M. Ruff and H. P. Singer, *et al.*, *Environ. Sci. Technol.*, 2014, **48**, 2097–2098.

110. J. Klingberg, B. Keen, A. Cawley, D. Pasin and S. Fu, *Arch. Toxicol.*, 2022, **96**, 949–967.
111. R. Bade, B. J. Tscharke, J. M. White, S. Grant, J. F. Mueller and J. O'Brien, *et al.*, *Sci. Total Environ.*, 2019, **650**, 2181–2187.
112. C. Cruz-Cruz, V. Yargeau, D. Vidaña-Perez, A. Schilmann, M. A. Pineda and M. Lobato, *et al.*, *Int. J. Drug Policy*, 2021, **88**, 103027.
113. M. Massano, A. Salomone, E. Gerace, E. Alladio, M. Vincenti and M. Minella, *J. Chromatogr. A*, 2023, **1693**, 463896.
114. L. Sabater-Liesa, N. Montemurro, A. Ginebreda, D. Barceló, P. Eichhorn and S. Pérez, *J. Hazard. Mater.*, 2021, **408**, 124984.
115. K. Chester, Written testimony to the Senate Committee on the Judiciary, 2019, p. 2, https://wwwjudiciarysenategov/imo/media/doc/Chester%20 Testimonypdf (accessed 11 May 2023).
116. L. P. Barron and G. L. McEneff, *Talanta*, 2016, **147**, 261–270.
117. R. Aalizadeh, N. S. Thomaidis, A. A. Bletsou and P. Gago-Ferrero, *J. Chem. Inf. Model.*, 2016, **56**, 1384–1398.
118. D. Pasin, C. B. Mollerup, B. S. Rasmussen, K. Linnet and P. W. Dalsgaard, *Anal. Chim. Acta*, 2021, **1184**, 339035.
119. F. Vaiano, E. Bertol, M. Mineo, L. Pietrosemoli, J. Rubicondo and C. T. Supuran, *et al.*, Development of a New LC-MS/MS Screening Method for Detection of 120 NPS and 43 Drugs in Blood, *Separations*, 2021, **8**(11), 221.
120. F. Vincenti, C. Montesano, S. Gobbi, M. Sergi, R. Curini and D. Compagnone, *J. Chromatogr. Open*, 2021, **1**, 100006.
121. V. Albergamo, R. Helmus and P. de Voogt, *J. Chromatogr. A*, 2018, **1569**, 53–61.
122. R. Bade, A. Abdelaziz, L. Nguyen, A. J. Pandopulos, J. M. White and C. Gerber, *Talanta*, 2020, **208**, 120479.
123. R. Bade, M. Ghetia, L. Nguyen, B. J. Tscharke, J. M. White and C. Gerber, *MethodsX*, 2019, **6**, 953–960.
124. G. Bertanza and R. Boiocchi, *Sci. Total Environ.*, 2022, **819**, 153236.
125. M. J. Wade, A. Lo Jacomo, E. Armenise, M. R. Brown, J. T. Bunce and G. J. Cameron, *et al.*, *J. Hazard. Mater.*, 2022, **424**, 127456.
126. X. Li, S. Zhang, J. Shi, S. P. Luby and G. Jiang, *Chem. Eng. J.*, 2021, **415**, 129039.
127. Q. Zheng, B. Tscharke, J. O'Brien, C. Gerber, R. Mackie and J. Gao, *et al.*, *Curr. Opin. Environ. Sci. Health*, 2019, **9**, 13–18.
128. B. Subedi, *ACS Symp. Ser.*, 2019, **1319**, 79–98.
129. H. E. Jones, M. Hickman, B. Kasprzyk-Hordern, N. J. Welton, D. R. Baker and A. E. Ades, *Sci. Total Environ.*, 2014, **487**, 642–650.
130. B. J. Tscharke, J. W. O'Brien, C. Ort, S. Grant, C. Gerber and R. Bade, *et al.*, *Environ. Sci. Technol.*, 2019, **53**, 10303–10311.
131. W. Sim, S. Park, J. Ha, D. Kim and J. E. Oh, *Sci. Total Environ.*, 2023, **857**, 159154.
132. K. V. Thomas, A. Amador, J. A. Baz-Lomba and M. Reid, *Environ. Sci. Technol.*, 2017, **51**, 11363–11370.

133. H. Yu, X. T. Shao, S. Y. Liu, W. Pei, X. P. Kong and Z. Wang, *et al.*, *Environ. Geochem. Health*, 2021, **43**, 4627–4635.
134. P. K. Thai, J. W. O'Brien, B. J. Tscharke and J. F. Mueller, *Environ. Sci. Technol. Lett.*, 2019, **6**, 265–269.
135. P. Ramin, A. Libonati Brock, F. Polesel, A. Causanilles, E. Emke and P. de Voogt, *et al.*, *Environ. Sci. Technol.*, 2016, **50**, 13397–13408.
136. F. Y. Lai, C. Ort, C. Gartner, S. Carter, J. Prichard and P. Kirkbride, *et al.*, *Water Res.*, 2011, **45**, 4437–4448.
137. J. Gao, B. J. Tscharke, P. M. Choi, J. W. O'Brien, T. Boogaerts and H. Jiang, *et al.*, *Environ. Sci. Technol.*, 2021, **55**, 7551–7560.
138. M. Huizer, T. L. ter Laak, P. de Voogt and A. P. van Wezel, *Water Res.*, 2021, **207**, 117789.
139. Australian Criminal Intelligence Commission, National Wastewater Drug Monitoring Program – Report 18, 2023. Available from: https://www.acic.gov.au/publications/national-wastewater-drug-monitoring-program-reports/report-18-national-wastewater-drug-monitoring-program.
140. M. Price and S. Trowsdale, *J. Hydrol.*, 2022, **61**, 59–75.
141. EMCDDA, European Monitoring Centre for Drugs and Drug Addiction, Ethical research guidelines for wastewater-based epidemiology and related fields, 2016. Available from: https://www.emcdda.europa.eu/drugs-library/ethical-research-guidelines-wastewater-based-epidemiology-and-related-fields_en.
142. M. R. Boleda, M. T. Galceran and F. Ventura, *J. Chromatogr. A*, 2007, **1175**, 38–48.
143. M. Huerta-Fontela, M. T. Galceran and F. Ventura, *Environ. Sci. Technol.*, 2008, **42**, 6809–6816.
144. F. Ahmed, J. W. O'Brien, A. Keshaviah, W. D. Hall, R. Bade and B. Tscharke, *et al.*, *Nat. Water.*, 2023, 401–404.
145. A. Keshaviah, The Potential of Wastewater Testing for Public Health and Safety (Mathematica Policy Research), 2017. Available from: https://www.mathematica.org/publications/the-potential-of-wastewater-testing-for-public-health-and-safety-special-report.
146. J. Prichard, F. Y. Lai, J. O'Brien, R. Bruno, P. Thai and W. Hall, *et al.*, *Curr. Issues Crim. Justice*, 2018, **29**, 195–208.
147. R. Bade, J. M. White, J. Chen, J. A. Baz-Lomba, F. Been and L. Bijlsma, *et al.*, *Water Res.*, 2021, **193**, 116891.
148. A. J. Gushgari, E. M. Driver, J. C. Steele and R. U. Halden, *J. Hazard. Mater.*, 2018, **359**, 437–444.
149. A. J. Gushgari, A. K. Venkatesan, J. Chen, J. C. Steele and R. U. Halden, *Water Res.*, 2019, **161**, 171–180.
150. A. J. Skees, K. S. Foppe, B. Loganathan and B. Subedi, *Sci. Total Environ.*, 2018, **631–632**, 1457–1464.
151. T. L. Croft, R. A. Huffines, M. Pathak and B. Subedi, *J. Hazard. Mater.*, 2020, **384**, 121306.
152. N. Centazzo, B. M. Frederick, A. Jacox, S. Y. Cheng and M. Concheiro-Guisan, *Forensic Sci. Res.*, 2019, **4**, 152–167.

153. Statistics Canada, Wastewater analysis suggests that consumption of fentanyl, cannabis and methamphetamine increased in the early pandemic period, 2021. Available from: https://www150.statcan.gc.ca/n1/daily-quotidien/210726/dq210726a-eng.htm.
154. P. Du, X. Liu, G. Zhong, Z. Zhou, M. W. Thomes and C. W. Lee, *et al.*, *Int. J. Environ. Res. Public Health*, 2020, **17**, 889.
155. P. Du, Z. Zhou, Z. Wang, Z. Xu, Q. Zheng and X. Li, *et al.*, *Sci. Total Environ.*, 2021, **795**, 148838.
156. A. Chappell, B. Armstrong, E. Jay, K. Phung, S. McCormick and S. Grigg, *et al.*, *Sci. Total Environ.*, 2022, **843**, 156956.
157. Commission ACI, National Wastewater Drug Monitoring Program reports. Available from: https://www.acic.gov.au/publications/national-wastewater-drug-monitoring-program-reports.
158. B. J. Tscharke, C. Chen, J. P. Gerber and J. M. White, *Sci. Total Environ.*, 2016, **565**, 384–391.
159. Commission ACI, National Wastewater Drug Monitoring Program – Report 18, 2023. Available from: https://www.acic.gov.au/sites/default/files/2023-03/NWDMP%20Report%2018_1.PDF.
160. F. Ahmed, B. Tscharke, J. W. O'Brien, W. D. Hall, P. J. Cabot and P. M. Sowa, *et al.*, *Environ. Sci. Technol.*, 2023, **57**, 1712–1720.

CHAPTER 10

Transferable Mass Spectrometry Methods: Examination of Authenticity in Artwork

C. BOUVIER,[a] A. BRUNELLE[b] AND S. VAN NUFFEL*[c]

[a] Scientific Laboratory of the Bibliothèque nationale de France, Bibliothèque nationale de France, Bussy-st-Georges, France; [b] Sorbonne Université, CNRS, Laboratoire d'Archéologie Moléculaire et Structurale (LAMS), Paris, France; [c] M4I, Faculty of Science and Engineering, Maastricht University, Maastricht, The Netherlands

*Email: s.vannuffel@maastrichtuniversity.nl

Abstract

This chapter discusses the use of mass spectrometry for the authentication of artwork and its transferability to forensics. The general process for art authentication is first discussed. An overview of the main mass spectrometry techniques that are currently used for art authentication is provided, along with commentary on the associated sample preparation and the data that can be obtained. Five examples of real authentication cases, where mass spectrometry techniques were used, are reported with the lessons learnt from them. Finally, the potential for the use of these analytical techniques during criminal prosecutions involving art is discussed, while also highlighting the numerous challenges that one may encounter.

New Developments in Mass Spectrometry No. 14
Applications of Mass Spectrometry for the Provision of Forensic Intelligence: State-of-the-art and Perspectives
Edited by Simona Francese and Stephen Bleay
© The Royal Society of Chemistry 2024
Published by the Royal Society of Chemistry, www.rsc.org

10.1 An Introduction to Art Authentication

The general approach to the authentication of art (with particular reference to paintings) generally consists of four steps, namely, (i) signature verification, (ii) review of the historical documentation, (iii) collection of new and/or review of existing scientific evidence and (iv) expert appraisal.

Although signature verification is usually the first step, the practice of the artist signing their work is fairly modern. It started around the 15th century as a signifier of completed work and sometimes details of time, place and medium were added too. It therefore started out as a practice for artists to keep track of their work. Signature verification is a seemingly simple process of reference matching using a loupe or microscope and a reference book with known and verified signatures. However, there are some additional aspects that need to match as well. Some artists have a tendency to sign all of their works in the same location, *e.g.*, lower left corner, cartellino style, hidden under the frame, on the back of the canvas, *etc.* Signatures also evolve over time so they can also be used to help date a work or establish a forgery if the painting style does not match the signature from that period. The materials used should ideally also match known references. Signature forgeries tend to fall under two categories: either an entire painting is a counterfeit or the signature of a more famous artist is added to an existing piece in order to increase its value. In the former case, an important feature to examine is the fluidity of the signature as it is surprisingly difficult to replicate someone else's signature with a smooth motion. In the latter case, the presence of craquelure can also be investigated, because it typically develops after the painting has been signed. In addition, pigments oxidise and fade over time so UV light can be used to check for a notable difference in pigmentation between the signature and the rest of the painting. A floating signature, *i.e.*, one placed on top of the varnish, is also an indication that a forgery has taken place, although some artists have been known to sign their works in this way. A forged signature does not necessarily mean the intent is criminal either. For example, in the case of "Children in the Breakers" (Jozef Israels, 1877), the signature of Bernard Blommers, one of Israels' students, was supposedly added during the Second World War in order to hide the fact that the painting was made by a Jewish artist and prevent it from being destroyed.

If no anomalies are discovered during the signature verification, the next step is an investigation of the provenance. Provenance is the record of ownership of a movable piece of art. In essence, the provenance accounts for the whereabouts of a work from the time of its creation to the present day. When buying from a present-day artist, one will normally receive a certificate of authenticity (COA) signed by the artist that will contain physical details of the painting such as style, subject, signature, materials, dimensions, frame, *etc.* Some artists are known to employ security printing technology for their COAs, similar to those used for banknotes, to prevent forgery. In the case of an older painting, where the artists would be deceased, the buyer will receive a sales receipt from the gallery at which they bought it. Apart from the physical details of the painting, a list of previous owners with supporting documentary proof is expected to be included, as well as any information regarding exhibitions where the work might have been displayed and a bibliography of catalogues it might have featured in. The nearer a work's pedigree approaches an unbroken provenance, the safer the purchase. A truly unbroken provenance is quite rare, but there are works of art that can boast it. One example of this can be the Ecce Homo, painted by the Venetian painter Titian, which he personally gifted to King Charles V in 1548. The painting has remained in Spain since then, becoming a part of the royal collection of the Prado Museum in Madrid in 1821. Documentary proof of ownership can be very varied in format and again needs to be corroborated.

Section 10.3.3 provides an excellent example of how mass spectrometry can be used for historical document authentication. Examples include bills of sale or sale receipts, auction records, wills or diaries, historic photographs, and also documents discussing the work such as inventories or *catalogues* raisonnés. In this regard, the back of a painting can also be very informative because it can feature exhibition marks, dealer stamps, gallery labels, shipping labels, *etc.* The reader can probably guess that there is one period in particular where provenance can be lacking, namely, the Nazi era and the Second World War. Looting, of course, also happened during many prior wars, but the Nazis did it systematically and on a larger scale than ever before.

After an investigation of the signature and provenance, collection of scientific evidence follows. Analytical techniques such as optical microscopy, various spectroscopy methods, X-ray fluorescence, *etc.* have all been successfully used for art research and authentication. Section 10.2 provides an overview of the mass spectrometry techniques used in art research along with commentary regarding sample preparation and modes of operation. Every technique has its own requirements, but the guiding principle is that any technique used needs to be as non-invasive and non-destructive as possible, because of the unique nature of the research specimen. Mass spectrometry is inherently destructive by its definition, but it delivers molecular information and does not require a lot of sample material. In the case of imaging mass spectrometry, it is also possible to sample small cross-sections of paintings from areas that are unseen, *e.g.*, underneath the frame, or areas already damaged and slated for modification during restoration. The decision to use any particular technique is always made following a cost-benefit analysis and the same applies for mass spectrometry. Section 10.3 consists of a collection of historical examples of the use of mass spectrometry to determine authenticity. In general, forgeries are identified by means of anomalies. When investigating the ingredients of a painting, one can, for example, identify anachronistic pigments. Section 10.3.4 tells the story of the famous forger Han van Meegeren who sold a fake Vermeer to Hermann Göring and later had to rely on scientific evidence in court to secure exoneration from charges of treason. He had used Bakelite, invented in 1907, to speed up the hardening of his paint layers, which, of course, should not be present in a 17th century painting. If the forger were to use the exact same materials as the original artist, a large age difference would be observed due to the absence of degradation products. In general, it would be correct to say that scientific evidence can be used to prove that something is not contemporary or that it contains ingredients not recognized to be used by the artist. The claim that a certain piece of art was made at a time by a certain artist can therefore be disproved by scientific evidence, but a positive proof can never be given.

Therefore, when all data have been collected, an expert will appraise the painting or artwork by examining the totality of the evidence. It is important that the expert is at least specialised in the style and period or, better yet, in the particular artist. Art authentication is unquestionably a difficult, multidisciplinary process. Ironically, the best forgers can be considered experts in their subject, rendering the authentication process a battle between specialists. Section 10.4 explores the potential relations between art and crime further and lays out the potential and challenges in the context of forensics.

10.2 Mass Spectrometry Techniques and Instrumentation Used to Investigate Authenticity in Artwork

10.2.1 Investigated Materials and Related Technique Specifications

Analytical science is a valuable tool in the authentication process, provided one understands its capabilities and limitations. To derive meaningful information about the original or non-original nature of a piece of art, an in-depth knowledge of the materials used by the artist must be combined with the assured identification of a given material in a complex mixture. The usual analytical uncertainties and the ambivalent results likely to be encountered in complex historical samples can make for a challenging interpretation of the data. The results are later used by conservation experts with varying knowledge levels of analytical science; as such, the analytical information that is disseminated should be as conclusive as possible in order to prevent any misunderstandings when all the evidence is pooled together.

Since the 1950s, the field of conservation science has exhibited growing interest in using analytical techniques to obtain reliable information regarding the composition, structure, and evolution over time for various historical materials.[1,2] These include structural materials including stone, wood, clay, marble, or substrate materials such as paper, canvas and fabrics, also metals, minerals like pigments or gems, organic dyes, organic materials like oils, resins, egg, glues, wax, used as cohesive binding material, and exogenous pollutants, biological contaminants, pests, and modern variants of these materials. The structure of paintings, one of the most familiar heritage objects, illustrates the diversity of materials employed in artworks. They are built by stacking paint layers, composed of organic binder(s) containing dyes and pigments, over a substrate, *e.g.*, wood and woven canvas made with vegetal fibres, which are then all covered with several coatings of resin-based or modern acrylic varnishes.

Due to their invasive nature, the application of analytical techniques to artwork is in fact a rare opportunity. Studies have highlighted good practices and limitations which should be considered when using analytical techniques to examine artwork, including sample preparation and data treatment. The analysed area should ideally be representative of the whole artwork, but this is seldom possible because of the value of the artwork. Often, sub-millimetre-sized samples are taken from historical objects of several square metres. Once the object is sampled, a combination of invasive or even destructive analytical techniques can be employed, but the excised sample must be kept as intact as possible to ensure that other analyses can still be performed afterwards. If no sample can be taken, for example, when the materials are too fragile, too valuable, or when no consent is given by the curator, only non-invasive techniques can be considered. Obtaining prior consent for sampling is very important. This is illustrated by the collections of the crown of Liechtenstein, none of which are allowed to be lent to exhibitions in France after the events occurring in 2016 during an exhibition in Aix-en-Provence (France), as reported in Section 10.3.1.

These small samples are often also spatially heterogeneous and composed of both organic and inorganic compounds. Mass spectrometry-based techniques have proved to be a powerful tool to identify and characterise many materials, in particular organic substances as shown in the recent literature.[3] The following subsection describes some of the most frequently used mass spectrometry techniques for the examination of authenticity of artworks, although it is important to remember that these are almost never the sole analytical technique employed, as illustrated in Section 10.3.5. Table 10.1 compares specifications for these techniques that are relevant for artwork authentication. Reporting times for these techniques applied to heritage samples are difficult to estimate, as they are not standardised and strongly sample-dependent. Each sample is unique and has its own unique issues of limited immediacy; acquiring and processing the data may take years.

10.2.2 Separation Methods Coupled with Mass Spectrometry

Separation techniques such as gas or liquid chromatography, coupled with mass spectrometry, can identify multiple organic compounds in complex samples. The analysed compounds are carried by a gas or liquid phase and separated based on their relative affinity to a stationary phase, before being identified by mass spectrometry.

Table 10.1 Commonly used MS techniques with their relevant features for artwork analysis.

	GC/MS	Py–GC/MS	LC/MS	AMS	MALDI-MS	SIMS
Sample type	Solid, liquid Microgram	Solid, liquid Microgram	Solid, liquid Microgram	Solid Milligram	Solid, liquid Microgram	Solid >mm
Sample preparation	Extraction Derivatization	None	Extraction Derivatization	Purification Selective extraction and treatments	Matrix coating Crystallised	Flat clean surface
Analysed compound type	Separated by polarity or T in inert gas	Pyrolysed + separated polarity or T in inert gas	Separated by polarity or T in solvent	^{14}C isolated from other isotopes	Desorbed ions protected by a matrix	All ions formed directly on the surface
Adapted for	Light and volatiles compounds	Large range of weights, including natural and synthetic polymers	Large range of weights, soluble compounds, proteins, polymers	Dating small samples of organic materials	Large bio-molecules: proteins, peptides, DNA, carbo-hydrates	Identifying and mapping inorganics and organics
Maximum mass range (Da)	<1000	<1000	<2000	—	<10 000	<1000
Prevalence in the field literature	++	+++	+	–	+	–

Gas chromatography–mass spectrometry (GC/MS) and its coupling with pyrolysis (Py–GC/MS) stands out as a widely used technique when analysing organic materials in historical artefacts. Its application has yielded a large amount of knowledge acquired over the last decades regarding the chemical composition, chemical structure and behaviour of organics found in art materials.[2,4,5] Both techniques can identify oils, carbohydrates, waxes, resins, synthetic paint and polymers, dyes, plant material, and even proteins.[3] These instruments are characterised by high mass resolution, sensitivity, reproducibility, speed and ease of operation. However, GC/MS requires extraction and derivatization steps, while Py–GC/MS allows one to analyse solid samples directly with minimal preparation. Liquid chromatography–mass spectrometry (LC/MS) offers similar advantages although controlling the liquid phase parameters adds more complexity. It also requires sample-specific preparation steps to analyse historical proteins, dyes, or plant extracts, and to characterise materials of natural origin.[6,7]

Because of the limited availability of historical samples and the variety of mixed compounds, GC/MS and LC/MS sample preparation steps have been specifically optimised to gather comprehensive data about important organic materials for heritage samples. In particular, to further characterise historical proteins, protocols have been developed for the identification of a selection of proteins of interest in their chemical environment, pigments, for instance, while considering their degradation state.[8–10]

In the case of paintings, to access the individual composition of layers with thicknesses of a few tens of micrometres, these layers must first be mechanically separated. A careful sampling procedure with a blade under a microscope ensures the collection of only the target material without residue from the surrounding layers. In this manner, localised chemical information can be obtained with sensitive techniques.

In addition to the molecular identification of the organic materials, these three techniques (GC/MS, Py–GC/MS and LC/MS) also yield quantitative information, which is relevant when investigating artistic techniques for authenticity research.[11] The relative quantity of a specific compound, combined with molecular and structural information, can help to trace back the original materials. Dyes or resins, for instance, contain molecules in different proportions depending on their nature and geographical origin. In the case of binding media (oils, resin, or glues), GC/MS and LC/MS data have been correlated with preparation processes, ageing and state of degradation.[12] In paintings, these results can be reviewed and compared with historical facts about the alleged artist's working technique, along with the alleged age of the binder and other observations made on the painting surface like the craquelure (cracking pattern) induced by the ageing process.

In summary, separation methods coupled with mass spectrometry are routinely used to achieve a good understanding of the organics found in artworks, specifically regarding their nature, state of degradation and relative proportions.

10.2.3 Accelerator Mass Spectrometry (AMS): Determining the Age of the Materials

Even with the necessary materials and expertise to imitate the technique of a past artist, hiding anachronism and harmonising the ages of each component found in the structure are difficult issues to tackle when forging an artwork. Therefore, dating techniques are often involved in authenticity investigations. An inconsistency in the ages of materials, supposed to have been prepared at the same date, offer incriminating evidence or clarifying information regarding the artist(s) hand(s). If a part of an artefact has a different look or composition from the rest of it, authenticity is doubted. This also occurs if the whole object has a different look from an expert eye compared to what is expected, given its alleged nature. The dating information helps to distinguish between several possibilities: a modern forgery, if a part is completely anachronistic, or a copy made by another workshop member at the time, or by another painter (or historical forger) much later, the latter's designation as "copy" or "forgery" being debatable.

Material analysis can help identify a specific ingredient that can be delimited in time, or that comes from a specific place.[13] This indirect way of placing the artwork in historical context is limited to time periods and relies on the known documentation about the use of artistic materials. Moreover, it cannot be excluded that a natural pigment from a specific period is retrieved and used in a forgery.

Simulating the age of historical organic material proves to be more challenging. Binders in paintings must be applied fresh and then cured. Whether the binder is a triglyceride-rich oil, triterpene resin, or a proteinaceous material, the processes through which these substances harden, dry and then age are chemically complex, leading to changes in their properties, optical behaviour and composition over time, often over centuries. Consequently, a binder applied in the 20th century cannot possibly have been taken from a historical painting and reused. Even if there are suspicions of artificial ageing through observations and chemical composition results, this will not provide the actual date at which the object was created, which is relevant information when investigating the source of a forgery. Radiocarbon ^{14}C (radiometric) dating can address this question, as it can provide absolute dates, even for objects realised in the 20th century. Calibrations according to specific conventions have been defined to homogenise the dating results over years.[14]

The sampling step is a crucial point, because the method is sensitive to any carbon-containing contaminant. Often for artworks, the underlying support frame has been used for dating purposes because of the large amount of matter needed for radiometric dating. An order of magnitude of grams was required until the 1980s. However, even if the artwork support frame has the alleged age, it could still have been taken from non-famous artworks of the correct period and re-used after removing other materials.

The support can therefore be older than the art composition on top of it, or younger in the case where a major undocumented restoration (like relining, moving of frescos, or reweaving) took place.

Identifying the age of smaller parts of an artwork has been a challenge over the last few decades. A breakthrough in the dating technique occurred around 1980. By separating ^{14}C from more abundant isobaric interferences (^{14}N or $^{12}CH_2$), thanks to high energy acceleration before mass detection, accelerator mass spectrometry (AMS) can detect low ^{14}C concentrations and $^{14}C/^{12}C$ ratios down to 10^{-15}.[15] AMS therefore requires much smaller quantities of material than radiometry, down to micrograms, which overcomes the limitation of radiocarbon dating for small samples.[1,14]

As described in the recent literature,[16,17] sample preparation for AMS is a challenging step to ensure that the measured proportion of ^{14}C is only correlated to the materials of interest. It involves identifying and removing sources of foreign carbon contaminants with different ^{14}C ratios, and then isolating the specific carbon fraction that represents the age of interest in the given sample. This depends on the characteristics of the original environment of creation for each material and on the entirety of its storage history. In the context of heterogeneous historical samples that enclose several diverse sources of carbon (like carbonates, carbohydrates, wax, varnishes, *etc.*), sample treatment must be adapted to accommodate selective extractions and washings. The obtained treated sample is ready for AMS analysis after combustion and/or graphitization.[18]

By combining the accuracy of the ^{14}C dating method with the increasingly smaller quantity of the required sample, dating paint layers separately becomes possible.[19] More precise time windows for the age of a painting, as well as for diverse types of artefacts can thus be obtained. Nowadays, determining the ages of various binder/pigment mixtures is an active field of research.[20]

10.2.4 MALDI-MS: Characterising Common Biomolecules and Polymers in Artworks

The principle of matrix-assisted laser desorption/ionisation mass spectrometry (MALDI-MS) relies on laser generated ions in the gas phase from a sample coated with a matrix, which aids the intact desorption and ionisation of molecules. The main advantage of MALDI-MS is the possibility to produce molecular ions and heavier fingerprint fragments for larger molecules, allowing for the identification of various polymers, triglycerides, phospholipids, carbohydrates, or proteins, all found in artworks. MALDI-MS has been widely used to characterise heritage materials, making it a valuable technique that can access detailed information about proteins and historical or modern polymers.[21,22]

MALDI-MS requires sample treatment and preparation, as the fraction of interest must be extracted with a procedure depending on its nature and chemical environment. The residue is mixed with a solution of the chosen matrix compound and deposited on an analysis plate where they crystallise together. Matrix molecules will be present in large excess, surrounding the analytes, thus protecting them from the direct impact of the laser and preventing fragmentation by absorbing the laser energy. The quality of spectral information will largely depend on the quality of the matrix–analyte co-crystallisation.

MALDI-MS imaging (MALDI-MSI) is a powerful technique combining the abovementioned detection/identification possibilities with surface imaging with a spatial resolution of around 10 μm. It is rarely applied on heritage samples, due to the necessary alteration of the matrix deposit and because the spatial resolution is often above the dimensions of interest in art samples, like paint layers. Recent developments push limits of resolution down to 5 μm, even with non-flat samples, providing MALDI-MSI with even more assets to analyse heritage samples.[23]

10.2.5 SIMS: Reaching the Spatial Chemical Information

The microscopic scale impacts the composition and structure of the whole object. When investigating artworks, small samples can be used in combination with GC/MS and AMS to provide somewhat localised data, but mass spectrometry imaging can chemically map a sample surface.

Secondary ion mass spectrometry (SIMS), more specifically time-of-flight secondary ion mass spectrometry (ToF-SIMS), has become an increasingly attractive technique to study heritage samples over the last few decades. The advantages of this surface imaging technique are the simultaneous detection of both molecular and inorganic compounds, a high sensitivity for all elements, and retention of the spatial information on the surface at the micrometric scale. The surface is impacted by a focussed beam of high energy (*e.g.*, 25 keV) polyatomic primary ions, and the produced secondary ions are identified according to their mass to charge ratio (m/z) using a time-of-flight mass analyser. Spatially resolved data are particularly relevant when analysing sub-millimetric cross-sectional samples taken from heritage samples, which are composed of a wide variety of complex heterogeneous materials. Current ToF-SIMS instrumentation also combines high mass resolution with high spatial resolution, thanks to delayed extraction.[24] The use of ToF-SIMS instrumentation to analyse heritage samples has been exhaustively described in a paper written by authors of this book chapter.[25] The limitations, and requirements of its application to historical, heterogenous, and unique samples are discussed with a special emphasis on instrument parameter optimization.[26]

Secondary ions produced from large molecules during a ToF-SIMS analysis are fragmented, limiting their unequivocal identification. MeV-SIMS, a modern version of the plasma desorption mass spectrometry (PDMS) technique developed during the 1970s and through the 1990s, allows ionisation with a MeV primary ion beam, yielding equivalent secondary ions.[27–29]

The number of MeV-SIMS applications in the context of artwork characterization has been increasing in recent years.[3]

The growing shared knowledge about the secondary ions expected for a given material at a given state of degradation will enable ToF-SIMS to become a more mainstream analysis technique. Having a dataset that contains the spatial distribution of pigments and associated molecular information about organics is very valuable to better understand how a painting was made, has aged, and has been preserved. In the case of artwork authentication, it offers an extensive overview of each part of the sample's microstructure at once.

ToF-SIMS and MeV-SIMS both have simple sample preparation procedures. The challenge is to obtain a surface that is flat and free from contamination. Several techniques can be applied on heritage samples, depending on the sample characteristics and the question to be addressed. Polishing or microtome cutting are the most used sample preparation techniques. Samples must have small dimensions and be in the solid state or be very viscous like a solidifying oil film.

Mass spectrometry imaging provides a deeper understanding of the identified chemical's distribution on the surface or across the material's bulk, down to the micrometric scale, including the localisation of degradation products, informing about the relative states of preservation. In the examination of authenticity of artwork, the spatial resolution, the high sensitivity and mass range, and the possibility to identify both inorganic and organic materials make ToF-SIMS imaging a powerful tool to detect, characterise, and localise incriminating impurities trapped inside a sample taken from a questioned artwork. As previously stated, to avoid hasty conclusions, a thorough examination and evaluation involving multidisciplinary expertise is then needed to define a possible origin for identified compounds.

10.3 Historical Examples of the Use of Mass Spectrometry in Art Investigation

The following section presents examples of authentication and legal cases where the application of mass spectrometry played a role. The interested reader can of course find other cases in the literature.[30,31] Investigations can take years or decades and are very complex so mass spectrometry sometimes covers only a small part of the investigation as a whole.[15,32]

10.3.1 A Complex Investigation Process: An Instructive Recent Case

Continuing advances in analytical science help tackle authentication problems, but the latest analytical capabilities are unlikely to be applied for every instance of artwork exchange. In parallel, available and routinely used analytical techniques become easier to deceive by educated forgers. Various experts may also deliver outright contradictory results, while analytical investigation can yield inconclusive data. This partly explains why identifying a forgery can take decades in some cases before reaching a formal and final conclusion.

This still-open case, further outlined in a recent book,[33] illustrates today's authentication process. It involves a masterpiece attributed to Cranach that has to date been proved to come from a series of alleged forgeries sold by art dealer Giuliano Ruffini. When the painting was first sold, both provenance and signature were sufficient to deliver an authentication certificate, even if some experts noted anomalies and expressed circumspection, suggesting the need for additional dating tests. Therefore, from the beginning of the validation process, there were some doubts concerning authenticity. Earlier, a reputable auction house was presented with the painting; however due to the presence of several questionable elements raising suspicion of an anomaly, the auction house demanded scientific examination before any transaction could take place. The investigation highlighted enough troubling facts to dissuade the house from putting it up for sale. At a further stage, imaging studies were suggested to examine the sub-layers and drawings, but these were never performed. Hence, the painting continued its journey supported by the positive assessment of renowned experts, until it was eventually acquired as part of the Royal Collection of the Crown of Liechtenstein. A second auction house, Sotheby's, eventually analysed the painting and, interestingly, the team stated that if they had not specifically sought to check if the painting was fake, they would have missed the analytical information. This case shows that the usual approach can prove to be insufficient, or that, if the artwork is presumed to be genuine, bias causes one to miss what reveals the forgery.

After forgery was suspected, the painting was taken away for analysis. Multiple analytical techniques were applied by both parties involved during the comprehensive additional investigations, either attempting to prove forgery or authenticity. At one point, GC/MS and FT-IR analysis of the binders from an alleged Brueghel also sold by Ruffini, showed the presence of 20th century materials in serval layers, such as polyurethane resin and acrylic paint.[33] One of the key points on which expertise strayed was dating the wooden panel below the painting. Dendrochronology studies led to a set of hypotheses ranging from 1550 to post-1980, keeping in mind that nothing proves a painting has the same age as its support. Because the collection owner prohibited further sampling, ^{14}C dating was not performed either, which could have allowed discrimination between an authentic past copy and a forgery (see Table 10.2). The disputed investigation so far concluded forgery, stating that the alleged Cranach is *"a fake intended to deceive, executed in the manner of Cranach, purposely using materials from the period and subjected to artificial ageing"*.[33]

Among the unusual features observed on the painting, lead sulphate protrusions were identified, a phenomenon known to be found in oil paintings although under precise conditions yet to be understood. Mass spectrometry has already been applied in the context of genuine painting examination studies aiming at elucidating the composition and structure of similar lead protrusions on paintings, and ToF-SIMS is especially suited for

Table 10.2 Short classification of situations in which the artwork was not executed by the person credited for it.

			Attribution correct: C; mistake: M		Further illustrative
Qualification	Intentional act	Signature	Seller	Expertise	ref.
Properly attributed original	None	Lost	None/C	C	ref. 51
		Original	None/C	C	ref. 42
Misattributed original		Lost	None/M	M	ref. 53
Original with past false authentication	Signature swap	Past alteration	None/M	M	ref. 36 p. 29 ref. 39
Misattributed original workshop copy	Replica	Master only	Partial/M	Partial	ref. 36 p. 34
Properly attributed copy	*None*	Copyist	None/C	C	ref. 37
Misattributed past copy		None	None/M	M	ref. 36 p. 34
Copy knowingly sold as original	Misattribution, artificial ageing	Forged	Falsified	M	ref. 36 p. 34
Past exposed forgery	Forgery	Past alteration	None/C	C	ref. 36 ref. 42
Past forgery		Past alteration	None/M	M	—
Exposed forgery		Forged	Falsified	C	ref. 33 and 38
Forgery		Forged	Falsified	M	—

investigating these organic–inorganic features and their distribution in the painting stratigraphy.[34] If a link could be established between the facies of these protrusions and their conditions of formation, this would help in detecting forgeries; in such cases, SIMS would be a valuable tool in this endeavour, given that microsamples are available.[35]

This illustrates how an exchange of all examination data between analytical scientists, art experts and authorities is needed for authentication investigations to be conclusive, and likewise, a difficult exchange process not only slows but also negatively affects the investigation by increasing the expert disagreement on which the fraud relies. Crooks on the art market are aware of how to take advantage of this system. This case also illustrates that analytical techniques like mass spectrometry bring formal results in authentication studies. To this end, fitting the techniques to the constraints of the artwork is mandatory to provide robust data. This case shows the need for developing these techniques as robust tools for routine art authentication.

10.3.2 Forgery or Copy? How can Analytical Science Disentangle the Intention of the Executive Hand?

The distinction between a forgery and a copy is discussed in the literature.[31] The intention defines the fraud of the forgeries, and Table 10.2 classifies some situations. These include instances such as a Raphael artwork decreed by the Pope to be given by one ruling family to another, which the former secretly refused and instead sent a copy, commissioned to Andrea del Sarto.[36] Hence, today two museums genuinely believe that they have an authentic Raphael piece, while one of them is actually a copy made by another Old Master. Fake items found clients in art enthusiasts as early as the 1st century. Nowadays, the buyer is sometimes aware of the fake, for example, when artists paint and sell "genuine fakes",[37] copies of masterpieces signed with their name, which are in fact reproductions. It is also important to isolate the case of replica(s) of one painting made in a single workshop during the same period, which had a commercial objective. Conversely, trying to sell an artwork that purposely matches the expectations of scholars eager to find a lost artefact or masterpiece constitutes an intention to deceive. Interestingly, a few famous artists experimented with forgery in their early days, and some, like Michelangelo (1475–1564), even received further commissions from the duped buyer, who was impressed by his skills. The legal consequences for forgers are equally diverse and rarely very severe, even today. After being exposed in the late 1980s, Eric Hebborn edited a few books and appeared in televised programs, as did John Myatt who later even helped the authorities in art fraud investigations.[37,38] Nowadays, the German forger Beltracchi regularly sets up successful exhibitions under his own name. Some forgers also reached such fame that their exposed forgeries get high prices on the market, so much so that forged forgeries are suspected to exist.

Proving the fraud in judicial cases requires finding and demonstrating who wanted to sell something while knowing it was not authentic. Finding the person who executed the forgery is not enough, because copying something is not illegal if no signature (monogram, watermark, *etc.*) is forged. In addition, someone else might have added the signature later (see Table 10.2).

Therefore, a careful characterization of signatures is important and benefits from gathering all available knowledge collected throughout the artwork's conservation history. A study related in the National Gallery Technical Bulletin (Vol. 21, 2000) illustrates how long-term expertise combined with analytical science provided decisive information about the genuineness of the signature on a painting.[39] The Madonna with the Iris (National Gallery, London) was attributed to Albrecht Dürer (1471–1528), with both date "1508" and signature "AD" visible. The signature was not mentioned in a thorough 19th century description in which the date was mentioned. The painting also has two sister versions without date or signature. Hence, its attribution has been questioned since 1945. Hypotheses are that either Dürer, or several hands from his workshop with or without his intervention, or a later copyist of Dürer executed it. The latter two options are supported by many motifs inspired from other Dürer's productions, variable under-drawing details, and the state of preservation depending on the area of the picture.

The "AD" inscription was declared fake in 1958, thanks to visual examination. The paint's look and texture distinguish it from the surrounding area and it is visibly above a later overpaint applied over scratched paint. In 2000, the analysis techniques that can be applied to paintings had evolved and the wooden support and pigments were proved to align with those of the 16th century. Microsamples taken around 1958 were still available, and the isolated overpaint layer under the signature could be analysed by both GC/MS and Py–GC/MS. It contained Manila copal, a resin never mentioned in Europe before the second part of the 18th century. Hence, the signature must have been added after 1750, more than two hundred years after Dürer's death, replacing a scratched signature that is now lost. This was perhaps a deceitful 19th century effort to increase the price of the painting. The hypothesis of other forged parts is excluded, as the paint of the date "1508" contains pigments used before 1750, suggesting authenticity. It was added, likely after the painting was finished, above a slow-drying varnish layer, which is original according to the Py–GC/MS identification. Hence, the artwork must be a genuine production of Dürer's workshop, executed by several hands at different times over a few years. Only the signature was forged later.

10.3.3 Historical Document Authentication with High-resolution Mass Spectrometry

Forging historical or legal paper documents has been a long-standing practice, motivated by many reasons. Forged documents sometimes had historical implications, like the text entitled "Donation of Constantine", originating allegedly from the 4th century and employed until the 15th century to defend the Pope's influence in Europe, until it was exposed as a forgery of unknown origin by a scholar around 1440.[36] The alteration of existing historical records was also illustrated in the context of art history by the tried method of trading forgery by altering archives to establish a false provenance. It indeed raises suspicion when one tries to sell a masterpiece by a renowned artist, which has no trace whatsoever in the past. John Drewe, a swindler, did this to sell the fakes he obtained from an unsuccessful painter named John Myatt, and the forged documents he executed flooded the archives of esteemed museums with many remaining undiscovered today, casting doubt on all that is mentioned.[36,40]

The motivation can also be direct profit by trading handwritten documents or drawings supposedly from famous artists or historical figures. Eric Hebborn was particularly gifted at this trick, as he sold a significant number of drawings inspired by existing artworks, which he let professional experts misattribute to genuine preparatory drawings by renowned historical artists, even though they were all his own work. He was eventually exposed because a similar touch (his own) was noticed across the drawings he "discovered" from supposedly different artists. Because he made a point to use historical materials, scientific investigations at the time were unable to reveal the forgeries in the 1970s, perhaps because MS techniques were not yet available for such samples.[36,38] Paper manuscripts executed with ink, charcoal or watercolour are historical artefacts whose nature complicates the authentication process. This was particularly well illustrated by the case of stolen and forged historical books involving Marino Massimo De Caro, the former director of the Biblioteca dei Girolamini in Naples. Only after he was exposed, it was proven that the manuscript edition of the *Sidereus Nuncius*, a text of major importance for the history of science, in which Galileo Galilei drew the moon for the first time, was a fake, just as it was about to be sold for several million euros.[41]

It requires an equally vast expertise to either forge or authenticate paper documents, which highlights the importance of the new analytical advances in the studies related to the actions of Alexander Holland Smith (1859–1913), who forged many writings from famous hands, including Robert Burns, the world-renowned Scottish poet (1759–1796). This has been known since the end of the 19th century, but most of his work had been dispatched and eluded the scrutiny of experts for over a century. Detected forgeries like those of Smith provide an opportunity to assess the ability of advanced analytical techniques to distinguish fakes from originals. A team from the University of Glasgow recently published an open study presenting an innovative mass spectrometry method efficiently distinguishing original Burns writings from Smith's forgeries.[42] They used a direct infusion mass spectrometry (DIMS) setup, which allows *in situ* minimally destructive sampling. Liquid extraction was performed in triplicates directly from the manuscript's surface with 2 μL of solvent using a pipetting tip hovering over the paper surface. The 8 μL samples were then transferred to a setup equipped with microfluidics chips (Advion Triversa nanomate), allowing analytes to be sent to an Orbitrap mass spectrometer (Thermo Fusion Orbitrap) through direct infusion nanospray with an injection time below 50 ms. Quadrupole isolation was used to collect fragmentation data at a HCD collision energy of 60% with a mass resolution of 15 000. This technique combines relatively high mass resolution, and an atmospheric working pressure. Their approach stands out through its simplicity in comparison to the usual ones involving LC/MS or MALDI-MS. They were able to isolate a selection of discriminatory peaks expressing significant chemical differences between the manuscripts (ink and paper) of the two writers (Figure 10.1). Specific ions associated with the forger Alexander Smith were identified as a plant metabolite ($C_{11}H_{10}O_4^+$, *m*/*z* 207.0654), bis(2-methoxyethyl) phthalate ($C_{14}H_{18}O_6^+$, *m*/*z* 283.1178), triphenyl phosphate ($C_{18}H_{15}O_4P^+$, *m*/*z* 327.0782) and $C_{14}H_{12}N_4O_5S^+$, *m*/*z* 349.0602. Specific ions associated with Robert Burns were citric acid adduct ($[C_6H_8O_7+K]^+$,

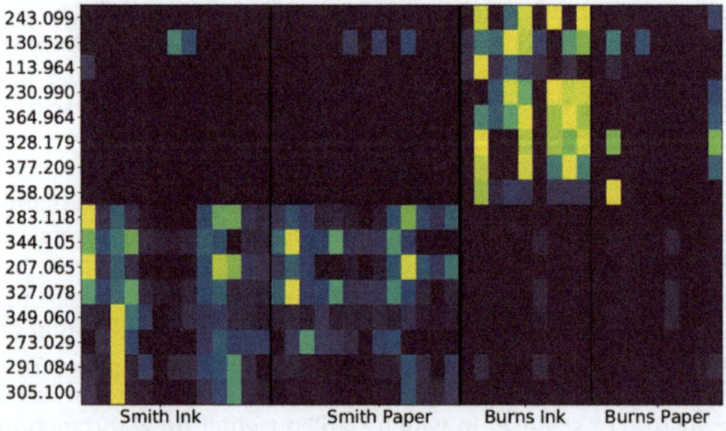

Figure 10.1 Heatmap of features found between Burns inks and paper and Smith's inks and paper. Reproduced from ref. 42, https://doi.org/10.1038/s41598-018-28810-2, under the terms of the CC BY 4.0 licence http://creativecommons.org/licenses/by/4.0/.

m/*z* 230.9904), $C_{11}H_{10}N_6O^+$, *m*/*z* 243.0993, $C_{14}H_{25}N_5O_2S^+$, *m*/*z* 328.1794 and $C_{21}H_{24}N_6O^+$, *m*/*z* 377.2091.

Most of the peaks associated with Smith's work were linked to the use of infused tea to artificially age the paper, a technique also used by abovementioned Hebborn,[38] and which is likely to be involved in many forged paper cases. According to the authors, their untargeted data analysis to compare both ink and paper opens the path to a wider range of applications in authentication of documents, or even a quick test available for buyer's experts themselves.

Mass spectrometry techniques provide solutions in the field of paper authentication research, where the stakes can be significant. We can also cite the contribution of ToF-SIMS in the characterization of banknote morphology,[43] as well as the development of proteomic analysis to further decipher the provenance and authorship of ancient documents and manuscripts through microbial, peptide and protein remains.[44] This could even help highlight the role of women in the production of illuminated manuscripts. Advances and perspectives of this proteomic approach have recently been described in the literature.[45] LC/MS, coupled with ESI and more recently with ion mobility spectrometry (IMS), constitutes one of the most explored paths, knowing that for ancient materials, nano-LC/MS presents the advantage of allowing a better sensitivity and peak separation than other separation setups. Thanks to the improved sensitivity of mass spectrometry, minimally invasive sampling methods are developed in parallel, for instance, with erasers or gels, to provide sufficient uncontaminated analytes. Approaches relying on MALDI ionisation are also explored for situations requiring limited fragmentation of the analytes.

10.3.4 Art Authenticity and Conflicts: Looted Art and Exonerating Forgery

Cultural sites and historical artefacts have highly symbolic significance and are among the first targets of war, while art trafficking is a significant resource for armed or criminal groups.[36] Artwork authentication thus goes beyond academic debates, as it tackles a challenge of international significance that is tracing artefacts looted from archaeological sites or during wars. In modern history, a vast amount of European art was stolen from personal owners, collectors, or museums by the Nazis during World War II. Such artworks entered the art market without apparent evidence of their criminal acquisition. Hence, verifications are routinely carried out when art auction houses intend to sell or buy.[46] Whenever a looted artwork is found, even unknowingly bought, it must be returned to its rightful owner. For this reason, the investigation focuses not only on the authenticity of the artwork, but also on the documentation supporting its provenance. Beyond conflicts, forging archives is also an effective strategy employed by forgers, as further detailed in Section 10.3.3.

The fate of the long career of Dutch forger Han van Meegeren (1889–1947) illustrates the importance of losing an artwork regarded as a national treasure. As most art forgers, he was thoroughly trained in old techniques and started out as a genuine painter. In the 1920s, he was no longer popular with the art critics and had to switch to painting restoration for a living, before eventually turning to forgery. He took advantage of the enthusiasm of scholars to find referenced yet lost artworks. He executed original work in the manner of Vermeer, for example, using historical materials, then subjecting it to artificial ageing, and finally having it authenticated by renowned experts who eagerly saw their hypotheses confirmed. His works were convincing enough to evade detection by the authentication techniques at the time, of which he had full knowledge. He developed a lucrative and prolific business from a workshop hidden in the basement of his villa. He painted a series of forgeries, most of them were attributed to Vermeer, and earned millions from their trade and sale. His success culminated during the Second World War.

In the aftermath of the German occupation in the summer of 1945, his case demonstrates the leniency of public opinion and the justice system towards forgers, who can even gain a favorable reputation based on how their motives are interpreted. When allies found Göring's war treasure in May 1945, "*The woman taken in adultery*" attributed to Vermeer was recovered, and an investigation to find out who sold a Dutch national treasure to the enemy led to Van Meegeren. He had lived a luxurious life during the war (he made about $2 million with various sales), and started out rather unpopular with the public, facing trial for treason and risking the death penalty. Even after stating that the painting was exchanged for over a hundred Dutch masterpieces looted by the Nazis, he was still guilty of surrendering a national treasure to the enemy. He had no choice but to reveal his long-kept secret that the painting was his own creation and a fake. Ironically, his work was so convincing that nobody believed him, so he had an idea to prove that he did it that went down in art forgery history. In front of witnesses, he painted a "brand-new" Vermeer in his cell that was convincing enough to motivate a full investigation of the Vermeer paintings he had sold.[47] He suddenly became the remarkable man who tricked Göring, even if this sale was almost incidental,[47] and was convicted of mere fraud in 1947 and received a very short prison sentence. He passed away before serving it. The interested reader may find in the literature further discussion about this case and its implications for authenticity research.[31,36]

Seven international experts during the physical and chemical examination of paintings gave their conclusions in 1947 after a series of invasive and destructive chemical tests performed on about 30 micro-samples taken from more than ten of his paintings, to reconstruct the working technique layer by layer. These trial experts revealed the use of an anachronistic material, a 20th century phenol formaldehyde resin called Bakelite, also found in Van Meegeren's workshop, recognized with chemical tests (Schiff reagent for aldehyde identification; red colour characteristic for phenolics after potash and diazo-*p*-nitroaniline addition; and blue colour after addition of 1% vanillin in sulfuric acid).[47] It has the property, verified by reconstructions made by experts, of hardening the paint when heated in an oven for a few hours, so that its surface exhibits the texture of a properly cured historical paint. Van Meegeren experimented a lot to achieve this result in his "alchemist's workshop",[36] drawing upon his knowledge of chemistry and paint examination. This not only proved that the Vermeer was not authentic but also that he was the forger. Following this case, described as a *"example of science applied in the service of art"*, experts insisted on the urgent need of applying physical and chemical sciences where authenticity is in doubt.[47] In the decades following his death, analytical science applied to the investigation of paintings improved, allowing the execution of new analyses (including MS) to seek definitive proof that the alleged Vermeers were forgeries, something not easily accepted by some owners who considered the existing analytical evidence insufficiently conclusive. In 1968, a study found that the isotope ratios in lead white pigment from these paintings indicated that they could not originate from Vermeer's time, providing solid scientific evidence.[48]

Subsequent analysis conducted in 1975, including pyrolysis gas chromatography, unambiguously identified the resin in the paintings, matching samples from six of the "Vermeers" with a sample of the resin found in Van Meegeren's workshop.[49] The combined studies provide certainty as to which painting was made by Van Meegeren. The range of MS techniques available today would be able to characterise the materials used by the forger in even greater detail, providing information which was impossible at that time; increasing the knowledge of the materials he used would help to expose others of his forgeries that might still be on the art market.

Currently, there is a growing discussion about the restitution of cultural artefacts stolen or looted from one country in the past or present, which are now being sold or exhibited in another country. Although not easily resolved, these concerns suggest that the nature of provenance is becoming increasingly significant. Regarding this question, mass spectrometry can play an ambivalent role, as discussed in the specific case of ^{14}C AMS dating.[50] Only artefacts of verified authenticity can be traded, making accurately dated pieces more valuable. AMS dating can even become a tool for looters, to identify archaeological objects granting them the most profitable transaction on the black market, on which authentic items also have great value.[50] Both examples show that mass spectrometry acts as a tool in art authentication cases, whether it be to expose forgeries or to trace historical materials for recovery or trade.

10.3.5 Identifying Artwork and Clarifying Provenance with Mass Spectrometry

The methodology to authenticate a lost Picasso painting again illustrates the role of mass spectrometry, this time in proving the authenticity of an artwork. Unlike other examples, the painting, entitled *Violon. Céret*, has no signature, which is not a rare thing for Picasso, but it also has an unknown collection history, no conservation history, and no provenance details. It was never studied, never exhibited, and has damage on its surface. Its style, however, matches that of Picasso at its supposed date of execution, 1912. As it was strikingly only documented once, along with a picture in a well-known catalogue of the work of Picasso edited in 1942, the suspicion of it being a forgery of a lost original painting is fairly understandable, considering the previously described cases. The team studying this painting discussed the methodological approach, combining historical, stylistic, scientific, and technical examination to verify its authenticity in a reliable and thorough way, knowing that *"no final outcome is ever wholly conclusive"*.[51] Indeed, when analysing artworks, no formal proof can indicate its origin with certitude, and only clues and the absence of suspicious elements can be confirmed.

A fluid and constructive interaction between disciplines ensures historical and technical relevance. The painting was visually examined before technical and scientific analyses were performed to gather multifaceted knowledge on the artwork and its constituting materials. Imaging with Vis/UV/IR, FORS, FTIR, XRF, SEM-EDS, Py–GC/MS, and HPLC/MS was performed. The visual examination showed paint damages due to the folding of the canvas and an ill-performed regluing of the original canvas on another support. The elements found in each part of the stratigraphy then made it possible to distinguish the non-original materials from the original. Mass spectrometry played an important role in the study. The material used between canvas layers was identified, using Py–GC coupled to a quadrupole mass spectrometer, to be a mixture of egg and animal glue, thanks to characteristic ions found in the recent literature regarding organic binders[52] (Figure 10.2).

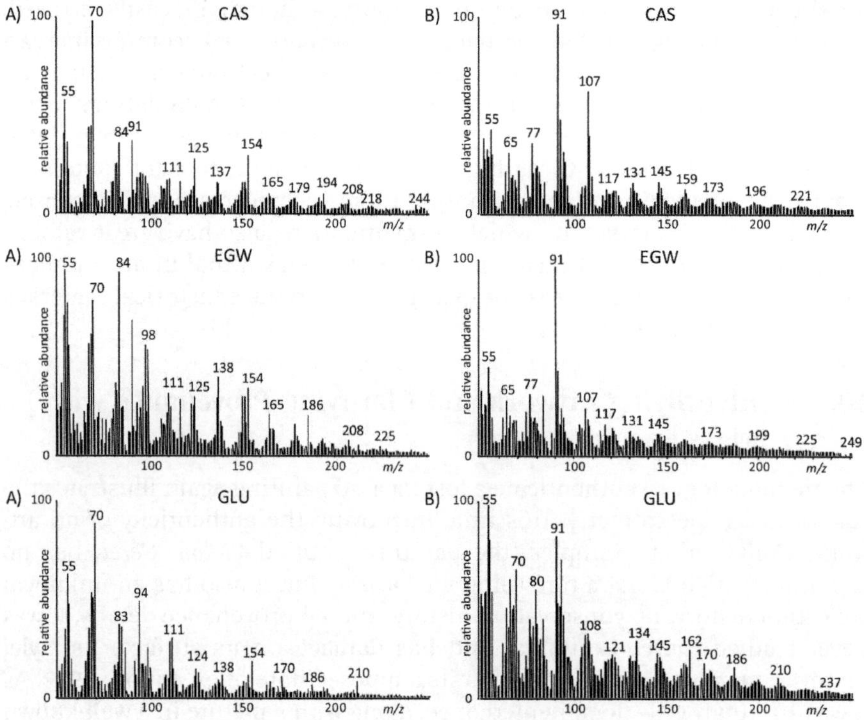

Figure 10.2 Average mass spectra of CAS (casein), EGW (egg white) and GLU (animal glue) obtained in the temperature ranges indicated in total ion thermograms (TITs) as areas A (310–320 °C, 300–310 °C and 320–330 °C, respectively) and B (425–435 °C). Reproduced from ref. 52 with permission from Elsevier, Copyright 2017.

An original triterpene resin was also identified using Py–GC/MS in the dark regions of the original paint surface.[51] This specific brown substance was available in 1912, and Picasso mentioned a brown pigment he was currently using in a letter to his art dealer when he was in Céret; the name of the painting is derived from Céret as it is suspected to be a part of a series he painted there.[51] Hence, an objective clue indicating that Picasso executed the painting was obtained by combining documentation and historical facts with analytical data that could be interpreted as a specific resin.

Authentication research is impacted by historical and modern artefacts. Painting materials underwent a commercial revolution during the 19th century, when tube paint became available to painters, replacing colour merchants or workshop production. In the case of modern paintings, materials are much more diverse, most of them being commercially purchased. In this study, HPLC-ESI-Q-ToF helped characterise the lipids found in the binder of the original painting. A series of oxidised triglycerides with unsaturated chains and hydroxyl groups were confidently identified, which correspond to the spectral profile of a highly oxidised linseed oil, a fat especially rich in unsaturated fatty acids. The possibility of it being modern house paint was ruled out thanks to knowledge about their distinctive look and pigment composition. The information gathered through the approach led to the official authentication of the painting as an original Picasso, dated 1912. Regarding the mysterious past of the painting, further studies are awaited.

Another aspect of historical artefact examination in which mass spectrometry plays a valuable role is isotope analysis, for instance, using radiocarbon dating, AMS or inductively coupled plasma mass spectrometry (ICP-MS). Depending on the fabrication and conservation of an artwork, or a material in the artwork such as a specific pigment like basic lead carbonate (lead white), the isotope ratios will vary. By accumulating background data about the ratios for relevant places, materials, and deposits, a sample can be matched with a given origin. This provides insightful information regarding provenance of antique or historical artefacts, or raw materials used in later objects, by combining the data with documentation and knowledge about history, the artist's life, trade of raw art materials and geological facts. This can help rectify misattributed artworks or artefacts such as decorated ceramics traded during the 16th century.[53]

10.4 Potential and Challenges for Forensic Applications

The Art Basel and UBS Global Art Market Report 2023 estimates that the international art and cultural property market generated a staggering $67.8 billion in 2022, roughly the GDP of Tanzania.[54] Due to the relative opacity of the global art market, which is subject to many different jurisdictions and often lacking robust regulation and certainly inadequately enforced, it is vulnerable to a host of illegal and lucrative activities. We will focus on crimes where art authentication can play an important role, which we can split into two categories: stolen goods and false goods.

Cultural artefacts are protected by international treaties that states sign and ratify such as the Hague Convention for the Protection of Cultural Property in the Event of Armed Conflict.[55] In total, 134 states ratified the 1954 Convention for the Protection of Cultural Property in the Event of Armed Conflict. Additionally, Article 8(2)(b)(ix) of the Rome Statute of the International Criminal Court makes it a war crime to deliberately target historic sites or buildings dedicated to art for destruction or defacement.[56] There is also the UNESCO 1970 Convention on the Means of Prohibiting and Preventing the Illicit Import, Export and Transfer of Ownership of Cultural Property, which gives member states the right to recover stolen or illegally exported cultural property from other member states and is currently ratified by 142 states.[57] It also led to the establishment of subsidiary bodies which help implement the treaty and settle disputes. Besides these international treaties, every country has their own national laws governing the sale, import and export of cultural heritage items. Additionally, most countries have laws regarding possession of stolen goods, which would also include cultural objects removed from a country without permission from its legitimate government. In general, the sale of stolen goods does not pass good title so an unsuspecting purchaser down the line would not legally possess the item: *"Nemo dat quod non habet"*. If one follows the laws of the country of origin regarding retrieval, transport and sale of archaeological artefacts and cultural heritage, the trade is generally legal. A merchant could prove this by providing a valid export certificate, for example.

As mentioned in the introduction, the Nazis looted priceless works of art and antique objects from museums and stole property of the people they murdered. These events were the impetus for the international treaties mentioned, but the reader should not be fooled in thinking this is all a matter of the past. In 2016, the International Criminal Court (ICC) first cultural destruction war crime trial was held upon the request of the State of Mali to pass judgement on the destruction of mausoleums in Timbuktu in 2012 and Ahmad al-Faqi, a member of the Islamist militia Ansar Dine, was sentenced to nine years' imprisonment.[58] In August 2015, the Islamic State destroyed the Temple of Bel in Palmyra. They can be tried under Syrian and Iraqi Law, but a body such as the ICC is much less amenable to corruption and intimidation. Unfortunately, neither Syria nor Iraq has accepted the jurisdiction of the ICC, which means it is unlikely that the perpetrators will stand trial for their acts of cultural genocide by destroying UNESCO World Heritage sites.[59] The Islamic State also ransacked historical sites and museums to steal and then sell valuable cultural heritage items in order to generate funds for their self-proclaimed government. This reportedly brought in an annual revenue from a few tens of millions up to $100 million.[60] Ukraine has accused Russia of violating the 1954 Convention for the Protection of Cultural Property in the Event of Armed Conflict during the ongoing Russian invasion of Ukraine. Ukraine's Ministry of Culture reports that Russia's forces have damaged hundreds of sites and objects important to Ukraine's history and cultural identity.[61] Reports have also surfaced that Ukrainian museums in occupied territory have been looted by Russian troops.[62] Again, the motivation appears to be acts of both cultural genocide and financial gain.

Other than theft in conflict areas, criminal organisations also target museums, galleries, churches, historic sites and private collectors, attracted by high-value items in not always secure environments. These items then find their way to the market and can fund other criminal activities. Famous works of art might be difficult to traffic and sell, however. There are databases one can consult such as the National Stolen Art File (NSAF), where law enforcement agencies in the U.S. and abroad submit stolen objects, and the Stolen Works of Art Database, where INTERPOL and specific international partner organisations such as UNESCO, and ICOM and ICCROM submit stolen objects.[63,64] The famous works can be ransomed or used as an alternative currency in the criminal milieu, but it has been suggested these might also be kept as collateral in exchange for reduced sentences once members of these criminal organisations are caught. It is important to bear in mind that some of these thefts could also be crimes of opportunity, wherein the thief likely does not know what to do with the stolen goods.

As mentioned, these stolen treasures often find their way to art markets and subsequently become part of private or museum collections. The complicity of actors in the art sector is an essential factor to launder these stolen goods, *e.g.*, by providing false export certificates and forging provenance documents. Once introduced to the legal art market, these objects become very difficult to identify, and art galleries and museums are caught having stolen artefacts in their possession with some regularity. The black market for art and antique objects is estimated to generate billions and is known to help fund wars, terrorism and various other activities such as the trade of illegal arms or drugs. It should be clear that there is a role for authentication and specifically the determination of the origin of art objects in the context of forensics.

Art forgeries are another area where authentication could play an important role in forensics. However, most criminal investigations and prosecutions focus on stolen art, because there is clear proof of a committed crime. Replicating art in and of itself is not a crime. Additionally, there must be proof of intent to commit larceny or fraud by deceiving another party into thinking the work is original. There are different ways a forger can go about it: one can recreate a lost painting or work of art, forge a hitherto undiscovered piece of art, modify a contemporary piece and/or falsify its provenance to increase its value. These forgers are experts in their chosen subject matter and often possess knowledge equivalent to or even surpassing that of authentication experts themselves, making it very difficult to identify these falsifications. Forgery is a longer-standing craft than scientific analysis of artwork, and forgers are skilled artisans. They use and develop ways to manipulate materials to create something matching the expected style, age, and technique, and try to make their forgeries pass expert examination. For this, they must be well-informed about the art history, the validation process, the techniques, the lives of the artists and any stylistic evolution. They also must stay up to date with the advances in scientific examination of the art they are forging. For example, Han Van Meegeren was acutely aware of the latest advances and literature in the field and carefully produced paintings for which he predicted forgery to be undetectable using the scientific techniques at that time. Beltracchi even created a fake painting and sent it to a specialised laboratory to find out what alterations were necessary before trying to sell one. Modern art in particular seems vulnerable to forgery, because its value is not so much derived from its intrinsic quality rather than the identity of the artist. Consequently, while researchers develop increasingly comprehensive analytical approaches, forgers learn about them and try to adapt. This raises the question of the necessity of discretion. In parallel, scientists explore exposed forgeries to develop identification approaches, as shown in Section 10.3.3, or apply several analytical techniques on it to highlight how each technique can contribute.[65]

The forged paintings then need to enter the art market, which means the forger needs to have established good connections with agents such as art dealers, auctioneers and art experts. This can be achieved through prior *bona fide* trade to build trust, or by seeking out individuals willing to partake in a conspiracy. In addition, things are often not very clear-cut. An art dealer or auction house might be too eager to get their commission of a sale and not do their due diligence. Alternatively, upon identification of a forgery, they may opt to withdraw the item without alerting the authorities in order to avoid damage to their reputation. It is not clear how many forgeries circulate in the art markets, but it likely makes up a larger share of art crime than theft. Even buyers might not want to identify their purchase as a forgery due to the negative impact on their investment. Experts who wrongly authenticated such forgeries might also be reluctant to admit mistakes.

In case a forgery is identified, criminal prosecution is possible under laws against fraud. In many cases, criminal prosecution does not take place due, in part, to high evidentiary burdens. There must be criminal intent beyond a doubt and many parties involved can claim they were unaware the piece was a forgery. The forger can even play it off as an elaborate joke or hoax. Instead of a criminal prosecution, the matter often ends up in a civil case where plaintiffs are trying to recuperate their losses. A final reason why criminal prosecution might not always take place in the case of forgeries is a lack of expertise. As mentioned, theft is usually clear-cut and law enforcement agencies know how to handle such cases. Nonetheless, there have been numerous successful criminal prosecutions involving art fraud.

10.5 Conclusion

Art authentication has evolved and now regularly relies on scientific evidence in addition to signature verification and provenance review. The evolution of the link between analytical science and art historians is fully discussed in the recent literature.[66] In the last few decades, mass spectrometry has proven particularly useful in authentication studies with different objectives and in various contexts as illustrated by the examples in Section 10.3. Its capability to obtain molecular information and identify organics along with its sensitivity, which limits the amount of sample needed, is particularly useful for the analysis of complex, precious samples such as art. With new developments in mass spectrometry such as AMS dating and notably advanced imaging capabilities which now allow the investigation of the microstructure of a cultural heritage sample, promising and wider applications can be expected. As exemplified in Section 10.3.5, a technique alone cannot authenticate an artwork and a multidisciplinary approach is necessary, including expert appraisal.

Given the large amounts of money involved in the global art trade and the current under-investigation of art forgeries, we believe this transferable expertise to be of utmost necessity. We therefore hope that this chapter has provided some insights into the role art authentication using mass spectrometry can play in forensics. It is just important to remember that identifying the forgery itself is not enough for a conviction, as intent also needs to be demonstrated beyond a reasonable doubt. There is probably also a need for added regulation in the art markets as well as international alignment, but that is something for legal experts and lawmakers to determine.

Acknowledgements

The authors acknowledge the support from their respective institutes during the writing of the manuscript.

References

1. J. Plesters, *Stud. Conserv.*, 1956, 2(3), 110.
2. J. S. Mills, *Stud. Conserv.*, 1966, **11**, 92.
3. E. Geddes da Filicaia, R. Evershed and D. Peggie, *Anal. Chim. Acta*, 2023, **1246**, 340575.
4. M. P. Colombini and F. Modugno, *Organic Mass Spectrometry in Art and Archaeology*, Wiley, Chichester, 2009.
5. A. Andreotti, J. La Nasa, F. Modugno and I. Bonaduce, in *Analytical Chemistry for the Study of Paintings and the Detection of Forgeries*, ed. M. P. Colombini, I. Degano and A. Nevin, Cultural Heritage Science, Springer, Cham, 2022, vol. 6, p. 157.
6. G. Favaro, D. Confortin, P. Pastore and M. Brustolon, *J. Mass Spectrom.*, 2012, **47**, 1660.
7. G. D. Smith, J. M. Esson, V. J. Chen and R. M. Hanson, *Forensic Sci. Int. Synergy*, 2021, **3**, 100130.
8. M. P. Colombini and G. Gautier, in *Organic Mass Spectrometry in Art and Archaeology*, ed. M. P. Colombini and F. Modugno, Wiley, Chichester, 2009, vol. 9, p. 237.
9. A. Lluveras, I. Bonaduce, A. Andreotti and M. P. Colombini, *Anal. Chem.*, 2010, **82**, 376.
10. M. Mackie, P. Rüther, D. Samodova, F. Di Gianvincenzo, C. Granzotto, D. Lyon, D. A. Peggie, H. Howard, L. Harrison, L. J. Jensen, J. V. Olsen and E. Cappellini, *Angew. Chem., Int. Ed.*, 2018, **57**, 7369.
11. S. Dallongeville, M. Richter, S. Schäfer, M. Kühlenthal, N. Garnier, C. Rolando and C. Tokarski, *Analyst*, 2013, **138**, 5357.
12. I. Bonaduce, L. Carlyle, M. P. Colombini, C. Duce, C. Ferrari, E. Ribechini, P. Selleri and M. R. Tiné, *PLoS One*, 2012, **7**, e49333.
13. J. E. Spangenberg, J. V Lavrič, N. Meisser and V. Serneels, *Rapid Commun. Mass Spectrom.*, 2010, **24**, 2812.
14. I. Hajdas, L. Calcagnile, M. Molnár, T. Varga and G. Quarta, *Forensic Sci. Int.*, 2022, **335**, 111292.
15. G. W. L. Hodgins, *Proc. Natl. Acad. Sci. U. S. A.*, 2019, **116**, 13158.
16. I. Hajdas, *E&G Quat. Sci. J.*, 2008, **57**, 2.

17. J. van der Plicht and I. Hajdas, in *Analytical Chemistry for the Study of Paintings and the Detection of Forgeries*, ed. M. P. Colombini, I. Degano and A. Nevin, Cultural Heritage Science, Springer, Cham, 2022, vol. 13, p. 421.
18. I. Hajdas, P. Ascough, M. H. Garnett, S. J. Fallon, C. L. Pearson, G. Quarta, K. L. Spalding, H. Yamaguchi and M. Yoneda, *Nat. Rev. Methods Primers*, 2021, **1**, 62.
19. L. Caforio, M.E. Fedi, P.A. Mandò, F. Minarelli, E. Peccenini, V. Pellicori, F. C. Petrucci, P. Schwartzbaum and F. Taccetti, *Eur. Phys. J. Plus*, 2014, **129**, 6.
20. L. Beck, *Physics*, 2022, **4**, 462.
21. C. D. Calvano, I. D. van der Werf, F. Palmisano and L. Sabbatini, *Anal. Bioanal. Chem.*, 2015, **407**, 1015.
22. S. Kuckova, I. C. A. Sandu, M. Crhova, R. Hynek, I. Fogas and S. Schafer, *J. Cult. Herit.*, 2013, **14**, 31.
23. M. Kompauer, S. Heiles and B. Spengler, *Nat. Methods*, 2017, **14**, 1156.
24. Q. P. Vanbellingen, N. Elie, M. J. Eller, S. Della-Negra, D. Touboul and A. Brunelle, *Rapid Commun. Mass Spectrom.*, 2015, **29**, 1187.
25. C. Bouvier, S. Van Nuffel, P. Walter and A. Brunelle, *J. Mass Spectrom.*, 2022, **57**, e4803.
26. C. Bouvier, H. Glanville, L. de Viguerie, C. Merucci, P. Walter and A. Brunelle, *Anal. Chem.*, 2021, **93**, 4463.
27. R. D. Macfarlane and D. F. Torgerson, *Science*, 1976, **191**, 920.
28. S. Della-Negra, A. Brunelle, Y. Le Beyec, J.M. Curaudeau, J.P. Mouffron, B. Waast, P. Håkansson, B.U.R. Sundqvist and E. Parilis, *Nucl. Instrum. Methods Phys. Res., Sect. B*, 1993, **74**, 453.
29. K. Boussofiane-Baudin, A. Brunelle, P. Chaurand, S. Della-Negra, J. Depauw, P. Håkansson and Y. Le Beyec, *Nucl. Instrum. Methods Phys. Res., Sect. B*, 1994, **88**, 61.
30. J. Fryc, Master thesis, State University of New York, 2021.
31. M. P. Colombini, I. Degano and A. Nevin, *Analytical Chemistry for the Study of Paintings and the Detection of Forgeries*, Cultural Heritage Science, Springer, Cham, 2022.
32. N. Khandekar, C. Mancusi-Ungaro, H. Cooper, C. Rosenberger, K. Eremin, K. Smith, J. Stenger and D. Kirby, *Stud. Conserv.*, 2010, **55**, 204.
33. V. Noce, *L'Affaire Ruffini: Enquête Sur Le plus Grand Mystère Du Monde de l'art*, Buchet Chastel, Paris, 2021.
34. J. Boon, K. Keune and J. Zucker, *Microsc. Microanal.*, 2005, **11**, 19.
35. S. Hageraats, M. Thoury, M. Cotte, L. Bertrand, K. Janssens and K. Keune, in *Analytical Chemistry for the Study of Paintings and the Detection of Forgeries*, ed. M. P. Colombini, I. Degano and A. Nevin, Cultural Heritage Science, Springer, Cham, 2022, vol. 12, p. 359.
36. H. Bellet, *Faussaires Illustres*, Actes Sud, Arles, 2018.
37. Genuine Fakes, https://www.genuine-fakes.com (accessed February 2023).
38. E. Hebborn, *The Art Forger's Handbook*, Cassell, London, 1st edn, 1997.

39. P. Ackroyd, S. Foister, M. Spring, R. White and R. Billinge, *Natl. Gall. Tech. Bull.*, 2000, **21**, 28.
40. L. Salisbury and A. Sujo, *Provenance: How a Con Man and a Forger Rewrote the History of Modern Art*, Penguin, London, 2009.
41. N. Wilding, *Faussaire de Lune: Autopsie d'une Imposture, Galilée et Ses Contrefacteurs*, Bibliothèque nationale de France, Paris, 2016.
42. J. Newton, G. Ramage, N. Gadegaard, W. Zachs, S. Rogers, M. P. Barrett, G. Carruthers and K. Burgess, *Sci. Rep.*, 2018, **8**, 10944.
43. A. Bejjani, M. Noun, S. Della-Negra, R. Tannous, G. Chalhoub, M. Hamdan and B. Nsouli, *Anal. Chem.*, 2019, **91**, 8864.
44. S. Fiddyment, M. D. Teasdale, J. Vnouček, E. Lévêque, A. Binois and M. J. Collins, *Heritage Sci.*, 2019, **7**, 35.
45. M. Creydt and M. Fischer, in *Exploring Written Artefacts*, ed. J.B. Quenzer, De Gruyter, Berlin, 2021, pp. 183–212.
46. The Nazi-Looted Art of Jakob Goldschmidt & Emanuël Vita Israel, https://www.sothebys.com/en/videos/the-restituted-art-of-jakob-goldschmidt-emanuel-vita-israel (accessed February 2023).
47. P. B. Coremans, *Van Meegeren's Faked Vermeers and de Hooghs: A Scientific Examination*, J.M. Meulenhoff, Amsterdam, 1949.
48. B. Keisch, *Science*, 1968, **160**, 413.
49. R. Breek and W. Froentjes, *Stud. Conserv.*, 1975, **20**, 183.
50. E. Huysecom, I. Hajdas, M. A. Renold, H. A. Synal and A. Mayor, *Radiocarbon*, 2017, **59**, 559.
51. E. M. Stella, S. Bracci, R. Iannaccone, J. La Nasa and M. P. Colombini, *J. Cult. Herit.*, 2019, **35**, 199.
52. S. Orsini, F. Parlanti and I. Bonaduce, *J. Anal. Appl. Pyrolysis*, 2017, **124**, 643.
53. S. Casale, N. Jaspers and L. Megens, presented in part at Authentication in Art 2018 Congress, The Hague, June, 2018.
54. C. McAndrew, *The Art Market 2023*, Art Basel and UBS, Basel, Zurich, 2023.
55. Convention for the Protection of Cultural Property in the Event of Armed Conflict, 1954, https://en.unesco.org/protecting-heritage/convention-and-protocols/1954-convention (accessed April 2023).
56. Rome Statue of the International Criminal Court, International Criminal Court, The Hague, 2011.
57. Convention on the Means of Prohibiting and Preventing the Illicit Import, Export and Transfer of Ownership of Cultural Property, https://en.unesco.org/about-us/legal-affairs/convention-means-prohibiting-and-preventing-illicit-import-export-and (accessed April 2023).
58. International Criminal Court, https://www.icc-cpi.int/mali/al-mahdi (accessed April 2023).
59. Why the Terrorist Who Destroyed Palmyra Won't Face Justice – The New York Times, https://www.nytimes.com/2016/09/30/world/africa/icc-hague-court-antiquties.html (accessed April 2023).
60. The Men Who Trade ISIS Loot – WSJ, https://www.wsj.com/articles/the-men-who-trade-isis-loot-1502017200 (accessed April 2023).

61. Destroyed Cultural Heritage of Ukraine, https://culturecrimes.mkip.gov.ua/ (accessed April 2023).
62. These Are Some Of The Most Famous Ukrainian Works Of Art Looted By Russia, https://www.forbes.com/sites/brianbushard/2023/01/14/these-are-some-of-the-most-famous-ukrainian-works-of-art-looted-by-russia/?sh=4e6014729b77 (accessed April 2023).
63. National Stolen Art File – FBI, https://www.fbi.gov/investigate/violent-crime/art-theft/national-stolen-art-file (accessed April 2023).
64. Stolen Works of Art Database, https://www.interpol.int/en/How-we-work/Databases/Stolen-Works-of-Art-Database (accessed April 2023).
65. G. D. Smith, J. F. Hamm, D. A. Kushel and C. E. Rogge, in *Collaborative Endeavors in the Chemical Analysis of Art and Cultural Heritage Materials*, ed. P. L. Lang and R. A. Armitage, American Chemical Society, New York, 2012, vol. 1103, ch. 1, p. 1.
66. A. Tummers and R. G. Erdmann, in *Analytical Chemistry for the Study of Paintings and the Detection of Forgeries*, ed. M. P. Colombini, I. Degano and A. Nevin, Cultural Heritage Science, Springer, Cham, 2022, vol. 1, p. 3.

CHAPTER 11

Transferable Mass Spectrometry Methods: Forensics in Time: Palaeoproteomics

A. BURNETT,[a,b] R. S. DENNIS,[a] S. DALED,[a] D. DEFORCE,[a] L. DE CLERCK[a] AND M. DHAENENS*[a]

[a] ProGenTomics, Laboratory of Pharmaceutical Biotechnology, Faculty of Pharmaceutical Science, Ghent University, Belgium; [b] ArcheOs Laboratory for Biological Anthropology, Archaeology Department, Ghent University, Belgium
*Email: maarten.dhaenens@ugent.be

Abstract

The information that defines all organisms, including humans, is encoded in their DNA, which altogether comprises the genotype. Throughout life, from initial organ formation to later lifestyle exposures, different genes are epigenetically activated and transcribed into mRNA, which is in turn translated into proteins that make up the final phenotype. Therefore, proteins are the most prominent source of phenotypic information and have consequently been exhaustively studied in different contexts, including clinical biomarker discovery for diseases. Despite the fact that proteins are extremely stable biomolecules that can hold phenotypic information over thousands of years, their potential in forensic and archaeological research has not yet been fully exploited. At least in part, this is due to the fact that the

New Developments in Mass Spectrometry No. 14
Applications of Mass Spectrometry for the Provision of Forensic Intelligence: State-of-the-art and Perspectives
Edited by Simona Francese and Stephen Bleay
© The Royal Society of Chemistry 2024
Published by the Royal Society of Chemistry, www.rsc.org

field of proteomics has mainly focused on fundamental molecular biology and medical applications, developing tools to maximize the number of peptide identifications in tryptic digests of (liquid) biopsies and cell cultures, which are very different from the sparse and heavily modified peptide mixtures often originating from unknown sources that need to be identified in forensic and archaeological contexts. Here, we describe the growing interest and recent developments in forensic proteomics and illustrate how palaeoproteomics, its equivalent for more ancient times, provides valuable methodological insights for future forensic applications.

11.1 Omics and the Information Content of Proteins

Hypothesis-driven biochemical assays have been the foundation of molecular biology for well over a century and have been carried out with great success. However, the lack of a more holistic view on the biomolecular complexity of systems has sent countless researchers down dead ends. Therefore, over the past few decades, large-scale omics approaches have been put forward for generating comprehensive hypotheses. These omics approaches initiated a shift towards an experimental design, wherein a broad biomolecular perspective of the biological sample is first generated in order to contextualise the hypothesis and the targeted biochemical assays beforehand.

Essentially, omics approaches were enabled by two relatively recent technical revolutions: the sequencing of nucleotides and the mass measurement of biomolecules by mass spectrometry (MS). Genomics was developed to create an image of the genotype, while epigenomics and transcriptomics were developed to study the activity of the genes. The phenotype arises only when the mRNA is translated into proteins, along with all the metabolites as inputs and outputs. The study of the phenotype through proteins is referred to as proteomics, while the study of metabolites in this manner is known as metabolomics. Unsurprisingly, such omics approaches have become a valued addition to forensic and archaeological toolboxes, because these fields intrinsically aim to maximise the information extracted from the samples under study.

At least in part due to the simpler makeup of DNA and mRNA (with only four nucleotide building blocks), the methods that sequence these biomolecules have developed most rapidly over the past two decades, with a plethora of software tools developed to suit and address innumerable different challenges that can be tackled through a nucleotide perspective. As a consequence, the (targeted) study of DNA has become one of the most adopted analytical techniques in forensics and it also had a high impact in archaeology (where the material under study is referred to as 'ancient' DNA, or aDNA). Since the conception of DNA profiling, rigorous curation of DNA databases and standardisation of DNA marker selections have been performed. Marker sets including 12, 17 or 20 different loci are selected if using the European Set of Standards, DNA17 (UK) or the CODIS system (US), respectively, as they demonstrate the greatest amount of variation between individuals. Statistical interpretation of these multiple marker profiles, *i.e.*, multiloci DNA profiles, involves match probability and likelihood ratio scores, which calculate the numeric possibility of two individuals sharing the same DNA profile. These numbers are often very large, and while they are treated with caution, they are sufficient to prove that two individuals having the same profile is statistically an extremely unlikely circumstance. As such, DNA profiling will remain a cornerstone in forensic investigations and subsequent court proceedings for the foreseeable future.

Intrinsically, however, the DNA sequence is identical in every cell of an individual, irrespective of the organ it originates from or the individual's phenotype. Therefore, except for epigenetic differences, DNA cannot identify the biological origin of a trace or the precise phenotypic characteristics of the donor, while such aspects can be of crucial importance and prove pivotal for a crime scene investigation or a deeper understanding of ancient life-ways. In addition, biological evidence can be degraded, containing minimal, damaged, or no DNA, especially at increasing time scales. This makes proteins specifically attractive information sources in archaeology, considering that the oldest DNA dates back to 1.2 million years old (Ma),[1] while proteins have been found in 80 Ma remains.[2] Finally, proteins also undergo modifications or degradation in ways that can themselves provide informative insights, even over shorter time periods, making them attractive biomarkers for estimating factors such as time since deposition (TSD). Excitingly, all these aspects have been explored to different degrees in both forensics and palaeoproteomics, but their full potential is yet to be uncovered.

11.2 Outstanding Analytical Challenges in Forensic Proteomics and Palaeoproteomics

Compared to traditional targeted biochemical methods, omics technologies focus more on increased throughput and analytical comprehensiveness. These traits make proteomics perfectly suited for analysing cell cultures and human biopsies in search for drug targets and biomarkers of disease. Briefly, in such studies, proteins are digested into peptides that are first separated by a reversed-phase liquid chromatography (LC) system before being ionised and analysed by mass spectrometry (MS). MS is not one technique, but a family of targeted and untargeted methods.[3] A targeted approach, such as multiple reaction monitoring (MRM) or parallel reaction monitoring (PRM), enables high-throughput and very sensitive detection of a limited number of predefined peptides representing the proteins of interest with high quantitative accuracy. Untargeted MS approaches aim to identify and quantify peptides in a sample with limited prior knowledge and these are described in more detail in the data record. Generally speaking, there are two main families of high-resolution MS instrumentation used in proteomics. The most common are Orbitrap mass spectrometers by Thermo Scientific™, followed by Q-TOF instruments, most popularly by Waters Corp™, AB Sciex™ and Bruker™ (more detail is provided under Section 11.2.1.3 'The Data Record'). Both styles of instruments perform comparably for most purposes, having adequate mass accuracy, sensitivity, speed and precision for standard proteomics applications.

Both forensic and archaeological samples differ greatly from the typically analysed cell cultures in several aspects: the tissue or biofluid is no longer "alive", has undergone taphonomic modifications, is derived from more than just one organism and often the samples are rare and irreplaceable. Therefore, a wide range of improvements are needed to minimise sample consumption and optimise data analysis, especially focusing on increasing the certainty of peptide and protein identifications in sparse data sets. Throughout the rest of this chapter, these analytical challenges are discussed in greater detail through concrete applications in the fields of forensic and palaeoproteomics.

11.2.1 Sample Preparation and Data Acquisition

11.2.1.1 Traceability and Chemical Contamination

Proteomics has not yet become a standard analytical technique for use on samples of recent or ancient age. Therefore, prior treatments and handling of these samples are not always well documented. This lack of traceability has been shown to lead to unexpected interferences, and documenting procedures should become part of the analytical process in the future.

For example, current forensic swabs used for trace sampling are still largely made of plastics. Polyethylene glycol (PEG) is the main plastic present in the tubes and shafts of these swabs. However, this PEG leaches into the samples and is a notoriously interfering compound for LC–MS analysis, not only impacting the retention of peptides on the LC column, but also inducing ion suppression and diverting the MS instrument to fragment plastics instead of peptides. While the use of swabs with wooden shafts could greatly alleviate this issue, many other surfaces still leach plastics into the sample (including the surfaces being swabbed, *e.g.* phones, toilet seats, and sex toys). Therefore, some studies have investigated the selective removal of PEG[4] and it will remain a very relevant area of research for the foreseeable future.

Additionally, for many archaeological finds excavated in previous decades, conservation efforts mainly focused on the physical integrity of artefacts rather than preservation of molecular content. Pottery fragments ('potsherds') and other archaeological remains have often been treated with conservation glues and varnishes such as Paraloid B72, in order to reconstruct or protect the artefacts, which may induce the complete loss of detectable peptide residues. Future sample screening efforts should investigate the suitability of existing solvent-based washing steps for removal of chemical treatments or aim to select artefacts or areas of artefacts that remain unaffected by conservation interventions. Chemical cleanup protocols could also be applied to reconstructed and varnished bones, making a greater number of collections available for proteomic analysis.

11.2.1.2 Clean Labs

In archaeology, more than in forensics, the wearing of gloves during sample manipulation – *e.g.*, during initial excavation, post-excavation processing, and curation – is not yet commonly established and this leads to the deposition of large quantities of keratins, other skin proteins, and cross-contamination from anything that was touched before (*e.g.*, food) onto these samples. The issue here is twofold: first, it becomes difficult or impossible to identify genuine ancient human proteins among modern contaminants, and second the modern (sometimes more abundant) peptides 'compete' with ancient peptides for selection by the MS. Therefore, implementation of forensic practices for sample handling throughout archaeological workflows can facilitate better data retrieval and future developments.

For both fields, specialised clean labs are universally established for DNA analysis. However, because the study of protein content is still relatively new, best practices for forensic and palaeoproteomics have been less well-established. In the same vein as early efforts to standardise and improve aDNA research practices,[5] recent efforts have now starting showing the way in palaeoproteomics[6] and these will also greatly improve forensic proteomics practice if adopted there. Briefly, it is important to use separate workspaces for the processing of archaeological, forensic, and other modern proteins to process blank extractions alongside samples during sample preparations and to avoid wearing PPE made of protein-rich materials such as latex or leather. Currently, laboratory practices in palaeoproteomics vary widely despite the increased urgency of good practice imparted by the comparatively more fragile and sparse protein remnants, while efforts to minimise laboratory contamination in forensic proteomics are peculiarly motivated by the need to provide legally defensible results.

Following extraction, the routine implementation of analytical blank injections before, between, and after sample injections in the LC–MS/MS system provides a means to assess issues such as peptide carry-over between runs – a precaution that is particularly relevant to samples containing collagens and other proteins with a propensity for sticking to the LC column. The abovementioned extraction blanks should be run alongside samples in the same manner and allow for the assessment of sample collection- and laboratory-related contamination including PEG traces, which are common not only in forensics but also in archaeological data due to the storage of artefacts and remains in plastics, as well as the use of plastic pipettes and sample tubes.

11.2.1.3 The Data Record

The completeness of the data record is even more crucial in forensics and archaeology than it is to fields that focus on fundamental biology. In large part, this is due to the huge informational potential that can be gained or lost from the identification of proteins that may be represented by very few peptides – for example, fluid biomarker proteins that could support a court conviction, or identify the earliest evidence of milking.[7,8] In the traditional untargeted (or 'shotgun') proteomics experiment, the mass spectrometer is operated in the data-dependent acquisition (DDA) mode, which involves automatic, yet *stochastic* selection of precursor ions for MS/MS fragmentation. Notably, a great deal of specificity can be achieved through the combinatorial complexity that is derived from only 20 different amino acids that form the basis for all functional units in the proteome. Most commonly, this is leveraged through database searches in proteomics, whereby each of the sometimes many thousand MS/MS spectra is mapped to the proteome under study.

More recently, data-independent acquisition (DIA) has gained a lot of interest. In DIA, peptides are continuously fragmented, *i.e.*, the mass spectrometer continuously cycles through predefined mass-to-charge (m/z) windows to equally and regularly measure the fragment ions throughout the LC gradient.[9,10] This generates more qualitative and quantitative data compared to DDA. In fact, a permanent record of the sample is generated, and the data can be interpreted in the same way as they are in targeted approaches – even if the targets are only defined post-acquisition. Until recently, the complexity of DIA data imposed the necessity of using previously identified peptides from DDA libraries as targets for peptide detection. Unfortunately, deriving targets from DDA data intrinsically transfers its limitations. With the introduction of machine learning and the capability to predict fragmentation patterns and peptide retention time on the LC system, DIA data analysis has now matured to yield tens of thousands of peptides from 15-minute LC gradients on fresh cell lysates or tissue samples. The potential implications of this total data capture for forensic and archaeological peptide mixtures remain unexplored but it may offer great advantages over DDA in some instances.

Different types of MS instrumentation used in proteomics produce slightly different but comparable data outcomes. Orbitrap mass spectrometers, which utilise a combined trapping analyser and detector, produce comparatively 'clean' data and have long outperformed other designs in the DDA mode, but this gap has gradually narrowed. Q-TOF instruments, on the other hand, produce a greater volume of raw data and are capable of performing a wider breadth of data-independent acquisition strategies because the TOF analyser is up to 100 times faster than an Orbitrap. This includes the use of full ion mobility separation, which adds an additional dimension to the data that allows further separation of ions based on their shape. Especially in the latest instrument designs, this is combined with advance acquisition settings on, *e.g.*, the quadrupole and the pusher to increase both the specificity and sensitivity in complex peptide mixtures. Both styles of instrument perform comparably for most purposes, although choice of instrument for laboratories performing forensic and palaeoproteomic analyses is expected to vary based on user expertise and analytical priorities.

11.2.2 Data Analysis: Challenges and Potential Solutions

When MS data are searched in a conventional proteomics setting, the experiment is performed *in silico* in order to define all possible target peptides that can be found. Therefore, a database needs to be selected for the organism under study, as well as the enzyme/s that was used for digesting proteins in the laboratory (most commonly trypsin), and the deliberate and potentially unwanted post-translational modifications (PTMs) that could have been chemically induced. It is also essential to select databases containing sequences for common laboratory contaminants (*e.g.*, the common Repository of Adventitious Proteins, or cRAP)[11] to avoid misattribution of peptides to the target organism. Together, the search engine then generates all possible peptides in the sample and scores the peptide-to-spectral matches (PSM) using a scoring algorithm. A decoy database of reversed or randomised peptide sequences is usually also searched, allowing a semi-supervised machine learning algorithm to weigh and re-score the target peptides to obtain a maximum of true targets at an empirically determined false discovery rate (FDR) using the so-called target–decoy approach. Today, state-of-the-art proteomics workflows allow users to run samples in less than 20 minutes, yielding well over 50 000 peptides, which collectively provide the basis for inferring of up to 6000–7000 proteins.

11.2.2.1 Sparsity

Compared to fresh tissues and cell lines that are usually subjected to proteomics analysis, forensic and archaeological samples tend to generate much more contaminated and sparse data, containing proteins from the sample surroundings and with a lot of modified and degraded peptides. Indeed, proteins can quickly start to decay when exposed to the elements and therefore the data are typically less rich. Most prominently, this sometimes does not allow for the application of a target-decoy approach to estimate the FDR or perform a semi-supervised machine learning algorithm to obtain more true targets. In the case of palaeoproteomics, sometimes only a few peptides are actually detectable, but their identity can have great impact on the archaeological understanding of the excavated sample. Therefore, probabilistic algorithms like Mascot and the derived Andromeda are usually used, as these calculate the chance of providing a random hit considering the number of spectra searched and the size of the database used, without requiring a decoy population for FDR estimation. Alternatively, because of the sparse nature of the data, much more computationally intensive *de novo* sequencing approaches can be used. However, to date, these approaches still struggle with significant ambiguity, because similar amino acid sequences can easily be matched to the same spectrum.

11.2.2.2 Database Size

The greater the number of options that are considered for database matching, the higher the chances of generating random matches. Therefore, increasing the database size is detrimental to most search engines. When degraded peptides are expected, a so-called "semi-tryptic" *in silico* digest needs to be performed in order to find the "frayed" ends of degraded peptides. This leads to an exponential increase in the size of the database and a corresponding increase in the number of potential random hits among candidates. The same is true for variable modifications which can either be present or absent. Especially when combining several PTMs in one search (as is often necessary for ancient samples with different modifications arising from decay and laboratory preparations), addition of each PTM increases the size of the database exponentially. To make matters worse, both forensic and palaeoproteomics samples are contaminated by the milieu, including potentially soil bacteria, plant residue, animal contamination, *etc.* Theoretically speaking, all these organisms need to be included in the database searching, again increasing the chances of a random hit considerably. In the case of palaeoproteomics, very old samples can even be derived from species that have point mutations compared to the reference sequence, therefore lacking a database to match the data altogether. Conventional proteomic data analyses were not designed for these challenging samples.

11.2.2.3 Computational Solutions

Fortunately, the recent addition of machine learning to the bioinformatics toolbox is opening doors for more robust and trustworthy peptide annotation. Most prominently, the retention time on an LC column as well as the fragment intensities in the spectrum of a peptide sequence can now be predicted with extremely high precision, even if these values were never measured before.[12–14] This can therefore be applied to increase the confidence in a (*de novo*) peptide annotation by plotting the measured peptide (Figure 11.1a) against the mirror spectrum that was predicted (Figure 11.1b) and calculating the Pearson correlation between the two. Visually, this comparison can be made with the Universal Spectrum Explorer (USE).[15] If the retention time equally matches the prediction, only minor differences could still underlie a false annotation. This additional security in peptide identifications is particularly important for the analysis of sparse datasets and limited proteomes, where the presence of key peptides can greatly influence the resulting interpretations.

Additionally, classification challenges, such as the routine identification of a tissue using diagnostic biomarker proteins, can now be performed using machine learning, provided a sufficiently large training dataset is used to find the proteins that hold the most significance for the classification task at hand. Together, it is clear that such tools will greatly boost the performance

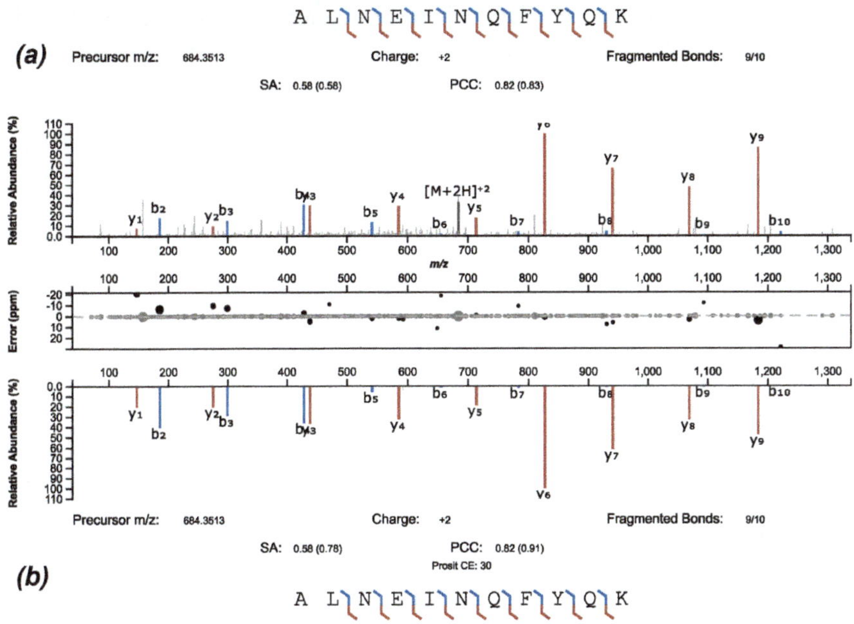

Figure 11.1 Spectral prediction for increased confidence in peptide annotations. The recent addition of spectral prediction to the bioinformatics toolbox is especially important for datasets where only few peptides are identified, but which can have a large impact. Here, this is exemplified with an alpha-S2-casein peptide from *Bos taurus* which could be interpreted as early evidence of milking. (a) Spectrum measured on a QTOF instrument and (b) predicted spectrum of this peptide using Prosit.[16] Picture created with the USE.

of forensic and palaeoproteomic research, hopefully contributing to the expansion of these fields in early development.

11.3 Applications in Forensics and Palaeoproteomics

The most prominent current proteomics applications are given below in reverse chronological order of impact from most recent (forensic) to most ancient (archaeological) and are summarized in Figure 11.2. Notably, this distinction is all but binary, and several approaches have shown to be relevant to both fields.

11.3.1 Tissue or Biofluid Identification

Forensic approaches for identification of the biological origin of trace residues traditionally employ the use of initial presumptive screening tests, which form part of standard forensic evidence examination practice.[17] Many of these tests rely on colorimetric changes induced *via* enzymatic activity, chemical reactions or antibody interactions.[18] However, presumptive screening tests suffer from a lack of specificity and sensitivity, resulting in false positive results. For example, the Kastle–Mayer test for blood determination additionally reacts with vegetable and plant peroxidases.[19,20] The assays also face difficulty in distinguishing body fluid types, *e.g.*, α-amylase is present in semen, vaginal, and nasal secretions and can

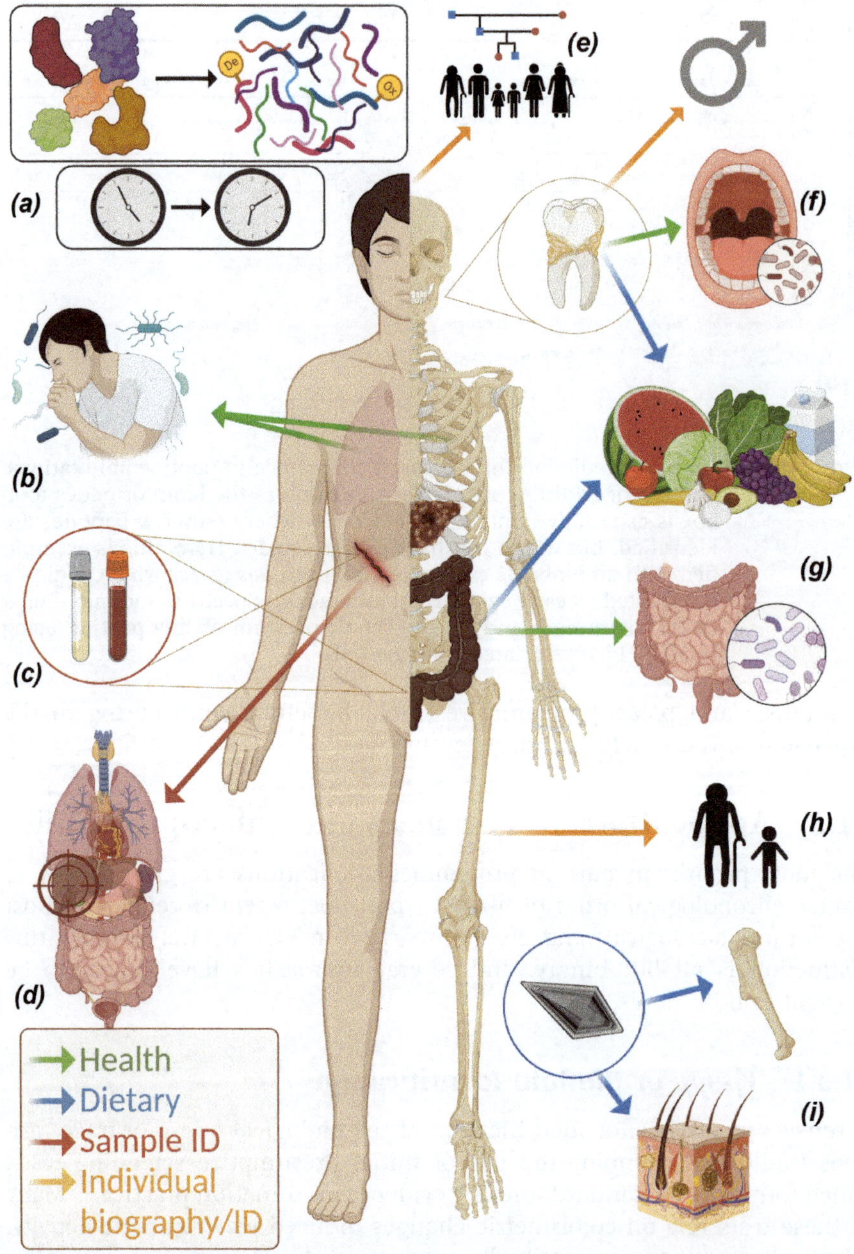

therefore react with the routine saliva test.[20] A similar issue can be observed for the acid phosphatase test for semen which also reacts with vaginal fluid. In addition, this semen test can react in the presence of plant originating phenols from field contamination.[21,22] Presumptive tests are only able to suggest the presence of a single target body fluid; consequently, they are unable to simultaneously identify individual fluids in a mixture and cannot discriminate species origin. For these specific details, cascade or serological testing is required, however, these are destructive in nature, which can be detrimental for subsequent DNA analysis.[23] These methods are also time consuming, expensive and do not exist for all biological fluids of forensic importance, *e.g.*, vaginal fluid, peripheral and menstrual blood.[17,24] Finally, while they have been applied to archaeological samples, results become more difficult to interpret with degraded protein residues. In light of these limitations, there is significant interest in finding alternative molecular genetic approaches based on messenger RNA, microRNA or DNA-methylation. However, DNA-methylation analysis requires a considerable amount of DNA. Quantitative polymerase chain reaction (qPCR) is widely considered as the gold standard for mRNA quantification and miRNA expression. Indeed, qPCR is limited to a small number of targets within a single reaction and needs a normalisation strategy to eliminate non-biological variance.[18]

Previously, untargeted forensic proteomics was performed in the ProGenTomics laboratory (Ghent University) for the determination of marker proteins within various biological fluids.[3] This DDA approach was applied for in-house investigations of multiple authentic forensic cases. For example, (i) a sex offender claimed that a stain on his garment, containing DNA from the victim, had originated solely from skin contact. However, we detected vaginal fluid proteins in that stain. (ii) A swab of a sex offender's glans penis was found to contain vaginal proteins, proving that penetration

---

Figure 11.2 Information potential of proteins from different sources in forensic and palaeoproteomic contexts. (a) Protein degradation and decay-induced modifications occur over time and may be linked to PMI and TSD; (b) respiratory infections leave traces of pathogens and immune responses in bone and soft tissues; (c) biofluid deposits contain biomarker proteins which can be used to identify blood, urine, semen, and more; (d) tissue biomarker proteins can be used to identify ancient organs and, when collected from a projectile or weapon, can be used to reconstruct trajectories of injury; (e) hair proteins contain genetically-variant peptides which correlate to ethnicity; (f) dental calculus is a rich repository of dietary and microbiome-derived peptides, while enamel peptides provide concrete proof of male sex; (g) vomit, coprolites, and gut contents analyses also present a bountiful source of dietary proteins and information on gut microbiota; (h) bone proteomes are thought to contain a number of biomarkers for phenotypic information including biological age; and (i) ancient proteins retrieved from artefacts including pottery impart dietary information and the ability to distinguish between different animal tissues. Created with Biorender.com.

had taken place in the 48 hours before the arrest. (iii) Blood stains on the car bumper of a traffic offender suspect turned out to be derived from a pigeon, explaining why the forensic lab could not obtain a DNA profile. Subsequent follow up studies have demonstrated the power of this approach, as detection of multiple protein markers increases sample specificity.[25,26]

In the field of molecular ballistics, backspatter and forward spatter can contain tissue residues that can help reconstruct the point of impact and therefore the course of events during a crime. Identifying tissues is intrinsically impossible using DNA sequencing because all cells in an organism contain the same DNA. Currently, approaches that aim to identify tissues rely on DNA methylation, mRNA, and miRNA analysis. Protein expression patterns have not yet been used to identify tissue of origin, despite their obvious utility in this regard, because machine learning computations are required to do so. Recently, however, a 'big data' approach has shown for the first time that classifying tissues purely based on annotated LC–MS/MS proteomics data is feasible. A human tissue proteome atlas was created through massive downloading and manual curation of public datasets, followed by a unified search on 35 467 775 spectra, yielding 15 108 proteins that were fed to a Random Forest model to classify samples into tissues and cell types.[27] Subsequently, a one-vs-all classification and feature importance were calculated and used to analyse the most discriminating protein abundances per class. This model is able to predict tissues with 98% accuracy, and cell types with 99% accuracy. Using such scoring algorithms, both biofluid and tissue classification could become commonplace in the future. Notably, the classification of tissues and biofluids could equally be trained on differential posttranslational modification abundances in future, which could be anticipated to increase its robustness.

Tissue classification capabilities are also of great interest in archaeology, both for identifying organs in complete mummies and for identifying isolated tissues in burial or apothecary jars. To the extent that can be stated in such an emerging field, mummified tissues are a particularly popular object of study in palaeoproteomics, which encompasses research on frozen,[28–30] anthropogenic,[31] temperate,[32] and bog bodies[33] from all over the world spanning thousands of years. Remarkably, in the case of Ötzi, the frozen Copper Age Alpine man, proteins thought to be linked to neurological disorders (including Alzheimer's and Parkinson's Disease) were recovered through an intensive and tailored approach designed to release the proteins from aggregates which are thought to have been an important factor in their preservation over 5300 years.[28] Palaeoproteomic analyses of mummified tissues have so far been restricted to recognisable or macroscopically identifiable tissues, including brain, skin, and muscle biopsies.[28,31,34] However, in a number of instances, the retrieval of proteins whose expression is specific to certain tissues (including cardiac muscle[31]) or is highly expressed in certain tissues (*e.g.*, lung[35]), is suggestive of the utility of proteomics to identify tissues that survive in an unrecognisable form. Indeed, extensive dissection campaigns carried out on Egyptian and South American mummies show that a large number of them retain tissues which cannot be identified through morphological, chemical, or radiological means, and which may represent ideal targets for proteomic investigations.[36] Tissue atlases, such as that created by Claeys *et al.*,[27] and Uhlén and colleagues' Human Protein Atlas[37] based on transcriptomics and immunohistochemistry, provide an ideal starting point for the identification of ancient and mummified tissues. However, tissue scoring approaches which rely on quantitative data may be less reliable for ancient proteomes or may require separate adaptations towards palaeoproteomic datasets. Additionally ancient tissue investigations should take into account the inevitable inclusion of biofluid proteins (*e.g.*, blood and tissue fluid) inside tissue proteomes and the likely mixed nature of tissue samples due to the movement and displacement during decomposition. The former is particularly relevant as there are no instances of singular biofluids preserved over archaeological timespans for comparison with ancient tissue proteome data. The characterisation of ancient tissue proteomes in this manner could provide useful data towards forensic tissue identifications in cases with extended PMIs, especially as protein preservation has repeatedly been shown to be dictated more by environmental conditions than by chronological age.

11.3.2 The Information Content of Hair

Hair is one of the most regularly encountered forms of trace evidence recovered from crime scenes, due to the easy transfer of hair as a result of natural shedding. Identification of the type of hair, whether human or animal, can provide critical information which can be used to link suspects, scenes and victims together.[38,39] The hair shaft itself is a poor provider of DNA; and while nuclear DNA can be obtained from an intact root bulb, severe degradation often renders it unsuitable for routine DNA profiling. In comparison, the proteins present in the hair shaft demonstrate a far higher intrinsic stability and contain genetically variant peptides (GVPs) as a source of genetic information.[40,41] Single amino acid polymorphisms (SAPs) are contained within these peptides and are translated from non-synonymous single nucleotide polymorphisms (nsSNPs). Recent deep proteome coverage studies have shown the potential of SAPs in enhancing forensic investigations. Random match probability scoring when calculated with nsSAP exome genotype data matching frequencies, demonstrate how the deep coverage approach could improve current SAP-based identification and establishment of the biogeographic origin of individuals for forensic purposes.[40–42] Shotgun proteomic analysis of hair shafts obtained from individuals of different ethnicities has previously uncovered notable variations in protein expressions of hair obtained from Korean, Kenyan, African-American and Caucasian individuals.[40,41] This untargeted technique has also revealed differences in keratin-related protein levels dependent on body hair location: successful differentiation of pubic, axillary, beard, and scalp hair shafts was achieved for Caucasian hair samples.[39] Subsequent studies have also found that keratin proteins, *e.g.*, K37 have notable differences in expression between scalp and pubic hair and are statistically more abundant in pubic hair in comparison to arm hair.[42,43] Histone H3.1, found in the hair cortex, also exhibits varying expression levels which are higher in arm and leg hair in comparison with pubic hair.[39] This could prove particularly useful for forensic investigations, especially in cases of violence and rape where there is a higher likelihood that the recovered hair samples will originate from different bodily regions. Importantly, however, hair and skin cells are both primarily composed of keratin, which is the most common laboratory contaminant protein and so great care needs to be taken when studying hair samples.

Remarkably, some of the same GVPs observed in modern hair persist for hundreds of years. Six British individuals' hair samples (*ca.* 1750–1853 AD) from Kent and London yielded 10 GVPs across 8 nsSNP loci, showing genetic and phenotypic consistency with modern European populations.[40] Recent studies focused on keratin proteins, the main structural component in hair, have demonstrated that keratins survive preferentially as compared to metabolic, membrane-related, and other modern hair proteins, and that keratins consequently represent the majority of protein recovered from ancient hair.[40] Analysis of 3880-year-old hair resulted in protein identifications using as little as 500 μg of hair. The proteins retrieved, including eight keratins, exhibited more extensive deamidation and cysteine oxidation in comparison to modern comparators; additionally the overall number of proteins retrieved was more limited. The destruction of such small quantities of an otherwise uninformative biomaterial demonstrates the promise of proteomic hair analysis as a complementary analysis, where sampling of tissues or bone is more destructive.

11.3.3 Time Since Deposition

Accurate estimation of the time since deposition (TSD) of a biological fluid trace, could provide forensic investigators pivotal information when establishing an approximate time frame for the commission of the criminal offence or when establishing the relevance of traces to crime under investigation. Over the past ten years, a number of studies have aimed to develop a means of estimating the TSD of bloodstains encountered at crime scenes. The methods and outcomes presented in these studies have confirmed that both the chemical and physical properties of blood stains change over time.[44,45] The majority of these studies have focused on identifying changes in enzymatic activity, changes in the spectral signal of haemoglobin and the time-related degradation levels and pattens present in RNA, and DNA.[44–47] However, despite showing great promise, these methods have not been able to achieve the level of repeatability required for forensic applications. In addition, these methods would also need prior identification of the recovered biological fluid, through various presumptive and confirmatory tests, which can prove detrimental for subsequent TSD estimations and DNA profiling when samples are of low volumes.

Proteins are known to degrade and undergo modification very shortly after deposition, *e.g.*, through oxidation, deamidation, and non-tryptic cleavages, making them ideal targets to search for signature modifications that correlate strongly with TSD. Previous research using colorimetrically-based assays for the TSD assessment of fingerprints found that samples exposed to Sakaguchi assay, which targets arginine, demonstrated a decrease in absorbance at 500 nm associated with increased TSD periods beyond seven days.[48] Although the Sakaguchi assay is not sensitive enough to estimate TSD with forensic accuracy, this work highlights the potential of TSD determinations based on a single amino acid. Sakaguchi assay, among others including Bradford and ninhydrin assays which are commonly used to test protein content prior to proteomic analyses and for the development of fingermarks on non-porous substrates, is also capable of providing sex estimations.[49–52]

It is already known that deamidation of asparagine, for example, occurs on shorter timescales than deamidation of glutamine, and as such the latter is more commonly seen in samples of archaeological provenance than in forensic samples.[53] Widespread oxidation is also commonly seen in degraded proteins. Current ongoing research is focused on identifying degradative TSD biomarkers under different conditions through machine learning approaches applied to differential ion abundances, which enables the targeting of fragment spectra of interest, followed by deeper manual curation to identify these ions.

Archaeological timescales extend beyond the applicability of forensic TSD estimation strategies, with ^{14}C radiocarbon dating most commonly applied to examine the antiquity of human remains or associated material. However, given the pervasive issue of the freshwater reservoir effect wherein waterlogging or significant consumption of freshwater dietary resources can artificially alter radiocarbon age estimations,[54] the implementation of a second biomolecular method for comparative or absolute assessment of antiquity is desirable. Hydroxyproline dating, whereby the ^{14}C content of a single amino acid in bone collagen is targeted to obtain more precise radiocarbon dates, has so far been proven promising for refining large date ranges, but does not qualify as a proteomic dating method. To date, establishing a protein modification 'clock' has not been possible over archaeological timescales, with rates and mechanisms of protein decay found to be more extensively linked to burial conditions than to chronological age. The variability of environmental conditions is so significant that it might eventually present a challenge that is too hard to overcome, despite ongoing efforts.

11.3.4 Sexing Through Tooth Enamel

Enamel provides a highly resistant environment for the preservation of proteins, shielding them from the degradative effects of groundwater, bioturbation, humic substances in soil, and bacterial agents of decay. In forensic anthropology and especially in archaeology, the detection through LC–MS/MS of amelogenin-Y (AMELY) has been used extensively to sex individuals. AMELY is an enamel protein which is only transcribed from the male Y chromosome; as such, it is present not only in adult males but also in children. This makes it an invaluable marker for defining the sex of children's remains, as their immature skeletons cannot be sexed osteologically, and it can be applied to large numbers of individuals.[55,56] This has been particularly impactful in population demography studies – including sexing 75 children from prehistoric cemetery sites[56] – and remarkably in the case of a suspected child homicide dating back ∼3950 years.[57] In this case, a 5–6-year-old was believed to have been violently killed on the basis of four perimortem impression fractures present on the skull, with the possible co-interment of a murder weapon in the form of a ceramic loom weight. The proteomic sexing of this child as male, alongside anthropological evidence, grant unprecedented insights to gender dynamics of prehistoric infanticide. Amelogenin-based sexing is also valuable to confirm osteological sexing of adult skeletons, having the greatest value for individuals whose skeletal sex is indeterminable, *e.g.*, where the pelvis and/or skull are incomplete. Sexing of isolated teeth can even be valuable in some contexts; from sites, cultures, or time spans with very limited human remains. However, because males express both amelogenin-X (AMELX) and AMELY proteins, while females express only AMELX, proving the absence of AMELY is necessary in order to identify a female donor, which is analytically very challenging: absence of evidence is not evidence of absence. Apparent lack of AMELY can, for instance, be caused by protein decay or by naturally low expression of this protein, especially given that AMELX expression is much higher than AMELY in males.[58] In order to enable bilateral sexing in a future-proof way, *i.e.*, its application to any tooth excavated in the future or enable future re-analysis of the current experimental data, a logistic regression model needs to be trained wherein one can define the chance of detecting AMELY, given the background of matrix proteins found in the enamel. The spiking of heavy labelled peptide standards will greatly improve the performance of such models; once these models are demonstrated to provide robust and reliable estimations, they will massively increase the utility of enamel sexing in forensic contexts.

11.3.5 Phenotyping Skeletal Remains

Collagens are the most abundant and intensely studied proteins in archaeological bones and are especially useful for taxonomic classification of fragmented remains. This is generally performed by employing a 'ZooMS' (zooarchaeology by mass spectrometry) workflow, wherein a tryptic digest of a bone collagen extract is analysed directly using MALDI–MS and the resulting peptide mass fingerprint (PMF) is used to identify the animal *via* spectral matching. Several groups have proposed to use peptide fragmentation through LC–MS/MS followed by a database search (the conventional proteomics workflow) for this, and both approaches have been shown to be very performant.[59] A combination of both techniques was used to identify a Denisovan mandible over 160 000 years old – the earliest fossil from this area of Tibet – which could not be securely distinguished from Neanderthals and other contemporary hominins by any other means.[60] This assignation was made on the basis of six collagen proteins, whose sequences differ between hominin species and have been used to create evolutionary phylogenies where DNA no longer survives.

Bone also contains a suite of non-collagenous proteins (NCPs), contained and preserved within the hydroxyapatite scaffold. These are slowly renewed during life as bone remodels, with the rate of remodelling varying in accordance with age, sex, illness, and biomechanical stress across the skeleton.[61] Consequently, certain bone proteins have been proven to hold potential as biomarkers for biological age determination[62] and ethnicity,[63] with sex-based differences also seen in animal bone elements.[64] With further refinements in practice, these bone biomarkers could constitute a powerful tool for building individual biographies and contribute to identification over shorter timescales. In forensic contexts, it has even been possible to explore relatedness through the identification of genetically variant peptides (GVPs) shared between individuals.[65] Applied to suitably preserved archaeological remains, GVP studies could increase the prevalence of relatedness studies and circumvent the greater expense and low success rates of aDNA-based efforts.

In both osteoarchaeology and forensic anthropology, and especially for bones recovered from mass or clandestine graves which often contain commingled or fragmented remains, the ability to recognise skeletal pathologies is essential to understand events and individual biographies. The recognition of traumatic injuries (*e.g.*, fractures) and their correct assignation as pre-, peri- or post-mortem is particularly important in this regard. Some pathological conditions result in changes to bone morphology in addition to affecting the soft tissue and are also important to recognise, including infectious diseases (*e.g.*, periostitis, osteomyelitis, syphilis, and tuberculosis), metabolic diseases (*e.g.*, iron deficiency anaemia, and rickets), or degenerative joint diseases (*e.g.*, DISH, rheumatoid arthritis, and gout).[66] Proteins indicative of health status can be present in skeletons, including proteins related to cancers and infectious disease.[62,67] Respiratory diseases are of great interest to archaeologists due to their causal links with animal domestication, urbanisation, and industrialisation[66] – as such, the proteomic profile of archaeological ribs is being investigated by archaeologists alongside markers for pathogenic disease and immune responses.[68,69]

However, existing data do not yet provide a sufficient basis for reliable and routine retrieval of physiological and pathological information from archaeological or forensic human bone proteins. Recent research efforts have begun to lay the foundations for a more comprehensive understanding of bone proteomes, involving state-of-the-art data acquisition and analysis methods to investigate entire mouse bone proteomes.[70] This approach, involving DIA MS and machine learning-based data analysis strategies to generate an exhaustive library of proteins, could equally be applied to human bones. However, sampling of a large and representative population subset, as well as sampling of multiple skeletal elements, may prove to be ethically and practically challenging. Nonetheless, such an endeavour would greatly increase the information yield from proteomic bone analyses regardless of post-mortem interval. Furthermore, the interim analysis of bones *via* DIA creates a permanent digital archive of the samples, enabling comprehensive reanalysis without the need for repeated sampling. This preparative 'data banking' approach could arguably accelerate the development of DIA-based analyses in both fields and maximise data gain especially from archaeological individuals, where the exploration of data is not limited by the necessity of legal relevance. In summary, bone holds significant potential for building increasingly more extensive individual biographies, especially in conjunction with protein data derived from dental samples.

11.3.6 Dietary Reconstruction

Following on from decades of successful explorations into ancient diets using lipid analysis, proteomic analysis of ancient pottery has been proven to be even more impactful and powerful in terms of taxonomic identification and potential for tissue specificity. Typically, lipid approaches are capable of classifying fats as aquatic, non-ruminant, or ruminant, but lack species and tissue specificity (*e.g.*, as seen in Northern European hunter-gatherer pottery[71]). In contrast, Anatolian pottery up to 8000 years old yielded peptides from numerous species of dairy products, meats, grains, and other plant organisms, demonstrating ability to differentiate components of complex food mixtures over long timescales through proteomics.[8] This is a direct consequence of the fact that a great deal of specificity can be achieved through the combinatorial complexity that is derived from 20 different amino acids that form the basis for all functional units in the proteome, as opposed to the more elemental composition of lipids built from CHO.

Very rich dietary data can also be retrieved though metaproteomic (shotgun proteomics of multi-species samples) analysis of human dental calculus, especially as metaproteomic analytical tools undergo rapid development.[72] Through sophisticated data analysis approaches it was possible, for instance, to carry out oral microbiome analysis and identification of human and dietary proteins from both modern and medieval (500–950-year-old) calculus samples.[73] This yielded information on oral and systemic health states, including distinct grouping of healthy and oral disease-susceptible individuals based on their oral microbiome compositions and host immune proteins, including the presence of so-called 'red complex' bacteria. The robust nature of this approach was improved through comparisons with modern calculus samples. Meanwhile, the trapping of oat and milk-derived dietary proteins also indicates that calculus could store information relevant to toxicological and other forensic investigations. However, numerous archaeological calculus investigations indicate a bias towards the preservation of beta-lactoglobulins and other dairy-related proteins over other foodstuffs,[74,75] which may be related to either preservational bias or the role of dairy proteins in dental biofilm formation. On a dietary level, calculus represents accumulation over time, while analyses of gut contents, whether expelled (vomit and coprolites) or retained in preserved remains (in fresh or mummified cadavers), provide information on consumption and digestive system microbiome composition in the hours before death or expulsion. Proteomic analysis was used to identify the presence of vomit at the scene of an alleged rape, validating the victim statement that they had been unwell at the time in question.[76] Crucially, the detection of meat-based proteinaceous constituents from the meal the victim claimed to have eaten, alongside the drinks she believed to have been spiked, was deemed corroborative evidence supporting the victim testimony and enabled continuation of the court case. Archaeological coprolite analyses[77] demonstrate the potential to extend this approach to faeces traces relevant in legal proceedings, while analyses of gut contents carried out on mummified remains[33] suggest that this approach could be applicable in cases of soft tissue preservation over shorter timescales, *e.g.*, in the case of so-called 'apartment mummies'.

11.4 Conclusions

This chapter provides an overview of the achievements and potential of proteomics as an analytical addition to the fields of forensics and archaeology. DNA genotyping will always be the gold standard for forensic person identification and proteomics can only define relatedness through genetically variant peptides. However, it is the wide range of phenotypic applications of protein analysis that makes proteomics such an appealing analytical technique. Proteomics can not only provide information on the biological origin of the sample, in terms of organs, biofluid and species, but it can also be applied to obtain more challenging information like time since deposition, biological age, and states of disease. MS-based proteomics, particularly state-of-the-art acquisition and data-analysis approaches, can make a huge difference in these fields. However, in order to fully harness this potential, specific challenges in sample preparation, data acquisition and especially data analysis still remain to be overcome. A recent tour de force in the form of data-independent acquisition has enabled a new way of acquiring a permanent record of everything, and the rise of machine learning during data analysis is enabling unprecedentedly accurate classification and deeper validation of sparse annotations through spectral and retention time prediction. Altogether, it is clear that the field of proteomics has matured to a level where it is ready to tackle the extremely challenging samples presented by forensics and archaeology.

References

1. T. van der Valk, P. Pečnerová, D. Díez-del-Molino, A. Bergström, J. Oppenheimer, S. Hartmann, G. Xenikoudakis, J. A. Thomas, M. Dehasque, E. Sağlıcan, F. R. Fidan, I. Barnes, S. Liu, M. Somel, P. D. Heintzman, P. Nikolskiy, B. Shapiro, P. Skoglund, M. Hofreiter, A. M. Lister, A. Götherström and L. Dalén, *Nature*, 2021, **591**, 265–269.
2. T. P. Cleland, E. R. Schroeter, L. Zamdborg, W. Zheng, J. E. Lee, J. C. Tran, M. Bern, M. B. Duncan, V. S. Lebleu, D. R. Ahlf, P. M. Thomas, R. Kalluri, N. L. Kelleher and M. H. Schweitzer, *J. Proteome Res.*, 2015, **14**, 5252–5262.
3. K. Van Steendam, M. De Ceuleneer, M. Dhaenens, D. Van Hoofstat and D. Deforce, *Int. J. Leg. Med.*, 2013, **127**, 287–298.
4. M. Waas, M. Pereckas, R. A. Jones Lipinski, C. Ashwood and R. L. Gundry, *J. Proteome Res.*, 2019, **18**, 1644–1656.
5. A. Cooper and H. N. Poinar, *Am. Assoc. Adv. Sci.*, 2001, **289**, 1139.
6. J. Hendy, F. Welker, B. Demarchi, C. Speller, C. Warinner and M. J. Collins, *Nat. Ecol. Evol.*, 2018, **2**, 791–799.
7. G. J. Parker, H. E. McKiernan, K. M. Legg and Z. C. Goecker, *Forensic Sci. Int.: Genet.*, 2021, **54**, 102529.
8. J. Hendy, A. C. Colonese, I. Franz, R. Fernandes, R. Fischer, D. Orton, A. Lucquin, L. Spindler, J. Anvari, E. Stroud, P. F. Biehl, C. Speller, N. Boivin, M. Mackie, R. R. Jersie-Christensen, J. V. Olsen, M. J. Collins, O. E. Craig and E. Rosenstock, *Nat. Commun.*, 2018, **9**, 4064.
9. C. B. Messner, V. Demichev, N. Bloomfield, J. S. L. Yu, M. White, M. Kreidl, A. S. Egger, A. Freiwald, G. Ivosev, F. Wasim, A. Zelezniak, L. Jürgens, N. Suttorp, L. E. Sander, F. Kurth, K. S. Lilley, M. Mülleder, S. Tate and M. Ralser, *Nat. Biotechnol.*, 2021, **39**, 846–854.
10. C. Ludwig, L. Gillet, G. Rosenberger, S. Amon, B. C. Collins and R. Aebersold, *Mol. Syst. Biol.*, 2018, **14**, e8126.
11. cRAP Protein Sequences, The Global Proteome Machine Organisation; 2012, Available from: https://www.thegpm.org/crap/.
12. R. Bouwmeester, R. Gabriels, N. Hulstaert, L. Martens and S. Degroeve, *Nat. Methods*, 2021, **18**, 1363–1369.
13. G. Varnavides, M. Madern, D. Anrather, N. Hartl, W. Reiter and M. Hartl, *J. Proteome Res.*, 2022, **21**, 2397–2411.
14. A. Declercq, R. Bouwmeester, C. Chiva, E. Sabidó, A. Hirschler, C. Carapito, L. Martens, S. Degroeve and R. Gabriels, *Nucleic Acids Res.*, 2023, W338–W342.
15. T. Schmidt, P. Samaras, V. Dorfer, C. Panse, T. Kockmann, L. Bichmann, B. Van Puyvelde, Y. Perez-Riverol, E. W. Deutsch, B. Kuster and M. Wilhelm, *J. Proteome Res.*, 2021, **20**, 3388–3394.
16. S. Gessulat, T. Schmidt, D. P. Zolg, P. Samaras, K. Schnatbaum, J. Zerweck, T. Knaute, J. Rechenberger, B. Delanghe, A. Huhmer, U. Reimer, H. C. Ehrlich, S. Aiche, B. Kuster and M. Wilhelm, *Nat. Methods*, 2019, **16**, 509–518.

17. C. M. Choung, J. W. Lee, J. H. Park, C. H. Kim, H. C. Park and S. K. Lim, *Leg. Med.*, 2021, **51**, 101872.
18. M. M. Foley and F. Oldoni, *Forensic Sci. Int.: Genet.*, 2023, **63**, 102819.
19. S. Harbison and R. Fleming, *Res. Rep. Forensic Med. Sci.*, 2016, 11.
20. J. Sloots, W. Lalonde, B. Reid and J. Millman, *Forensic Sci. Int.*, 2017, **281**, 141–146.
21. D. J. Wornes, S. J. Speers and J. A. Murakami, *Forensic Sci. Int.*, 2018, **288**, 81–88.
22. N. Dawnay and K. Sheppard, *Sci. Justice*, 2023, **63**, 206–228.
23. E. Mistek-Morabito and I. K. Lednev, *Commun. Chem.*, 2020, **3**, 178.
24. A. Aliferi, S. Sundaram, D. Ballard, A. Freire-Aradas, C. Phillips, M. V. Lareu and D. S. Court, *Forensic Sci. Int.: Genet.*, 2022, **57**, 102637.
25. K. M. Legg, R. Powell, N. Reisdorph, R. Reisdorph and P. B. Danielson, *Electrophoresis*, 2017, **38**, 833–845.
26. N. Procopio, A. T. Chamberlain and M. Buckley, *J. Proteome Res.*, 2018, **17**, 1000–1013.
27. T. Claeys, M. Menu, R. Bouwmeester, K. Gevaert and L. Martens, *J. Proteome Res.*, 2023, **4**, 1181–1192.
28. F. Maixner, T. Overath, D. Linke, M. Janko, G. Guerriero, B. H. J. Van Den Berg, B. Stade, P. Leidinger, C. Backes, M. Jaremek, B. Kneissl, B. Meder, A. Franke, E. Egarter-Vigl, E. Meese, A. Schwarz, A. Tholey, A. Zink and A. Keller, *Cell. Mol. Life Sci.*, 2013, **70**, 3709–3722.
29. A. Petzold, C. H. Lu, M. Groves, J. Gobom, H. Zetterberg, G. Shaw and S. O'Connor, *J. R. Soc., Interface*, 2020, **17**, 20190775.
30. A. Corthals, A. Koller, D. W. Martin, R. Rieger, E. I. Chen, M. Bernaski, G. Recagno and L. M. Dávalos, *PLoS One*, 2012, **7**, e41244.
31. J. Jones, M. Mirzaei, P. Ravishankar, D. Xavier, D. S. Lim, D. H. Shin, R. Bianucci and P. A. Haynes, *Philos. Trans. R. Soc., A*, 2016, **374**, 20150373.
32. A. Bonicelli, A. Di Nunzio, C. Di Nunzio and N. Procopio, *J. Proteome Res.*, 2022, **21**, 1285–1298.
33. N. H. Nielsen, P. S. Henriksen, M. F. Mortensen, R. Enevold, M. N. Mortensen, C. Scavenius and J. J. Enghild, *Antiquity*, 2021, **95**, 1195–1212.
34. B. Demarchi, R. Boano, A. Ceron, F. D. Bello, S. E. Favero-Longo, S. Fiddyment, E. F. Marochetti, G. Mangiapane, M. Mattonai, C. Pennacini, E. Ribechini, J. Woolley, G. Zilberstein and P. G. Righetti, *J. Archaeol. Sci.*, 2020, **119**, 105145.
35. J. Hendy, M. Collins, K. Y. Teoh, D. A. Ashford, J. Thomas-Oates, H. D. Donoghue, I. Pap, D. E. Minnikin, M. Spigelman and M. Buckley, *J. Archaeol. Sci.*, 2016, **66**, 146–153.
36. A. C. Aufderheide, *The Scientific Study of Mummies*, Cambridge University Press, Cambridge, 2010.
37. M. Uhlén, L. Fagerberg, B. M. Hallström, C. Lindskog, P. Oksvold, A. Mardinoglu, Å. Sivertsson, C. Kampf, E. Sjöstedt, A. Asplund, I. M. Olsson, K. Edlund, E. Lundberg, S. Navani, C. A. K. Szigyarto, J. Odeberg, D. Djureinovic, J. O. Takanen, S. Hober, T. Alm, P. H. Edqvist, H. Berling, H. Tegel, J. Mulder, J. Rockberg, P. Nilsson, J. M. Schwenk, M. Hamsten,

K. Von Feilitzen, M. Forsberg, L. Persson, F. Johansson, M. Zwahlen, G. Von Heijne, J. Nielsen and F. Pontén, *Science*, 2015, **347**, 6220.
38. Y. Xu, N. Wang, S. Gao, C. Li, P. Ma, S. Yang, H. Jiang, S. Shi, Y. Wu, Q. Zhang and Y. Cui, *Int. J. Leg. Med.*, 2023, **137**, 319–327.
39. F. Chu, K. E. Mason, D. S. Anex, A. D. Jones and B. R. Hart, *Sci. Rep.*, 2019, **9**, 7641.
40. G. J. Parker, T. Leppert, D. S. Anex, J. K. Hilmer, N. Matsunami, L. Baird, J. Stevens, K. Parsawar, B. P. Durbin-Johnson, D. M. Rocke, C. Nelson, D. J. Fairbanks, A. S. Wilson, R. H. Rice, S. R. Woodward, B. Bothner, B. R. Hart and M. Leppert, *PLoS One*, 2016, **11**(9), e0160653.
41. Z. C. Goecker, M. R. Salemi, N. Karim, B. S. Phinney, R. H. Rice and G. J. Parker, *Forensic Sci. Int.: Genet.*, 2020, **47**, 102314.
42. C. N. Laatsch, B. P. Durbin-Johnson, D. M. Rocke, S. Mukwana, A. B. Newland, M. J. Flagler, M. G. Davis, R. A. Eigenheer, B. S. Phinney and R. H. Rice, *PeerJ*, 2014, **2**, e505.
43. J. A. Milan, P. W. Wu, M. R. Salemi, B. P. Durbin-Johnson, D. M. Rocke, B. S. Phinney, R. H. Rice and G. J. Parker, *Forensic Sci. Int.: Genet.*, 2019, **41**, 19–23.
44. A. Marrone, D. La Russa, A. Montesanto, V. Lagani, M. F. La Russa and D. Pellegrino, *Molecules*, 2021, **26**(20), 6272.
45. K. Krassowski, *J. Mod. Sci.*, 2018, **38**, 209–219.
46. M. Bauer, S. Polzin and D. Patzelt, *Forensic Sci. Int.*, 2003, **138**, 94–103.
47. R. H. Bremmer, D. M. de Bruin, M. de Joode, W. J. Buma, T. G. van Leeuwen and M. C. G. Aalders, *PLoS One*, 2011, **6**(7), e21845.
48. E. Brunelle, M. Eldridge and J. Halámek, *ACS Omega*, 2021, **6**, 12898–12903.
49. E. Brunelle, C. Huynh, E. Alin, M. Eldridge, A. M. Le, L. Halámková and J. Halámek, *Anal. Chem.*, 2018, **90**, 980–987.
50. V.-A. Duong, J.-M. Park, H.-J. Lim and H. Lee, *Appl. Sci.*, 2021, **11**(18), 3393.
51. L. Toma, G. Vignali, E. Maffioli, S. Tambuzzi, R. Giaccari, M. Mattarozzi, S. Nonnis, M. Milioli, L. Franceschetti, G. Paredi, A. Negri, B. Riccardi, C. Cattaneo, M. Careri, G. Tedeschi and S. Bruno, *Sci. Rep.*, 2023, **13**, 6116.
52. F. Zampa, M. Hilgert, J. Malmborg, M. Svensson, L. Schwarz and A. Mattei, *Sci. Justice*, 2020, **60**, 191–200.
53. M. Pal Chowdhury and M. Buckley, *Methods*, 2022, **200**, 67–79.
54. B. Philippsen, *Heritage Sci.*, 2013, **1**, 24.
55. R. Gowland, N. A. Stewart, K. D. Crowder, C. Hodson, H. Shaw, K. J. Gron and J. Montgomery, *Am. J. Phys. Anthropol.*, 2021, **174**, 859–869.
56. K. Rebay-Salisbury, P. Bortel, L. Janker, M. Bas, D. Pany-Kucera, R. B. Salisbury, C. Gerner and F. Kanz, *J. Archaeol. Sci.*, 2022, **139**, 105549.
57. K. Rebay-Salisbury, L. Janker, D. Pany-Kucera, D. Schuster, M. Spannagl-Steiner, L. Waltenberger, R. B. Salisbury and F. Kanz, *Archaeol. Anthropol. Sci.*, 2020, **12**, 265.

58. J. D. Bartlett, *ISRN Dent.*, 2013, **2013**, 1–24.
59. P. L. Rüther, I. M. Husic, P. Bangsgaard, K. M. Gregersen, P. Pantmann, M. Carvalho, R. M. Godinho, L. Friedl, J. Cascalheira, A. J. Taurozzi, M. L. S. Jørkov, M. M. Benedetti, J. Haws, N. Bicho, F. Welker, E. Cappellini and J. V. Olsen, *Nat. Commun.*, 2022, **13**(1), 2458.
60. F. Chen, F. Welker, C. C. Shen, S. E. Bailey, I. Bergmann, S. Davis, H. Xia, H. Wang, R. Fischer, S. E. Freidline, T. L. Yu, M. M. Skinner, S. Stelzer, G. Dong, Q. Fu, G. Dong, J. Wang, D. Zhang and J. J. Hublin, *Nature*, 2019, **569**, 409–412.
61. N. Procopio, A. T. Chamberlain and M. Buckley, *J. Proteome Res.*, 2017, **16**, 2016–2029.
62. R. Sawafuji, E. Cappellini, T. Nagaoka, A. K. Fotakis, R. R. Jersie-Christensen, J. V. Olsen, K. Hirata and S. Ueda, *R. Soc. Open Sci.*, 2017, **4**, 161004.
63. E. D. Merkley, D. S. Wunschel, K. L. Wahl and K. H. Jarman, *Forensic Sci. Int.*, 2019, **297**, 350–363.
64. E. Johnston and M. Buckley, *J. Proteome Res.*, 2021, **20**, 538–548.
65. K. E. Mason, D. Anex, T. Grey, B. Hart and G. Parker, *Forensic Sci. Int.*, 2018, **288**, 89–96.
66. C. Roberts and K. Manchester, *The Archaeology of Disease*, The History Press, Stroud, Gloucestershire, 3rd edn, 2010.
67. A. Bona, Z. Papai, G. Maasz, G. A. Toth, E. Jambor, J. Schmidt, C. Toth, C. Farkas and L. Mark, *PLoS One*, 2014, **9**, e87215.
68. A. Boros-Major, A. Bona, G. Lovasz, E. Molnar, A. Marcsik, G. Palfi and L. Mark, *J. Archaeol. Sci.*, 2011, **38**, 197–201.
69. R. Kendall, J. Hendy, M. J. Collins, A. R. Millard and R. L. Gowland, *Sci. Technol. Archaeol. Res.*, 2016, **2**, 15–24.
70. J. P. Rose, C. A. Schurman, C. D. King, J. Bons, J. B. Burton, S. K. Patel, A. O'Broin, T. Alliston and B. Schilling, bioRxiv, 2022, 7, 192016.
71. B. Courel, H. K. Robson, A. Lucquin, E. Dolbunova, E. Oras, K. Adamczak, S. H. Andersen, P. M. Astrup, M. Charniauski, A. Czekaj-Zastawny, I. Ezepenko, S. Hartz, J. Kabacinski, A. Kotula, S. Kukawka, I. Loze, A. Mazurkevich, H. Piezonka, G. Piliciauskas, S. A. Sørensen, H. M. Talbot, A. Tkachou, M. Tkachova, A. Wawrusiewicz, J. Meadows, C. P. Heron and O. E. Craig, *R. Soc. Open Sci.*, 2020, **7**, 192016.
72. T. Van Den Bossche, B. J. Kunath, K. Schallert, S. S. Schäpe, P. E. Abraham, J. Armengaud, M. Arntzen, A. Bassignani, D. Benndorf, S. Fuchs, R. J. Giannone, T. J. Griffin, L. H. Hagen, R. Halder, C. Henry, R. L. Hettich, R. Heyer, P. Jagtap, N. Jehmlich, M. Jensen, C. Juste, M. Kleiner, O. Langella, T. Lehmann, E. Leith, P. May, B. Mesuere, G. Miotello, S. L. Peters, O. Pible, P. T. Queiros, U. Reichl, B. Y. Renard, H. Schiebenhoefer, A. Sczyrba, A. Tanca, K. Trappe, J. P. Trezzi, S. Uzzau, P. Verschaffelt, M. von Bergen, P. Wilmes, M. Wolf, L. Martens and T. Muth, *Nat. Commun.*, 2021, **12**, 7305.
73. R. R. Jersie-Christensen, L. T. Lanigan, D. Lyon, M. Mackie, D. Belstrøm, C. D. Kelstrup, A. K. Fotakis, E. Willerslev, N. Lynnerup, L. J. Jensen, E. Cappellini and J. V. Olsen, *Nat. Commun.*, 2018, **9**, 4744.

74. S. Wilkin, A. Ventresca Miller, W. T. T. Taylor, B. K. Miller, R. W. Hagan, M. Bleasdale, A. Scott, S. Gankhuyg, A. Ramsøe, S. Uliziibayar, C. Trachsel, P. Nanni, J. Grossmann, L. Orlando, M. Horton, P. W. Stockhammer, E. Myagmar, N. Boivin, C. Warinner and J. Hendy, *Nat. Ecol. Evol.*, 2020, **4**, 346–355.
75. S. Wilkin, A. Ventresca Miller, R. Fernandes, R. Spengler, W. T. T. Taylor, D. R. Brown, D. Reich, D. J. Kennett, B. J. Culleton, L. Kunz, C. Fortes, A. Kitova, P. Kuznetsov, A. Epimakhov, V. F. Zaibert, A. K. Outram, E. Kitov, A. Khokhlov, D. Anthony and N. Boivin, *Nature*, 2021, **598**, 629–633.
76. M. Pieri, A. Silvestre, M. De Cicco, G. Mamone, E. Capasso, F. Addeo and G. Picariello, *J. Proteomics*, 2019, **209**, 103524.
77. A. K. W. Runge, J. Hendy, K. K. Richter, E. Masson-Maclean, K. Britton, M. MacKie, K. McGrath, M. Collins, E. Cappellini and C. Speller, *Proc. R. Soc. B*, 2021, **288**, 20210020.

Subject Index